MW00615140

WINGWALKERS

WINGWALKERS

A History of Canadian Airlines International

PETER PIGOTT

Harbour Publishing

Copyright © 1998 by Peter Pigott
Revised edition copyright © 2003 by Peter Pigott

All rights reserved. No part of this publication may be reproduced, stored in a
retrieval system or transmitted, in any form or by any means, without prior
permission of the publisher or, in the case of photocopying or other reprographic
copying, a licence from Access Copyright, the Canadian Copyright Licensing
Agency, 1 Yonge Street, Suite 1900, Toronto, Ontario, M5E 1E5,
www.accesscopyright.ca, 1-800-893-5777, info@accesscopyright.ca.

Harbour Publishing
P.O. Box 219
Madeira Park, BC Canada V0N 2H0
www.harbourpublishing.com

Harbour Publishing acknowledges financial support from the Government of Canada
through the Book Publishing Industry Development Program and the Canada Council for
the Arts; and from the Province of British Columbia through the British Columbia Arts
Council and the Book Publisher's Tax Credit through the Ministry of Provincial Revenue and
the British Columbia Heritage Trust.

THE CANADA COUNCIL | LE CONSEIL DES ARTS
FOR THE ARTS | DU CANADA
SINCE 1957 | DEPUIS 1957

BRITISH
COLUMBIA
ARTS COUNCIL
Supported by the Province of British Columbia

Dustjacket design by Martin Nichols, Lionheart Graphics.
Photograph sources: ACA—Air Canada Archives; CAA—Canadian Airlines
Archives; CPA—Canadian Pacific Archives; DND—Department of National
Defence; SAB—Saskatchewan Archives Board; WCAM—Western Canada Aviation
Museum.

Printed and bound in Canada

National Library of Canada Cataloguing in Publication Data

Pigott, Peter
  Wingwalkers : the rise and fall of Canada's other airline / Peter
Pigott. — Rev. ed.

Includes index.
ISBN 1-55017-292-1

1. Canadian Airlines International—History. 2. Canadian Airlines
International—Biography. I. Title.
HE9815.C33P53 2003   387.7'06'571   C2003-910743-4

For Holly and Jade

Contents

Introduction

The first edition of *Wingwalkers*, while documenting the serious problems besetting Canadian Airlines International, did so in the present tense. No more. Following its takeover by Air Canada, Canadian faded away entirely in 2002. Its passing marked the end of the wild and woolly era of western Canadian aviation, dominated by such larger-than-life characters as Wop May, Jim Spilsbury, Grant McConachie and Max Ward.

This revised edition of *Wingwalkers* caps that colourful story with a new final chapter chronicling the airline's desperate manoeuvres to survive—and the fiscal and political forces that thwarted those efforts. As it turned out, the machinations behind the demise of Canadian were as dramatic as anything in its checkered history.

This book takes its title from the name given to a group of extraordinary men and women in the 1920s. Royal Flying Corps veterans returning to Canada from World War I found that the airplane was treated as little more than a fairground toy. Faced with a choice between indifference and starvation, the early aviators, some of whom would later fly for Canadian Pacific Airlines, took to travelling the rural circuits in weary Curtiss JN-4 "Canuck" biplanes, performing as barnstormers, selling rides or looping the loop to thrill the crowd below. The more foolhardy among them left their cockpits to walk between the aircraft's wings. The public soon tired of that, and to bolster gate receipts the wingwalkers clambered onto the top wing, then stepped cautiously along its length, between the bracing wires and the king posts and, for a finale, waved to the crowd or sometimes stood on their heads.

The wingwalker was always aware that he was completely at the mercy of the elements and the pilot's skills. The audience felt cheated if parachutes and safety wires were used, and because stunts had to be performed at a low level to be within sight of the crowd, such lifesaving devices would have been useless anyway. It was an exhilarating, hair-raising way to make a living but, until commercial aviation began, the only one open to aviators.

The archives trace Canadian Airlines' ancestry through a hodge-podge of bush outfits to the year 1927. Despite its vicissitudes, the company clawed its way to legitimacy, not content to play second fiddle to Trans-Canada Air Lines. For three decades it was a pariah within the rigid, government-imposed air transportation system. Because of this, in best wing-walker mode, the company periodically performed in a series of heart-stopping, life-and-death dramas, alternately evoking public sympathy and admiration. In every stage of its metamorphosis, whether as Canadian Airways, Canadian Pacific Airlines, CP Air or Canadian Airlines International, as a victim of circumstances or just poorly managed, the airline's directors, employees and investors must have empathized with those barnstormers: this, too, was an exhilarating way to make a living.

The ignominious role of the federal government in all of this warrants some explanation. To paraphrase Hugh Keenleyside, the history of modern Canada can be divided into two eras: before aircraft and after aircraft. The country's climate and geography required a safe, reliable and efficient system of air transport to carry the mail and link the main cities, and it was necessary to do this before the Americans did. However, so traumatized was the government of the day by the burden of two transcontinental railways in a nation that hadn't the population to support even one, that Prime Minister Mackenzie King and his minister of transport, C.D. Howe, ensured that the new mode of transport was kept out of private hands. Until the Diefenbaker years, Ottawa was dogmatic about regulating commercial aviation: small carriers had their uses in the North or in the bush, but transcontinental air transport was a state monopoly. A government-owned airline was indivisible from the nation's development, and a private interest like Canadian Pacific Airlines competing against it would not only affect the public treasury but would be deemed almost unpatriotic. As a result, when they outgrew their provincial limits, Western Canada Airways, Canadian Airways Ltd. and Canadian Pacific Airlines were treated as saboteurs of the grand design. Like Prometheus, by giving the travelling public an alternative, they had stolen what belonged to the gods. And like Prometheus, they were punished.

Scholars have ignored the emotional aspects of commercial aviation, but working for Canadian Airlines and predecessors, when compared with the security of Trans-Canada Air Lines, must have required extreme devotion. Like the unlamented Vickers Vanguard airliner, Canadian in its various incarnations was a collection of spare parts flying in close formation. Each of the bush outfits or "regionals" bought up and merged into the company considered itself the glue that kept its piece of Canada together. During World War II, all of them did their patriotic duty by training thousands of airmen and maintaining and ferrying aircraft overseas. But until recently they were run with the innocent enthusiasm of the returning war veteran. To James Richardson, W. Leigh Brintnell, Russ Baker and Grant McConachie, the airline business was a religion; it promised a utopian future that always seemed to be just within reach, if only the next stage of redemption—mail contracts, transcontinental routes or financial aid—could be attained.

This narrative documents my affection for and interest in the airline and its genealogical roots. My own experiences with Canadian Airlines are comparatively recent, as I grew up in the Montreal suburb of Dorval, within sight of the main Air Canada base. My earliest memories of Canadian extend no further back than a recent posting to Hong Kong with the Department of Foreign Affairs. Most of my duties seemed to consist of waiting each day on the tarmac at Kai Tak Airport to meet the diplomatic courier aboard the Canadian Airlines flight from Vancouver. I still do not know if it was the sight of that DC-10 threading its way through the rows of Kowloon tenements or the fumes from the sewage drain at Kai Tak, but the experience never failed to take my breath away and has remained with me ever since.

The Fokker Universal G-CAGD was one of two bought by "Doc" Oaks in 1927 for the Churchill airlift. Christened *The City of Toronto*, it is shown here in 1931, having just been sold to Grant McConachie. CAA

CHAPTER ONE

James Richardson and the Creation of Commercial Aviation in Canada

It began as a lot of Canadian dreams do—on a wet summer's day at the lake. Among the dripping pines and the call of the loon at Minaki Lake in northwestern Ontario, a Liberty engine coughed and spluttered into action, to the relief of the engineer sweating over the hand crank. Soon, as cumbersome as a rigged sailing ship, the Curtiss HS-2L seaplane lumbered tentatively away from its moorings into the centre of the lake.

The pilot knew that he had a clear run for the takeoff because the usable runway area of the lake extended to about 6 miles (10 km). Designed during the Great War by Glenn Curtiss to search the coastal waters for German submarines, the HS-2L, with its seventy-four-foot wings held together by high-tensile steel wires, was less manoeuvrable on inland waterways, and because of its ungainly shape Air Board personnel had nicknamed it "The Pregnant Pelican." To help manoeuvre the clumsy biplane, the engineer had clambered to the edge of a wing and sat down, his weight allowing the pilot to turn into the wind. On a hand signal from the pilot—the straining engine made shouting impossible—the engineer raised himself up, dodged the wires once more and made for the cramped cockpit. Hardly waiting for him to get settled, the pilot adjusted his goggles and opened up the throttle. Ignoring the

13

protestations of the heron near the bank, the Curtiss gathered speed in the green water and lifted into the air.

The performance was keenly watched by a vacationer on shore. Although he had often seen the flying boat take off and land on Minaki Lake, James Armstrong Richardson remained fascinated. The latest Winnipeg newspapers were full of Alan Cobham's record-setting air journey from London to Sydney, Australia, that month. It was difficult to believe that in this summer of 1926 the miracle of powered flight was less than twenty years old in the British empire. The previous year, Donald MacLaren, a former air ace, had bought a Curtiss HS-2L for his British Columbia company, Pacific Airways Ltd. He then persuaded the government to award him a contract for fishery patrols from Swanson Bay, near Princess Royal Island. But since government air regulations had become law in Canada only six years before (when the first commercial pilot's licence was issued), few paying passengers and little air freight had been carried aloft. Richardson was aware that aircraft had been used to carry mail from Remi Lake to Moose Factory in Ontario, to photograph remote parts of Labrador and to carry out forest fire patrols in Quebec. Even as Richardson watched the flying boat lift off Minaki Lake, in Montreal a wealthy American sportsman, J. Dalzell McKee, and Squadron Leader Earl Godfrey were preparing to make the first seaplane flight across Canada.

In comparison with Europe and the United States, Canada was a country with a huge frontier, which was largely unexplored north of the city of Edmonton. The nation's largely rural population was too scattered and the physical barriers too daunting, and air companies could not hope to make up even their operating expenses. The only aircraft available were war-surplus JN-4 "Canuck" trainers and the Curtiss flying boats, neither of which had been designed to carry freight or passengers, or indeed to remain in active service so long after the Great War had ended. The primary reason for the stagnation of commercial aviation in Canada was the lack of public interest and private investment. Since air travel was unknown to the Fathers of Confederation, it had not been mentioned in the British North America Act of 1867. Thus it was left to the courts to decide on the division of powers concerning international and interprovincial aeronautics. As one historian observed, Confederation was an instrument of steam power, entirely unsuited to the limitless uses of hydroelectricity—and aviation. Canada was still paying for an overbuilt railway network, a legacy of the

Several of the original pilots for Canadian Pacific Airlines began their careers as barnstormers in the 1920s, including Mel Knox, Ernie Boffa, Russ Baker, Bob Randall, Sheldon Luck and Grant McConachie. SAB S-B

Edwardian era. In 1923, when the country was faced with a breakdown in regional rail communications as a result of the 1920–22 economic depression, only the federal government had the power and the wealth to merge a collection of the bankrupt rail lines into Canadian National Railways. From that moment on, Ottawa's control over all forms of transportation was absolute.

The economic depression had the added effect that no sane politician would contemplate investing in what was at that time more a form of amusement than a mode of transport. In the private sphere, until James Richardson thought of taking an interest, most aerial entrepreneurs were enthusiastic Royal Flying Corps veterans who—deprived of duelling with the Hun over the Western Front—travelled the country as aerial gypsies, thrilling audiences with barnstorming and wingwalking until their tattered "Canucks" fell apart.

In contrast, as the ink on the Treaty of Versailles was drying, European governments seized on the new mode of transportation as an extension of their national ambitions. In 1922, seven British air companies operated between Croydon airport, in the suburbs of London, and Paris. Unfortunately they undercut each other to the verge of bankruptcy, and on March 31, 1924, the British government brought four of the largest companies under its control to form Imperial Airways. Keenly watched by other governments, Whitehall then designated Imperial Airways as Britain's "chosen instrument" in the air. The director of British civil aviation, Sir Sefton Branckner, influenced a generation of British and Canadian politicians with his belief that unfettered competition in commercial aviation led to disastrous consequences.[1]

Rather than allowing Imperial Airways to become a flying bus service between European capitals, thereby enabling it to make a profit, Whitehall moulded it into a patriotic glue with which to join the scattered colonies of the empire to Mother England.[2] Imperial Airways was subsidized to the tune of £750,000 annually on the condition it buy only British-made aircraft and fly to distant Australia, India and Africa—whatever the cost. The idea of Imperial being run as a business was probably the last thing its board of directors wanted. As a result, in the very summer that Richardson watched the seaplane at Minaki Lake, Imperial Airways launched its inaugural service from London to India.

The British concept of a state monopoly in air travel was followed in Holland, Belgium and France,[3] but nowhere more so than in Weimar Germany. Bereft of its merchant marine and colonies, and forbidden an air force, the German government saw in Luft Hansa (the name remained two words until 1934) a means of recapturing some of its prewar overseas prestige. German aviation engineers solicited government subsidies to build airships and the first all-metal airliner, the Junkers F-13, so that by the summer of 1926 Luft Hansa was carrying more passengers in Europe and to South America than all other air companies combined.[4]

Like most Canadians of his day, James Richardson accepted aviation as he did other new ideas that were flooding into postwar North America. Despite inflation and the increasing unemployment of the 1920s, in the unbridled optimism of the time it must have seemed to him that the country's hour had struck. There were mineral strikes in the Kootenays, the Klondike, Cobalt and Sudbury. The long bull market of the 1920s, the flapper girls, wingwalkers and the Jay Gatsby-type of instant millionaires all reflected the mood rather than creating it. In this supercharged climate anything seemed possible, and shrewd investors capitalized on the new technology of the automobile, aircraft, radio and talking pictures to make their fortunes.[5] The railways and the telegraph, those solid foundations of the Victorian era, had been displaced by more ethereal mediums that offered a tempting vision of the future.

James Richardson's interest in mining led him to aviation. He risked his fortune to establish Western Canada Airways and later Canadian Airways Ltd. before an American incursion.
James Richardson & Sons (UNN)

Freed from the physical and political restrictions of the pre-1914 world, these innovations seemed to promise an end to what had caused the Great War: political frontiers and economic rivalries.

With a perspicacity inherited from his grandfather, James Richardson believed that flying had the potential to be profitable. Five years earlier Canada's first airline, Laurentide Air Services, financed by Montreal businessman Thomas Hall, had carried prospectors and trappers between Haileybury, Ontario and Angliers, Quebec. But the mining communities that Laurentide serviced could never generate enough traffic to pay more than its operating costs, and Hall was always out of pocket—a lesson not lost on Richardson. When the Ontario government launched its own air service in 1924 Laurentide folded, never having made any money.

In one of its last moments, Laurentide did buy thirteen Curtiss HS-2Ls from the United States to overhaul before handing them to the Ontario Provincial Air Service (OPAS). From its Sault Ste. Marie base, OPAS blossomed beyond forestry patrol duties to become the first large-scale aviation company in Canada. It hired Laurentide staff, including H.A. "Doc" Oaks, "Al" Cheesman, W. Leigh Brintnell and Romeo Vachon, who through the summer months transported teams of firefighters in flying boats across the province. In 1925, when a gold rush began at Jack Howey's mine in Red Lake, Ontario, the OPAS HS-2Ls carried men and supplies into the area before the winter froze the waterways, thus participating in the first such airlift in history.

In the United States, commercial aviation had progressed beyond the pioneering stage and the glimmerings of future airlines were making themselves seen. The business lobby in Washington saw the potential of transcontinental air mail, and the world's first night "airway" between

Chicago and Cheyenne, Wyoming, opened in 1923 for mail flights. The 885-mile (1416-km) route was equipped with lighted airports, which were connected across the continent by using Sperry electric beacons as navigation markers. The advantages of delivering lightweight, expensive commodities like film or bank drafts by air gave birth to air freight.[6] Within three years, 45 million letters and parcels had been flown from coast to coast in the United States. On April 10, 1926, when President Calvin Coolidge signed the Kelly Act, authorizing the transfer of air mail delivery from the US Post Office to private companies, he unknowingly began the rise in American industrial aviation. By his action, talented aeronautical engineers like Bill Boeing, Donald Douglas and Allen Lockheed were assured of a market for their aircraft from airline entrepreneurs like C.M. Keys, Eddie Rickenbacker and Juan Trippe.

As a businessman Richardson appreciated that to compete with the railways, an air carrier had to gain the public's confidence by proving its dependability. An aircraft's single advantage over a train was speed, and to be truly effective against surface transportation in Canada, a pilot had to be able to fly at night, in poor weather and throughout the long Canadian winter. Rather than searching for farmers' fields to land on, pilots needed well-mapped routes, paved and lighted airports, rotating beacons and the security of radio and meteorological services. Aircraft rugged enough for use in all seasons and terrains, that is, with enclosed cabins and interchangeable undercarriages—floats for water, wheels for airports and skis for the winter—were essential. Finally, a carrier relying on mining, with its "boom and bust" cycles, was inviting bankruptcy. Nothing less than the regular carriage of mail, freight and passengers would attract investors. On that wet summer's day at Minaki Lake, James Richardson decided that these conditions would be the basis of any air company he began.

Born in Kingston, Ontario, on August 21, 1885, James Richardson was named after his grandfather, who had emigrated to Kingston from Ireland more than half a century before. Orphaned at the age of four, within months of arriving in Canada, in best Horatio Alger fashion the young Irishman began life as a tailor's apprentice, but soon ran his own tailor shop. His introduction to the grain industry might have begun when he sometimes accepted payment in grain from rural customers. By 1857, Grandfather Richardson had given up tailoring. With his two sons, George and Henry, he went into the grain business full time. Within a few years he had acquired a fleet of schooners on the Great

Lakes to carry his grain, and in 1882 he built the first grain elevator in Ontario. Taking advantage of the recently completed Canadian Pacific Railway and the settling of the prairies, his son George expanded into western Canada. Soon the Richardson family's grain elevators dotted the new provinces of Manitoba, Saskatchewan and Alberta. In 1912 James Richardson & Sons moved its offices to Winnipeg, a city that touted itself as "The Gateway to the Canadian West." The company expanded rapidly as it reaped the rewards of selling wheat on the open market.

In 1918 George's son, James, at the age of thirty-three became president of the family firm, then the largest private company in Canada. By 1923, Canada was the world's largest grain exporter and the Richardson family firm the most powerful in the industry. Like influential men everywhere, James was sought after to serve on boards of companies like the Canadian Bank of Commerce, Canadian General Electric and, significantly, the Canadian Pacific Railway. It came as no surprise that when he was elected president of the Winnipeg Grain Exchange, Richardson championed free enterprise, contending that the government had no place in the regulation of a prosperous industry. "Success in the grain export business," he said, "requires the constitution of an ox, an enormous capacity for hard work and the peculiar qualifications which belong to a trader."

Knowing that government regulation of the grain industry was inevitable, Richardson diversified the company into radio and aviation. In 1926, six years after Westinghouse opened the world's first radio station to create a market for its ham radios, Richardson obtained a licence to broadcast on the prairies and opened six radio stations. Ostensibly created to beam crop prices to farmers, the stations soon provided music and entertainment.

It was Richardson's interest in mining that led him into flying. The family had been involved in mining for decades, and James's younger brother was a mining engineer. It followed that his own "interest in Airways was inspired entirely by a desire to accomplish the early opening up . . . of the mineral wealth of Canada." James was intrigued that a mining claim had been staked from the air as early as 1920 by the world's first bush pilot, Stuart Graham, while on forestry patrol over the St. Maurice River area, and that in 1922 the Canadian Aerial Services had flown exploratory flights around Moose Factory, looking for evidence of oil.

Central Canada Airlines was owned by former Royal Flying Corps pilot Jack Clarke. With this Curtiss HS-2L, Clarke ferried supplies from Minaki Lake, Ontario, to the gold mines in the Red Lake area. WCAM 9400

The flying boat at Minaki Lake belonged to Central Canada Airlines, a one-aircraft outfit servicing gold mines in the Red Lake area. Richardson introduced himself to its owner, Jack Clarke. Like many military pilots, Clarke had dreamt of owning his own air company but could barely afford even the HS-2L. The two talked of using aviation to exploit the mineral potential of Ontario and Manitoba, and on August 27, 1926, the grain entrepreneur agreed to invest $25,000 in Central Canada Airlines and assume control of it.

Much later, Richardson would set down his conditions for his involvement in commercial aviation. "You know my view," he wrote to Doc Oaks, "I want to see every possible precaution taken to provide every safeguard for our work and no flying done except when our equipment is in absolutely perfect condition and no one put in charge of a plane who is not thoroughly competent and experienced."

The decision to pay $60,000 for a Fokker Universal landplane, the company's second aircraft, was Richardson's. Anthony Fokker's masterpiece, the highwing Universal was the first Fokker to be built in the United States specifically for the North American market. Like Dr. Hugo Junkers, Fokker designed the best aircraft then available because of his experience in designing aircraft for Germany during the Great War. Throughout the 1920s, his products were constantly in the headlines: first airplane across the United States non-stop (1923), first airliner to fly from Amsterdam to Batavia (1924), first to fly over the North Pole (1926) and, soon, first flight across the Pacific (1928). In Canada, the RCAF used Fokker Universals for its Hudson Strait expedition in July 1927.

Powered by the Wright J-4B 200-horsepower engine, the Universal was of another generation from the decade-old Curtiss HS-2L. The flying boat, with its watercooled 400-horsepower engine and unwieldy shape, was limited to use in the summer months and could take off and land only on water. The Fokker was constructed of a welded steel fuselage with a "one piece" plywood wing, and had a maximum speed of 118 miles per hour (189 kph) and a range of 535 miles (856 km). Most important, it had an interchangeable undercarriage of skis, floats or wheels. As with the HS-2L, the pilot still sat in the open but his four passengers were in an enclosed cabin.

Sent to pick up the aircraft at the Fokker factory in New Jersey, Jack Clarke succumbed to the lure of flying in warmer climates, leaving Richardson to dissolve Central Canada Airlines on October 23, 1926. Now the owner of a new Fokker and an old HS-2L, he must have been relieved when Doc Oaks paged him on the floor of the Winnipeg Grain Exchange. Oaks had trained as a pilot in the Royal Flying Corps, then flew for the OPAS. Before both jobs, he had been a mining engineer, sharing Richardson's enthusiasm for aerial exploration. Called "Doc" because he was the son of a doctor, he came well recommended to Richardson by J.A. Wilson, the federal controller of civil aviation. Oaks had begun his own air company, Patricia Airways, to serve the Red Lake mining area but, unable to raise investment, was bankrupt within a year.

In later years both men would claim that each had thought of the name of the new company. Whoever was responsible, on December 10, 1926, Western Canada Airways (WCA) was given life, with Richardson as its president, his wife Muriel its vice-president, John Hunter as secretary and Oaks as managing director and chief pilot. The head office and maintenance base were to be at Winnipeg, and its only shareholder, James Richardson, invested $200,000 in capital stock into the company. (All the money he used to finance his aviation companies came from his private fortune, not the family firm.)

Commercial aviation was not specialized in the 1920s. Besides

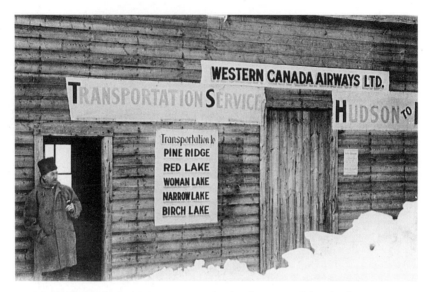

J.A. McDougall, the WCA treasurer, in the doorway of the shed at Hudson, Ontario, 1927. The fact that James Richardson employed an accountant distinguished his operation from any that had come before.
WCAM

providing transportation, WCA was chartered to manufacture and repair aircraft, operate flying schools and engage in aerial mapping and photography. Richardson also diversified, becoming a director of the Montreal shipbuilder Canadian Vickers, which was then building Vedette flying boats as a sideline. With other bush pilots, Oaks suggested several improvements to "Canadianize" the Fokker design, and in August 1927 Richardson acquired the rights to build Fokker aircraft in Winnipeg. He even purchased land for the facilities in the suburb of Norwood, but there were problems with a golf course nearby and the plan never materialized.

That Christmas, Oaks ferried Clarke's purchase, the Fokker Universal G-CAFU, christened *The City of Winnipeg*, from New York to Hudson, Ontario. When the American skis that the aircraft came with were found unsuitable for Canadian conditions, he replaced them with a pair of his own design, made of ash, brass-shod and weighted—a design that became standard on all WCA aircraft and was later used by Admiral Richard Byrd on his Antarctic flight.

The new air carrier began its first revenue flight on December 27 from Hudson to Lac Seul on to Narrow Lake, a distance of 217 miles

Left to right: J.R. "Rod" Ross, Bernt Balchen, S.A. "Al" Cheesman and Fred Stevenson on their return from the Churchill airlift, April 22, 1927.
WCAM/H.A. Oaks Collection

(347 km), with a shipment of 700 pounds (315 kg) of supplies. By the end of the year it had carried three passengers and 850 pounds (383 kg) of freight, earning Richardson $180. The company's first big contract came the following month, when the federal government asked if WCA could fly supplies and men to Churchill, Manitoba.[7] Earlier, as part of a government scheme to ship grain and cattle to Europe through Hudson Bay, the construction of a railway line had begun in 1912. It had been discontinued during the war, but with the help of British engineers the railhead had reached Cache Lake, 350 miles (560 km) from the port, by 1926. Exploratory drilling to deepen the harbour at Churchill could only be carried out while it was still frozen, hence the urgency.

The air force thought the airlift a mad endeavour, and it is a sign of Richardson's desperation to break into the market that he accepted the challenge. Oaks was sent down to New Jersey to buy two more Universals: G-CAGD, *The City of Toronto*, and G-CAGE, *Fort Churchill*. Three more pilots, J.R. Ross, Fred J. Stevenson and Bernt Balchen,[8] were hired. The initial shipment to Churchill expanded to 17,894 pounds (8052 kg) of freight and fourteen passengers. From March 22

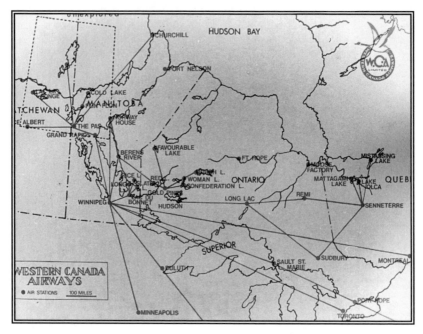

The Western Canada Airways network. On June 1, 1927, WCA's first scheduled air service began between Winnipeg and Long Lake via Lac du Bonnet. The routes to Minneapolis, Toronto and Montreal were wishful thinking. WCAM (UNN)

to April 17, 1927, using a boxcar as sleeping accommodation at Cache Lake, WCA's pilots made twenty-seven round trips in the three Fokkers, each time landing on the frozen Churchill waterfront.

On the heels of the first large-scale airlift in North American history was news of the completion of Charles Lindbergh's solo flight across the Atlantic on May 29. Both events publicized air transport as never before. The Churchill airlift resulted in a steady stream of orders: a contract to fly prospectors from Montreal to Senneterre, Quebec; a large air freight contract for Sherritt Gordon Mines at Cold Lake, north of The Pas; and, on June 1, 1927, a weekly air service from Winnipeg to Long Lake, Manitoba.

More aircraft were bought, among them the first of the Fairchild FC-2s. Powered by the Wright Whirlwind J-5, the FC-2 was more popular than the Fokker because its pilot was enclosed with the passengers and its fabric-covered wing was easier to repair. In November 1927, the increasing number of its aircraft prompted WCA to open a maintenance base to service them at the foot of Brandon Avenue, on the banks of the Red River.[9] Since it had no runway to land wheel-equipped aircraft on, the machine shop was moved to a small airfield in the suburb of St. Charles.

Grant MacDonald had recently arrived in Winnipeg from Nova Scotia and enjoyed watching the WCA aircraft take off and land on the river. He and his two brothers owned a sheet metal company on Aikins Street off Dufferin Avenue. MacDonald became friendly with the pilots and soon learned that WCA preferred floats built by the Edo Aircraft Corporation of Long Island, New York. MacDonald negotiated the licence to manufacture Edo floats in Canada, and soon after MacDonald Brothers Aircraft Ltd. opened across the lane from WCA on Robinson Street. The business now included fuselage welding repairs and propeller straightening. With investment from Richardson the company grew, helping to make Winnipeg the centre of early bush aviation.

In December, Richardson decided to start a flying school at Kirkfield Park, in the St. Charles suburb of Winnipeg. Using Avro Avians, instructors Andrew Cruickshank and W.B. Burchall guided former OPAS and RFC pilots through refresher courses. Many of the school's graduates were employed by WCA. F. "Russ" Baker, an office boy for WCA, signed up for flying lessons on August 29, 1928, paying $300 initially and later $756 for solo time. He obtained his commercial licence in 1929, but, unable to get a job with WCA, Russ went off to fly for Ginger Coote Airways and later Grant McConachie's United Air Transport. His destiny and that of WCA would intertwine far in the future.

Fred Stevenson was awarded the international Harmon Trophy for 1927 for his role in the Churchill airlift. (The other winner was Charles Lindbergh.) Sadly, Stevenson died on January 5, 1928, when he stalled and crashed one of the three Fokkers while coming in to land at The Pas, Manitoba. W. Leigh Brintnell took his place, becoming general manager at the same time as Tommy Siers was hired as chief mechanic. Winnipeg named its new airport, then being built in the suburbs, Stevenson Aerodrome in Fred Stevenson's memory. The city, conscious of its place in aviation, painted a navigation aid on the tank on the roof of the Winnipeg Electric Company. The marker consisted of a chrome-yellow arrow, 115 feet (35 m) long, with AIRPORT appearing in letters 15 feet (4.5 m) high, pointing toward the Stevenson Aerodrome.

Doc Oaks's ability was recognized as well. In 1927 he became the first recipient of the McKee Trophy, an annual prize for outstanding service in the advancement of aviation in Canada. But his first love was mineral exploration, and he left WCA soon after to join Jack Hammill and form Northern Aerial Minerals Exploration Ltd., another company that Richardson would invest in.

In all flights, the pilot and his engineer in the 1920s worked as a team, particularly in winter. The evening routine was synchronized. After the last flight of the day, one of them would immediately drag out the big engine cover while the other ducked under the engine to drain the oil into a five-gallon pail before it thickened in the cold weather. Next came the refuelling routine, with one manning the handpump and the other up on the wing holding the hose and strainer. There was the getting of poles to put under the skis to prevent them from freezing to the ground, then a discussion on how to tie the aircraft down so that it would not blow away or be damaged if a big wind came up, and finally the actual tying down when a decision had been reached.

All this had to take place after a heavy day of loading and unloading as the pair landed, refuelled and took off from several locations. Some days they went without a break for lunch because the winter days were short and the idea was to go while there was still light. A WCA engineer earned an average of $25 a month, with room and board thrown in.

By the end of 1927 WCA had flown 145,834 miles (233,334 km) in 1,905 hours, carrying 420,730 pounds (189,329 kg) of freight and 1,141 passengers. The increased business demanded larger aircraft, and in 1928 Richardson purchased fourteen Super Universals, in which the pilot and six passengers were enclosed. Richardson's pilots saw that they needed aircraft better designed for Canadian climate and bush conditions, with more reliable engines, metal fuselages and greater cargo-carrying capacity. Built by Canadian Vickers in Montreal under licence from Fokker, the Super Universal had a range of 675 miles (1080 km), a higher ceiling than the Universal (18,000 feet, compared with 11,500 feet) and a Pratt & Whitney Wasp engine of 420 horsepower—double the power and endurance of its predecessor.

Other businesses were also realizing the potential of commercial aviation. On May 5, 1928, the Canadian Pacific Railway started the first air express service in the country. It promised delivery of packages taken from a ship in the St. Lawrence River to Montreal, Ottawa and

Toronto—on the same day. At noon, goods were transferred from a CPR delivery van in Toronto to an aircraft bound for Montreal's St. Hubert Airport. For the inaugural run, part of the shipment was put aboard another plane for Rimouski, where it would be picked up by the outgoing Canadian Pacific liner *Empress of Scotland* for Britain. The remainder of the shipment was displayed in the windows of Montreal's T. Eaton store—before the store had closed for the day! On May 6, the first air express arrived in Toronto with packages that had been taken from the Canadian Pacific ship *Empress of Britain* that very day. The beginning of the courier industry in Canada gave the CPR executives at Montreal's Windsor Station some food for thought.

On May 1, 1928, Richardson expanded WCA's reach from Winnipeg to the West Coast by buying Donald MacLaren's Pacific Airways and its Boeing and Vickers flying boats. More importantly, however, WCA took over his federal fishery-patrol contracts. More pilots were hired in the summer, including C.H. "Punch" Dickins from the Royal Canadian Air Force, G.A. "Tommy" Thompson[10] from the OPAS, Harry Marlowe Kennedy from the Manitoba Government Air Services, and a number of men from bush outfits—Matt Berry, Walt Fowler, Herbert Hollick-Kenyon and C.M.G. Farrell. Others, like "Wop" May, Stan McMillan, Archie McMullen and Jack Moar, would join from competing carriers. Like most pilots of their generation, they had seen action with the Canadian army or the Royal Flying Corps; the aircraft engineers Tommy Siers and Sammy Tomlinson had also learned their skills in the military.

Walter Gilbert's first aerial experience predated them all. As a boy in Cornwall, Ontario, he had been taken to the famous Aerial Meet held at Lakeside, near Montreal, in 1910 and had gawked at J.A.D. McCurdy, Count Jacques de Lesseps and Walter Brookins. At a time when few believed in commercial aviation, all were given an opportunity by James Richardson to earn a living doing what they loved: flying or maintaining aircraft. All of those named would be awarded the McKee Trophy[11] for their contributions to Canadian aviation.

The greatest bush pilot of them all was Punch Dickins. He reported for duty at the WCA office at Hudson, Ontario, on January 3, 1928. Told to take a passenger to Gold Pines and Red Lake, he overheard the passenger say, "Well, I'm not flying with a young bugger who has never been to Red Lake in his life and looks as though he isn't dry behind the ears." With that the speaker, a heavy-set man, took a bottle out of his parka and tilted it up for two large gulps. Dickins replied, "I'm just as

The historic WCA Fokker Super Universal G-CASK in 1928. Beloved by Punch Dickins, Leigh Brintnell, Walter Gilbert and Francis Roy Brown, it was eventually destroyed in 1933 at Fort McMurray by an engine fire.
Schade family photo

anxious to get to Red Lake as you are. In fact, I want to make my living doing so. Now get in and let's go." Expecting an alcohol-flavoured blast, he was surprised to hear the infamous Jack Howey, owner of the Howey mine at Red Lake, say meekly, "Well, I'll be damned," and get in.

Initially busy distributing cargoes to various settlements throughout the Gold Pines area, Dickins was sent to the Fokker factory in New Jersey to pick up a Super Universal G-CASK, an aircraft destined to serve WCA and Canadian aviation well. With it, he started a series of exploratory flights from Fort McMurray, Alberta, down the length of the Mackenzie River, with the idea of drumming up business along the way. On March 6, 1929, he became the first commercial pilot to cross the Arctic Circle, 10 miles (16 km) north of Good Hope. Appropriately, on Dominion Day he flew the length of the river, landing at Aklavik.[12]

Dickins later wrote:

I recall vividly the end of one day's flying on an operation which had taken me to Aklavik. Standing on the muddy bank of the Mackenzie River in the light of the midnight sun, tired but happy, I thought of that doughty explorer, Alexander Mackenzie. He and his party of six Indian paddlers in two birchbark canoes had taken months to reach the Arctic Delta that now bears his name. Had he

Pilot C.H. "Punch" Dickins, wearing gloves, and engineer Mickey Sutherland, kneeling, on their return to Edmonton from Fort Good Hope, NWT, bringing the first airborne bales of fur out of the North, March 1929. In the same aircraft but equipped with floats, Dickins flew to Aklavik. He arrived on July 1, 1929, becoming the first airman to reach that Arctic post. WCAM/F. Ellis Collection

stood there, on the very bank on which I was standing, 140 years before? My feelings were a mixture of humility and pride as I contrasted . . . his ordeal with my flight from Edmonton, covering the nearly 1,800 miles in only two days. I was convinced that air service would be a permanent part of the future of this north country, changing forever the lives of the people who lived and worked here.[13]

In the summer of 1928, several WCA pilots pioneered flights into the far North. On August 12, A.H. Farrington was ordered to take his Fokker from Winnipeg, fly along the west coast of Hudson Bay to Eskimo Point and pick up a prospecting party. The trip of more than 1,000 miles (1600 km) was the first in the Hudson Bay area.

True to his ambition of using aircraft to explore the inaccessible North for minerals, in 1929 Richardson was persuaded by a prospector,

Gilbert Labine, that there were large mineral deposits in the Great Bear Lake region of the Northwest Territories. W. Leigh Brintnell was flying G-CASK to the Yukon, and Labine was allowed to hitch a ride with him to Great Bear Lake. When Dickins picked him up three weeks later, Labine was very excited about the discolouration of rocks in the area, recognizing them to be pitchblende. He returned in March 1930, confirming that he had discovered the richest deposit of pitchblende in history. His mine produced silver, and later radium and uranium. The Great Bear Mine, as it came to be known (and the claims of the other four thousand prospectors who followed) was 800 miles (1280 km) from the nearest railhead. Since all supplies had to be flown in, WCA was provided with a steady business for years to come.

As Richardson had hoped, mining and bush flying complemented each other. During the development of a strike, while it was uncertain whether a "find" would become a paying proposition, the cost and delay of inroad or railway construction were prohibitive and not justified. This was an occasion when an aircraft was ideal. There was little "start-up" investment, and if the mine proved non-productive the plane could be withdrawn immediately. Freight, in the form of mining equipment and supplies for the camps, became the carrier's main source of revenue. WCA's freight hauling, for example, increased from 800 pounds (360 kg) in December 1926 to almost 40,000 pounds (18,000 kg) three years later.

No one wrote more eloquently of the effect of the introduction of WCA air service to the North than C.M. Bourget, the Indian agent at Resolution in 1929:

> I beg to express my appreciation and feel in doing so I am the interpreter of all the residents of the north, missionaries, traders, Government officials, trappers and even natives. We had the pleasure of receiving our winter mail accumulation which was previously held at McMurray or transported by horses as far as Smith but no good to us because it lay there until open water . . . without result . . . We received more mail in January than we had in the last four winters combined. Every winter I have to visit only once . . . Providence and Hay River, a 150-mile [240-km] trip with dogs at a cost of $400 with misery during the 12 days the journey takes of travel. This year I made it in three hours, round trip at a cost of $135 including excess baggage . . .

The real test of the company's ability to function as a carrier in the Far North came in 1929 with the MacAlpine expedition. The president of Dominion Explorers, Lieut. Col. C.D.H. MacAlpine, planned an overly ambitious mineral exploration through the Canadian Arctic. The year before, Dickins had flown MacAlpine and Richard Pearce, the editor of *The Northern Miner*, using G-CASK on a similar trip, covering 4,000 miles (6400 km) in twelve days. This time, to accompany his own Fairchild FC-2W-2, CF-AAO, MacAlpine leased G-CASK from WCA. "Tommy" Thompson served as pilot and A.D. Goodwin as engineer. Fuel caches were vital; the average range of an aircraft was 600 miles (960 km). Here, however, they were non-existent, so MacAlpine had arranged to rendezvous with the supply schooner *Morso* in Churchill harbour before proceeding to the Arctic coast to resupply.[14] Neither aircraft was radio-equipped and both were on floats because it was thought that the trip would be completed before "freeze-up," which customarily began in October. The seven-man party consisted of mining engineers and, once again, Pearce, who kept a diary of the ordeal.

The two aircraft left Winnipeg at 10 a.m. on August 24, flying north to Hudson Bay. Smoke from forest fires on the way lessened the visibility, and both aircraft arrived at Norway House only after becoming lost and flying in circles for some time. Fuel ran dangerously low, and it must have been with some relief that they bumped the aircraft down at Churchill harbour two days later and moored offshore to await the *Morso*. The next day, however, two of the schooner's lifeboats were rowed in, filled with the *Morso*'s crew. They had abandoned ship when it caught fire and blew up (it had been carrying a cargo of dynamite).

Hardly had MacAlpine and his party come to terms with this when, that night, the Fokker dragged its anchor and was swept into the sea by the tide. Its floats were torn open by jagged ice, and the aircraft sank below the surface, damaged beyond repair. The WCA office in Winnipeg was radioed, and Francis Roy Brown flew G-CASK out on September 6. Both machines were loaded up and continued on to a fuel cache at Baker Lake, where another passenger was taken on. Here it was discovered that condensation in G-CASK's fuel tank resulted in frost particles in its radiator, and this again delayed the party. They eventually flew onward and had passed Pelly Lake when ice storms developed and the temperature began falling alarmingly. The pilots realized that the lakes that were to be their landing areas would start to freeze over earlier than expected. Faced with the prospect of being trapped overnight in the ice, they decided to make for the open sea of the Arctic coast and land at one of the coastal settlements, from where a ship could be hailed.

There was less than two or three hours' fuel in the aircraft when the coast appeared. When a low ceiling forced both aircraft down at Dease Point, on the mouth of the Koolgaryuk River, less than three pints of fuel remained in each machine—and there was no fuel cache within hundreds of miles.

The party realized they were marooned until help arrived and prayed that it would come before winter set in. But they were reassured that it was still early in September, with a month remaining before the freezing temperatures were to begin. While waiting to be rescued, the crew built a small house out of stones, moss and parts of the Fairchild's wing. An engine cowling served as a stove. On September 12 they tried to get G-CASK into the air with the remaining fuel, but once more its carburetor gave trouble. The party's rations were supplemented by fish and caribou from curious Native hunting parties who also went off to find help. They advised against travelling on foot to the nearest settlement before freeze-up, when the ice would be strong enough to support the men's weight over the shoreline.

By October, when the first of the snowstorms struck, there were still no rescue aircraft to be seen. But the temperature had not dropped below freezing long enough for ice to form on the sea. Their sorry predicament caused Thompson to marvel: "Through this raging blizzard two Eskimos travelled for five hours to get back to us with fresh caribou meat . . . This incident gives an insight into the sterling qualities typical of the Eskimo character, the obligation of humanity and a promise to fulfill—blizzard or no blizzard." The nearest settlement was a Hudson's Bay Company post at Cambridge Bay, estimated by the Natives to be three days' and two nights' walking distance—but only if the shoreline ice was strong enough to bear their weight. Food rations were almost gone, and the Native diet of dried fish, while generously shared, gave the white men severe cramps. Yet the weather refused to turn colder, and all took to bed in the hut to conserve their energy.

Not until October 18 was the ice judged thick enough by the Native guides for the party to cross along the coast. Two days later the men and their Native guides began the trek to Cambridge Bay. On October 24 they made it to a point on the Kent Peninsula where, across the Queen Maud Gulf, lay the outpost of Cambridge Bay. By now they had

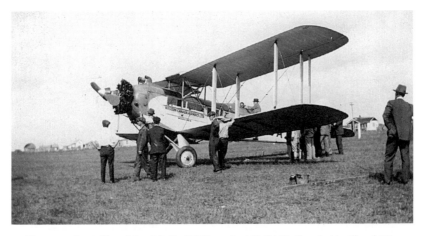

The very short life of the Moth: WCA rented G-CAJT, the de Havilland 61 "Giant Moth," from Canadian Vickers in September 1928, only to have it crash and burn at Calgary the following month. Schade family photo

run out of food, so they rested while some of the Native guides returned to Dease Point for more. When they returned with news that an aircraft had flown over the hut, the MacAlpine party was encouraged.

Both WCA and Dominion Explorers had mounted aircraft searches for the party, despite the imminence of the fall freeze-up, when flying was not usually possible. Floats could be used early in the search, but it was understood they would have to be replaced by skis in October. Two teams were organized: a main one, with five floatplanes to operate as far north as they could, and a second group that would ferry fuel and supplies to the main group. Brintnell was put in charge of the overall search. He, Dickins, Francis Roy Brown,[15] Andy Cruickshank, Bill Spence and Herbert Hollick-Kenyon[16] searched the vast Mackenzie District, flying as far west as Fort Norman and Coppermine, and as far east as Baker Lake. Combing the wilderness through October and into November took its toll on men and machines, and two aircraft were temporarily abandoned until the following spring, when one was written off.

It was a race against the winter as more of the inland water froze below the searchers daily. The crews had to be very careful not to be trapped in it wherever they set down overnight. By October 13, all the lakes were frozen and the conversion to skis had begun. Tommy Siers, the WCA maintenance manager, used Baker Lake as a base. All five aircraft were hauled up on shore and fitted with skis. A severe gale lashed the thin ice against the beach on October 17, damaging the aircraft, and

WCA's Winnipeg base, around 1930. The original buildings on the right date back to 1927; the building in front with the crane was added later. The King George Hospital is recognizable in the background. Because of its size, the airfield was rarely used for landings and takeoffs; the Red River served better with floats and skis. In 1928 Stevenson Aerodrome, named after WCA pilot Fred Stevenson, opened in the municipality of St. James. WCAM (UNN)

not until October 24 were four serviceable enough for the search to continue. This time the searchers reached the coast and landed their aircraft at a fuel cache on the frozen Burnside River. Three made it down safely, but Cruickshank, in G-CASQ, misjudged the strength of the ice and plunged through, settling the aircraft's nose down in the water with only its wings showing. The plane was abandoned until the MacAlpine party could be found, and the other aircraft pressed to the Arctic Ocean, flying over the now-empty hut at Dease Point.

Back at Burnside, in subzero conditions and during a cruel blizzard, Siers and his crew had G-CASQ raised by block and tackle in order to work on the engine. Siers kept it suspended until the water had refrozen, then overhauled the engine. The aircraft was ready to resume the search when, on November 5, a dog team arrived with the news that the MacAlpine party had been found.

The schooner *Fort James* at the magnetic North Pole had intercepted a message from Roald Amundsen's old ship, the *Bay Maud*, which was being used as a Hudson's Bay Company supply depot at Cambridge Bay. On November 3, in –27°F weather, the MacAlpine party had staggered into the post, starving and suffering from frostbite. The four rescue aircraft landed alongside the *Bay Maud*, and the weakened men were flown out in relays. Because of the ferocity of the winter, not until December 4 was the entire MacAlpine expedition party landed at the railway station at Cranberry Portage for the train to Winnipeg. MacAlpine, knowing that the party owed their lives to the Native guides, made sure that all the guides they had met were rewarded with money, rifles, ammunition, tobacco and sewing machines.

Although the sunken G-CASP was insured, WCA made little money on these epic flights. The extensive use of its aircraft and personnel over the four-month period of the expedition meant a loss of revenue elsewhere. Financial considerations apart, however, much of northern Canada had been flown over for the first time, and the heroism of WCA's pilots and engineers—who, as Brown said, "were just doing their jobs"—is worthy of note. No lives had been lost. As with bush pilots before and since, Hollick-Kenyon and Dickins survived by thinking well ahead of their airplanes at all times and by seldom venturing into a situation without first devising a plan to get out.

The pilots employed by WCA and later Canadian Airways Ltd. altered the significance of Canada's geography. They navigated uncharted regions in a harsh, cruel climate for little recognition and less profit, and they transformed the sheer novelty of flying into a profession. Their reticence was never better illustrated than in the following excerpt from a WCA file:

> While flying an injured man to hospital, which was over fourteen hundred miles [2240 km] away, one of our pilots had the misfortune to be forced down out of gas forty miles [64 km] from his destination. Adverse winds and a temperature of –50 degrees F. didn't help. A newspaper reporter in the area later wrote: "Sustained only by a bar of chocolate and using a pair of snowshoes, the WCA pilot walked the forty miles in 24 hours, loaded some gas on a dog sleigh and brought his machine out.

The pilot's log of the same incident read: "Forced down out of gas— found emergency rations not up to standard, especially when the passenger was helpless. Took appropriate action."

Unfortunately, piloting ability and navigational skills were no match for the upheaval to come. For the enemy this time was not the weather or frozen engines or unmapped wilderness, but that root of all business problems: money.

Like the DC-3 two decades later, the Junkers played a pivotal role in Canadian bush aviation. It could carry anything—from bales of hay, sled dogs, canoes (on top and under) to fuel caches—and was equally at home on pontoons or skis. Schade family photos

CHAPTER TWO

"In Years Gone By
I Used to Fly a Fairchild Eighty-Two... "[1]

July 26th: Charter trip to Pagua Lake, ENE of Pickle via Badesdawaga and Kakagiwizia Lakes, Ozheska River and north to Ishanakie and Pagua—camped at Paaonga Lake. —Jarvis.

July 27th: Pakwash Lakes very shallow making it difficult for water transportation. —Farrington.

July 30th: First plane to land at Great Bear Lake from Fort Norman, landing at Bolands Trading Post at the head of Bear River and going on to Deer Pass. The Lake is large and roughly shaped like a star fish. There is practically no shelter on the north shore, and the shoreline is rocky. —Dickins.

August 4th: Between McMurray and Lac La Biche, on the Athabasca River, poor places to land due to rocks and rapids. Telegraph wires across river about 20 ft. off water at Pelican Settlement and opposite Calling River; at each end they are 60–100 ft. high. —Farrell.

August 9th: Taking Dr. Watson to hospital at Grand Prairie accompanied by Indian Agent. Hudson's Hope to Saskatoon Lake and return—300 miles. —Calder.

August 25th: Caution—Shoal Areas, Emma Lake, marked red buoys. —Kenyon.

The WCA log books for 1928 provide a dramatic window on aviation in

A pair of WCA Fokker Super Universals at an unknown lake: G-CASN
would be one of the aircraft burned in a fire at the Stevenson Field
March 4, 1931 and a week later "G-CAIX" crashed at Elk Lake, Ontario.
Schade family photo.

remote bush communities. In an era when the brief aerial odysseys of
Charles and Anne Lindbergh, Wiley Post and Harold Gatty over the
Canadian North grabbed the headlines, the day-to-day flights by WCA
bush pilots were equally dramatic, not only because "of their immediate purpose, but also . . . for the light they cast on the venturesome spirit of the period."[2]

But the records of such flights ignore the co-operative effort that
went into preparing for them. Purchasing maps, calculating distance,
estimating the hours of daylight to fly in, obtaining weather forecasts
by radio or telegraph and loading the aircraft all had to be accomplished before the pilot could take off. Most important was the placing
of fuel caches in forty-five-gallon drums at strategic locations. The few
maps of the Arctic coast that were available had been drawn for sailors,
so the maps' geographic features petered out a few miles inland. The
proximity of the magnetic north made magnetic compasses unreliable,
and sun compasses were useless in low cloud.

Such flights, providing services to trappers, missionaries and
prospectors, were insufficient for James Richardson's finances—especially during the winter months, when prospecting dropped off.
Depending on these flights for much longer would have bankrupted
WCA as surely as it had Laurentide. The revenue generated by them

never paid for more than the company's operating expenses, allowing nothing for the upgrading of aircraft or route expansion. When Col. R.H. Mulock,[3] a fellow board member at Canadian Vickers, let Richardson know that "friends in the Post Office" were looking for a company to run an air mail service on the prairies, he saw a chance for financial stability. Rural mail delivery of any kind in Canada dated from 1908, the year that the first aircraft was flown in the country.[4] Before passengers became commercial aviation's mainstay, air mail contracts were much sought after. Unlike the cyclical traffic generated by carrying frontier passengers, transporting the King's Mail provided a steady income for air carriers without the bother of looking after passengers.

For Richardson and WCA's other investors, the timing was propitious. In the summer of 1927, Postmaster General P.J. Veniot had been caught up in the aviation hysteria that had swept the world after Charles Lindbergh's solo flight across the Atlantic. Ottawa could hardly be immune; not only did "Lucky Lindy" visit the nation's capital to inaugurate the opening of the Parliament buildings—captivating everyone from the prime minister down—but the Imperial Conference was to be held in the city that fall. With the eyes of the world on Canada, the private and public sectors clamoured for air mail services, and Veniot awarded the first contracts.

The most imaginative of the contracts was given to Canadian Airways. Using an HS-2L, H.S. Quigley was to collect mail from transatlantic ships as they entered the St. Lawrence River off Farther Point. When the transfer of bags from ship to seaplane proved too unwieldy, the bags were thrown onto the pilot boat, which took them to a government-built airfield at the Rimouski pier. From there a Canadian Transcontinental Airways wheeled aircraft flew the bags to St. Hubert, the only other federal airport, near Montreal. The mail was then distributed for flights to Ottawa, Toronto and Albany, New York. When the aircraft returned, the transatlantic mail was collected at Toronto, Montreal, Ottawa and Quebec to catch the ship before it reached Rimouski. This flurry of activity to deliver first-class and Imperial Conference mail cut almost a week off the usual delivery time from Britain.[5]

Once they were flown to Toronto, however, the letters destined for points farther west went either by rail across Canada or across the border to Buffalo, where the US air mail system picked them up. Although the Post Office's decision to help the young commercial aviation industry by spreading contracts among as many air companies as possible was well intentioned, critics claimed that it took advantage of the carriers' desperation for business by playing them off against each other and forcing them to offer the lowest rates. In the Post Office's defence, it should be noted that there was no official policy regarding the use of aircraft to deliver mail. So experimental was the whole service that the Post Office had neither issued air mail stamps nor negotiated an air mail treaty with the United States.

The success of the maritime air mail service for the Imperial Conference prompted Veniot to ask several air companies to submit bids to carry the mail over the winter of 1927–28, most notably from Moncton to the Magdalen Islands, La Malbaie to Anticosti, and Leamington to Pelee Island. A number of firms—Canadian Transcontinental Airways, Western Canada Airways, National Air Transport, Canadian Colonial Airways, General Airways and Compagnie Aérienne Franco-Canadienne—tendered for contracts, knowing that their futures depended on being awarded one. WCA did receive a contract, and on October 6, 1927, Fred Stevenson flew WCA's first air mail from Lac du Bonnet to Long Lac, Manitoba, a two-hour journey that previously took four days on foot.

Thus encouraged, Richardson applied to the Post Office on March 28, 1928, for a contract to fly the mail at night between Winnipeg, Regina, Moose Jaw, Medicine Hat, Lethbridge and Calgary, with a branch line flying to North Battleford and Edmonton. To strengthen his case he wrote to certain members of Prime Minister Mackenzie King's cabinet, asking them to exert pressure on his behalf. The heightened interest in the awarding of postal delivery contracts, always a contentious item in Parliament, led to an investigation of the Post Office's methods. In June the leader of the Opposition, R.J. Manion, questioned Veniot in the House of Commons as to how the tendering was done: although Canadian Transcontinental Airways had bid $57,000 against Fairchild Air Transport's tender of $40,500, it had been awarded the Montreal–Rimouski route. Established in Canada since 1923, Fairchild had done sterling aerial photography for the government, photographing much of the country. Nevertheless, the federal government claimed that since owner Sherman Fairchild was an American, his company was ineligible to bid for federal contracts.

As one of Fairchild's investors, James Richardson was disappointed. He did, however, find some solace in the news that on June 25 WCA

Much was expected of the ungainly Fokker F.14s bought for the mail contracts. CF-AIJ, seen here in 1929 at Blatchford Field, Edmonton, before partially built Hangar No. 1, was wrecked in a forced landing in the fog later that year. CAA 988-136-4

had been awarded the prairie air mail contract for an unprecedented period of four years. This was a commercial coup; although air companies in eastern Canada enjoyed higher rates for carrying the mail, their contracts were negotiated annually. But, to Richardson's disappointment, mail delivery along the Mackenzie River was given to Wilfrid Reid "Wop" May's Commercial Airways Ltd. of Edmonton. May would achieve fame on January 3, 1929, for flying antitoxin in bitter cold from Edmonton 600 miles (960 km) north to Little Red River during a diphtheria outbreak.

On September 13, Richardson incurred the Post Office's disfavour when he announced that WCA would carry passengers as well as mail between Regina, Calgary and Winnipeg. Post Office officials felt that the pilot might be unwilling to take necessary risks if passengers were aboard. However, since it could not back out, on November 16 the department stipulated that the WCA run was to be on a trial basis for the 1928 Christmas season, from December 10 to 29. It was a bittersweet victory for Richardson; he knew that all the other carriers had been awarded contracts without having to perform trial flights. In trying to force the Post Office's hand, Richardson had made enemies in that department, something he would later regret.

The indestructible, all-metal Junkers W-33/34 was the best aircraft that Canadian Airways Ltd. ever owned. It achieved fame in 1928, when W-33 *Bremen* became the first aircraft to make an east–west crossing of the Atlantic Ocean. All four of these aircraft remained in service from 1931 to 1946. CAA (UNN)

For the inaugural air mail flight on December 10, from Winnipeg to Regina and back, pilot W.J. Buchanan had Premier John Bracken of Manitoba along for the ride. The event was held in daylight so that it could be recorded by movie cameras. Bad weather hampered G-CASK from getting through the whole route for ten of those December days, but in all a creditable 3,152 pounds (1418 kg) of mail were carried. The Post Office seemed satisfied with the results and tentatively agreed that WCA be given its four-year contract, to begin on December 2, 1929.

For this and future mail runs, Richardson placed an order for six F.14s from Fokker and hired more staff. The ungainly F.14s were the first non–bush aircraft that WCA had ever bought. Conceived by Anthony Fokker to compete against Ford's Trimotor for a US Army requirement for ambulance aircraft, the F.14 betrayed his European roots. The pilot, still in an open cockpit, sat at the rear of the fuselage (current thinking was that he was more likely to survive a crash in this position), and the wing was raised in parasol style to give him some visibility. Almost as fast as the Super Universal, the F.14 suffered in performance and reliability in comparison with WCA's Boeings and Junkers. But it could take eight passengers in the comfort of a closed cabin—with a small washroom in the rear. As W. Leigh Brintnell noted, the high winds made it necessary to provide for flights of up to seven hours' duration; thus the toilet was essential.

By the end of 1928 WCA owned twenty-seven aircraft, a great increase over the six it had started the year with. The company's planes had flown 545,193 miles (872,309 km) in 6,873 hours and carried 1,192,646 pounds (542,112 kg) of freight and 9,648 passengers.

The federal government, which had confined its investment in civil aviation to the building of two airports—Rimouski and St. Hubert—had in 1927 begun to subsidize flying clubs. Within two years there were twenty-four such clubs across Canada, each charged with maintaining its own airfield. Since flying clubs were usually near urban areas, the next step was to designate them as municipal airports and link them with an "airway," which was equipped with revolving acetylene beacons, radio aids and meteorological services.

On February 22, 1929, the Department of National Defence was ordered to construct a network of rural airports that would be linked by beacons from coast to coast. The Trans-Canada Airway, the nation's most extensive construction project until the building of the Distant Early Warning (DEW) line two decades later, was the brainchild of J.A. Wilson, the controller of civil aviation, and the army chief of general staff, Maj. Gen. A.G.L. McNaughton. McNaughton justified the expense to the government by maintaining that not only did the Airway serve civil interests, if there were ever to be a war in the Pacific, aircraft could be flown from the main RCAF base at Trenton, Ontario, along the Trans-Canada Airway to reinforce squadrons in British Columbia. Whatever McNaughton's agenda, nothing gave more impetus to commercial aviation than the ten-year construction of the Airway. Ironically, its speedy completion was catalyzed by unemployment relief programs during the Great Depression.

In a portent of future relations with Richardson, the Post Office attempted to delay the prairie air mail contract. Richardson suspected that Sir Henry Thornton,[6] president of Canadian National Railways (CNR), was behind the attempt. Since Richardson was also a director of the Canadian Pacific Railway (CPR), Thornton knew that it was only a matter of time before his rival, already involved with its air express service, would be further drawn into aviation. By far the most powerful, wealthiest entity in the Dominion, the CPR in 1919 had Parliament alter its charter to allow for the operation of aircraft "for the carriage and hire of passengers, mails, express and freight between such points and places, within or without the Dominion of Canada." Once this imaginative step was taken, however, the company ignored the potential of commercial aviation for another decade.

Thornton's alleged involvement notwithstanding, Veniot may have regretted awarding so large a contract to a single carrier. The actual signing was delayed until October 18; the reason given was that because it involved flying the mail at night, Wilson's lighting beacons and emergency landing strips had to be operational and then the routes had to be checked over by air force survey flights. Now paying for an idle staff, extended insurance coverage and new aircraft, Richardson offered to have WCA do the survey flights without charge. This offer was refused and only a portion of the route was offered. By late August 1929, there were lighted airfields at 30-mile (48-km) intervals between Winnipeg, Edmonton and Calgary. Although the flat terrain of the prairies made this the easiest portion of the whole Airway to build, federal bureaucrats dragged out the granting of permission until March 3, 1930, when WCA was allowed to fly nightly air mail runs between Winnipeg and Edmonton, with stops at Regina, Moose Jaw, Medicine Hat and Calgary. It had been a long, acrimonious fight to get the prairie air mail running, but at least now Richardson could look forward to a steady income from the mail runs for the next four years.

For an air company to demonstrate to the world that it was more than a rough bush operation, an official emblem was *de rigueur*. The company colours had evolved gradually: the first three WCA Fokkers, named after Toronto, Winnipeg and Churchill, were finished in aluminum, with their cowlings painted dark blue. In 1928 the colours were standardized throughout the fleet with royal blue fuselages and yellow wings, elevators and stabilizers. The aircraft also had the manufacturer's signature on the rudder. Usually it was "Fokker"; "Western Canada Airways, Winnipeg, Manitoba" appeared on the fuselage.

Aircraft logos are the heraldry of aviation, symbolizing an airline's values and aspirations. Juan Trippe's Pan American Airways, for example,

James Richardson chose a Canada goose as a logo because "it represented nationality, strength of wing, regularity and organized flight." The design is attributed to Leonard Watson, the father of a future Pacific Western Airlines president, Don Watson. CAA 989-169-6

sported a blue globe, although at that time it flew only as far as the Caribbean. KLM used the "Flying Dutchman" logo, and Luft Hansa had a black crane on a yellow background.[7] The CPR trademark was a beaver, chosen because it was amphibian—neatly tying in the shipping branch—but also because of its industriousness and uniquely Canadian origin. (Some in the Grand Trunk Railway said it looked like a rat.)

To symbolize his airline, Richardson wanted a bird native to Canada. Not only would it reassure WCA's customers with its airborne endurance, but when WCA's international routes began it would do for Canada what Pan American's globe or Luft Hansa's crane had done for their countries. He sought the advice of a professor of biology at the University of Manitoba, who mentioned that the Canada goose might suit: it flew vast distances and had an infallible sense of finding its way back home. It was "noble, game, organized, recognized a mile high by its orderly formation, its steady progress and its northern destiny. To the understanding, it represents nationality, strength of wing, regularity and organized flight." These were exactly the qualities that Richardson wanted the public to identify when thinking of his airline, so a Canada goose in flight was chosen.

The first issue of the company magazine *The Bulletin* (July 1929) featured a flying Canada goose on its cover. The design is attributed to Leonard Watson, a wildlife artist, who was the father of Don Watson, who later became president of Pacific Western Airlines. Serving a staff of 120 (25 pilots and 95 ground personnel) scattered at offices, air harbours and airfields from the British Columbia coast to western Ontario, from the mouth of the Mackenzie River to Winnipeg, the company

In 1928, Wilfred Leigh Brintnell was made superintendent of WCA's base at Hudson, Ontario. He left in 1931 to form Mackenzie Air Services.
WCAM 17082

magazine kept everyone informed of official policies and personnel movements and would later (naturally) be renamed *The Honker*.

The company ended the decade with fifty-one aircraft, which included Universals, Fairchilds and Boeing flying boats (based in British Columbia) as well as the all-metal Junkers W-34s. The W-34s were rugged, cantilever low-winged monoplanes, suited to rough terrain because of their corrugated metal outer skin. When a Junkers W-33, *The Bremen*, made the first east–west crossing of the Atlantic on April 13, 1928, Richardson was impressed enough to buy eight W-34s (which were similar to the W-33s), despite a heavy import tax imposed on them by the government. By WCA's (and later Canadian Airways') own account, the Junkers were the best machines they had ever owned.[8] One benefit of such low-winged aircraft, pilots wryly noted, was that when they fell through the ice, as they usually did during the spring thaw, no one got wet.

WCA's first nightflying attempt was inadvertently carried out on a cold, clear night in January 1929 by Art Schade, recently graduated from WCA's flying school. Ferrying an unheated de Havilland floatplane

The Fairchild 82 was flown by Canadian Airways Ltd., Wings Limited and Starratt Airways. The only survivor, CF-AXL, was refurbished in 1965 by Canadian Pacific Airlines employees in Vancouver and presented to the National Aviation Museum in Ottawa. Although veteran CPA pilot Bob Randall, shown here, offered to fly it to the museum, the Fairchild was delivered by flatcar. CAA (UNN)

from Emerson to Winnipeg, he delayed at Portage la Prairie for a cup of coffee and took off after dark but in the light of a bright moon. Schade could not recall anyone ever having flown at night but did not see why he should not, particularly in such fine weather. He followed the silver ribbon of the Assiniboine River to Winnipeg and arrived over the city at about 9:30 p.m. He was so enthralled by the street lights below that he circled the city a few times before setting down at the WCA docks at the foot of Brandon Avenue. There he was greeted by a crowd of people who, having heard the aircraft overhead at night, had come out to see the crash. Among them was Leigh Brintnell, who was not at all amused and suspended Schade for three days for flying after dark.

When regular nightflying did begin, the WCA pilots discovered that the Post Office was correct to have been cautious. Westbound flights left Winnipeg at 9 p.m., with twenty-minute stops at Regina, Moose Jaw and Medicine Hat, and arrived in Calgary at 5 o'clock the next morning.

This photograph of Windsor Station, Montreal in the 1950s conveys the Canadian Pacific Railway's baronial control of all forms of transportation and communication in Canada. CPA NS. 8416

Eastbound flights left Calgary at 2:15 a.m. and reached Winnipeg at noon. No one in Canada had ever flown regularly after dark; at a time when pilots navigated by sighting landmarks like water towers and silos or following the "iron compass" (the intercity railway line) before rural electrification, this meant venturing into total blackness. Even when the Trans-Canada Airway's new revolving acetylene beacons worked and emergency airfields could be found, it was a dangerous and nerve-wracking experience.[9]

The pilots complained that the beacons and runway lights were useless in blizzards, when their need was the greatest—maintenance crews couldn't get to them because the access roads were blocked by snow—so they came to rely on city street lights for navigation, bitterly comparing the whole experience to "flying in a black bag." Unable to depend on the Department of National Defence's navigational aids, carriers began building their own. Within six years, Canadian Airways Ltd. had twelve ground stations; when necessary, its pilots used American agencies for weather reports.

Since there were no government nightflying regulations, WCA

personnel worked out their own. This was prompted by the fear of a mid-air collision. In 1930, G.A. "Tommy" Thompson noted:

> With regard to machines passing each other between Medicine Hat and Calgary without seeing each other, I have only once seen the machine going in the opposite way, even in fine weather. This is taken care of . . . when the beacons can be seen by the machines keeping a few miles to the right of the beacons. In bad weather we have an arrangement whereby the Westbound pilot flies low and the Eastbound pilot flies high.[10]

Not until 1931 did pilots have the comfort of flying at night on a radio beam. By the end of the year, five radio beam stations had been installed by the military at Forrest, Regina, Maple Creek, Lethbridge and Red Deer. Despite the hazards, of the 4,864 scheduled air mail trips on the prairie route, 4,512, or 92.8 percent, were completed—a creditable score, considering that in the United States, with its established "airways," 93.9 percent of air mail trips were completed.

Calling WCA an "airline" conjures up visions of passengers enjoying meals and the attention of stewardesses, amenities that were familiar to anyone who flew commercially south of the border. There, Eastern Airlines recruited its first twenty-two female cabin attendants in 1930 and equipped its Curtiss Condor biplanes with upholstered reclining seats, toilets with hot and cold running water, carpeting and— a big plus—soundproofing.[11] In Canadian skies, however, apart from hardy Native trappers and prospectors, there were few passengers because WCA was primarily a freight hauler; its machines were considered "flying trucks." Although WCA carried 5,685 passengers in 1930, its aircraft were conspicuously lacking the comforts offered on American and European airliners. In the cabin, there were no upholstered seats or individual reading lights. Sometimes there were no seats, and when dog teams were carried, their owners crammed themselves into the cockpit with the pilot. And, as if the ear-shattering engine noise and the freezing, grimy cabins were not bad enough, the Post Office insisted that His Majesty's Mail be locked in the F.14's toilet compartment for the whole flight.

The company hadn't the facilities to cater to passengers in the air or on the ground. Its main base at Stevenson airport was still more clapboard than concrete, and the "passenger terminal" was a lean-to beside

Art Schade, recently graduated from WCA's flying school at St. Charles, with one of the two de Havilland 60 Cirrus Moths he would have trained on. Schade family photo

Nightflying, essential for mail delivery, was a hazardous occupation. This Canadian Airways Ltd. Boeing 40B-4, CF-AIN, bought on January 23, 1930, crashed at Southesk, Alberta, on September 25, killing the pilot and two passengers. WCAM 309

the maintenance hangar. "Punch" Dickins recalled that there were no paved runways at any of the airports he used. At Calgary the field used as an airport was near a park on the south side of the Bow River—on a 5° slope running toward the river. In the days when aircraft had no brakes, it was essential for the pilot to get the tail skid on the ground right after touchdown and then to hold the stick hard into his stomach, hoping that the drag would stop the plane before it went over the river-bank. Edmonton's airfield wasn't quite as bad, but Dickins remembered small clumps of willows and poplars between the runways that kept the pilot on his toes.

It was while trying to secure the prairie mail route that James Richardson reluctantly looked into the commercial aviation market in eastern Canada. Here, eleven small companies battled it out. The two largest firms, Canadian Transcontinental Airways and International Airways, had barely survived the prosperous 1920s. Canadian Transcontinental Airways and its subsidiary, Canadian Airways, flew between Rimouski and Montreal with mail from the pilot boats. Its other routes were from Rimouski to Anticosti Island and between Moncton and the Magdalen Islands. International Airways, located in Hamilton, had the route between Montreal, Toronto and Detroit, and its board of directors was well connected both politically and socially.

Even with the steady mail contracts, the airlines made no profit risking lives and aircraft over the stormy lower St. Lawrence or in southern Ontario, which was extensively served by railways and now automobiles. All of the eastern air carriers looked to the future with grim despair. Except one . . .

Despite its name, Canadian Colonial Airways (CCA) was owned by Aviation Corporation of Delaware and flew the mail between Montreal and Albany, New York. Although it had barred Sherman Fairchild from bidding for contracts, Ottawa treated CCA as a special case because it needed the access CCA gave to the US air mail system. To CCA's president, Maj. Gen. J.F. O'Ryan, the Montreal–Albany route meant that his airline was well placed to expand into the Maritimes and eventually to secure a "jumping off" spot for Atlantic air travel—which was, no doubt, very near. O'Ryan was not alone in his hopes.

In the last years of the free-flowing 1920s, when money seemed to be plentiful, Charles Lindbergh's flight across the Atlantic had the indirect effect of making the destitute central Canadian airlines look increasingly attractive to American aviation entrepreneurs. The maturing airline industry in the United States was going through a period of consolidation, fighting "turf" wars as ferocious as those in the bootlegging industry. By dangling air mail contracts before these companies, Washington forced several small one-aircraft outfits to merge, giving birth to future giants like American Airlines, United Airlines and Transcontinental and Western Air. O'Ryan had little doubt that the pickings in the Canadian aviation market would be easy. He saw a possible point of entry in the failing International Airways, and in December 1928 he had approached one of the directors, Victor Montague Drury. The brother-in-law of Max Aitken (the future Lord Beaverbrook), Drury was also a director of Canadian Car and Foundry, a railway-carriage maker that like Canadian Vickers wanted to diversify into aircraft manufacturing. An American connection like O'Ryan could open the doors to loans and shared technology, and Drury had purchased shares in International Airways to attract foreign investors such as O'Ryan.

It did not take Col. R.H. Mulock at Vickers long to hear of this and sound the alarm. He raised the spectre of commercial aviation in Canada being controlled from a boardroom in Delaware, and Drury was persuaded to hold out against the American offer until a local rescue package could be put together. The mood in Montreal financial circles was tense. If nothing was done soon, said J.A. Wilson, it would not be long before the Americans could "tap traffic in the main centres of population in Canada, all of which lie adjacent to the international boundary and feed it into the United States airway system." Wilson suggested that perhaps the time was ripe for the railway companies to be brought in to help with air transport.

Several businessmen from Hamilton who had shares in International Airways were also uneasy about the possibility of their air carrier falling into American hands. They persuaded James Richardson to join them. It wasn't just a matter of patriotism but the sure knowledge that if American entrepreneurs gained a foothold in Canadian aviation, these shareholders would be squeezed out of the market.

O'Ryan knew that Canadian Transcontinental Airways (CTA), with its St. Lawrence routes, was the key to establishing his company on the East Coast, so he began negotiating with CTA as well. Drury also realized CTA's importance, and on June 19, 1929, he merged it with Interprovincial Airways. A month later, the Aviation Corporation of Canada was formed as a holding company to purchase controlling

To give the airline a corporate identity, all CAL aircraft were painted with a dark blue fuselage and orange wingtips and were adorned with the flying goose insignia. CAA 989-169-7

interests in other small companies in eastern Canada. The air companies would keep their own names in order to retain their air mail, aerial photography and training contracts, but they now belonged to the syndicate. Drury then offered Richardson two fully paid shares in the corporation, and he and John Hunter were made directors. The Aviation Corporation of Canada bought 26,000 International Airways shares at $25 each, and Richardson took 60 percent of the outstanding CTA shares. Colonel Mulock was appointed managing director. His accounts show that the cash investment in the new company was $2,190,544.12, and that in 1929 the combined revenue of all of CTA's eastern air companies was $703,575.49. (By way of comparison, WCA's revenue in the same period was $793,451.76.) By the end of the

Sir Edward Beatty, the CPR's first Canadian-born president, began the railway's investment in commercial aviation. Early in World War II, he organized an air ferry service to Britain and tried unsuccessfully to interest the government in creating a single airline. CPA NS 17837

decade, besides owning flying schools and aerial survey companies, the Aviation Corporation of Canada controlled air mail contracts in the east, effectively thwarting an American incursion—for the time being.

Among the corporation's directors was the CPR's Edward Beatty. There is some speculation that Beatty entered the field of aerial transport because both his predecessors had carved out their own starry trails in the national transportation firmament: William Van Horne had joined Montreal and Port Moody, British Columbia with a band of iron, and Thomas Shaughnessy had expanded the CPR's operations into shipping. It is reputed that on his deathbed Shaughnessy spoke his last words not to members of his family in attendance but to Beatty, his chosen successor: "Maintain the property."

With such deep shoes to fill, the young president sought out his own destiny. He chose to invest not only in shipping but in his friend James Richardson's air company. But—perhaps fearful of what his shareholders might say—he kept his investment anonymous, asking Richardson to control his shares in the syndicate. Neither man could have guessed that this was to be the start of the CPR's six-decade relationship with aviation in Canada.

Although Beatty's wish was respected, his participation in WCA soon reached the ears of Sir Henry Thornton, who called on Richardson in Winnipeg to tell him that Canadian National Railways would like to participate in any future expansion of WCA's air mail activities. The railway president was revelling in the recent success of the CNR's shipping service to the West Indies; in 1925 he had single-handedly modernized the government merchant marine (which was part of the CNR) by building a fleet of "Lady" ships[12] that would, like Trans-Canada Air Lines' DC-8s forty years later, add considerable prestige to the railway.[13] He had no desire, he pointedly informed Richardson, to begin his own air company but, like the CPR, he wanted to be part of the twentieth century's most innovative form of transport.

Aware that refusing Thornton might force him to organize his own air company—which, as a government-owned corporation, would be entitled to take over all the mail contracts—Richardson agreed. He was pleased that both railways were equal partners in the corporation. For the government-owned CNR to participate, however, Thornton required the approval of the governor-in-council, and the bureaucratic process necessary to grant this dragged on through the summer and into the new year.

The effects of the stock market crash on October 24, 1929, meant that money for propping up risky ventures like airlines evaporated overnight. With more established industries shutting down, investors looked to unload their stock in the Aviation Corporation of Canada as soon as they could. Drury, who now must have regretted not accepting O'Ryan's offer, feared for his aviation investments and asked Richardson to sell WCA to the syndicate. Since WCA was the largest, most promising air company in Canada, it was felt that the company would survive what was then considered a temporary downturn in the economy and that it could carry the eastern airlines with it. Richardson, who had also been approached by American syndicates, declined; he was waiting for Thornton to get federal approval. Before the effects of the stock market crash could be properly judged, he had every reason to feel optimistic.

Four years after seeing Jack Clarke's flying boat take off at Minaki Lake, the Winnipeg grain entrepreneur could look back on his achievements with some pride. He had put together the second-largest airline in the British empire. Only Britain's Imperial Airways had more aircraft, staff or routes, but Imperial was Britain's "chosen instrument"; it had a monopoly on air travel to the colonies, and its annual deficits were forgiven—something that other air carriers could only envy.

Buoyant with ambition, Richardson sent the company secretary, John Hunter, to London in September. His "right-hand man" was to study the infrastructure of Imperial Airways and learn from its general manager, George Woods-Humphrey, how WCA might assume respectability as Canada's "chosen instrument in the air."[14]

Some hint of the amalgamation of the Aviation Corporation of

Wireless communication, which became more common by 1930, was a mixed blessing that allowed bush companies to eavesdrop on each other's frequencies and steal traffic meant for competitors. CAA 988-10-4j

Canada and WCA appeared in the press in the spring, but since Thornton had not received permission, nothing happened until November. The name "Canadian Airways, Ltd." was chosen on June 27; the comma was added to distinguish the name from the old "Canadian Airways," which became "Canadian Airways Ltd." (CAL in common usage). Richardson sold WCA to Canadian Airways Ltd. on November 19. He received 60,000 shares, which, together with his stock in the Aviation Corporation, gave him a controlling interest of 72,672 shares out of the total 127,088. The CNR and CPR each put in $250,000, and the company was capitalized at $3,177,200.

In order to demonstrate to Ottawa and American financiers that the new company was well connected and in a sound financial position, it chose a powerful board of directors. Richardson was appointed president of CAL, and Thornton of the CNR and Beatty of the CPR became vice-presidents. The presidents of the Royal Bank, Bank of Montreal, Banque Canadienne Nationale, Consolidated Mining, Willys Overland and E.B. Eddy were made directors of the new company. CAL's territory was divided along the country's 85th meridian (about thirty-five miles [56km] west of Sault Ste. Marie) into Eastern Lines, headquartered in Montreal, and Western Lines, based in Winnipeg. Colonel Mulock and C.G. Drury, Victor Drury's brother, looked after the company's interests from the head office at 920 University Street and Richardson from the Trust & Loan Building in Winnipeg.

In its first annual report, Canadian Airways Ltd. was, its accountant calculated, "in splendid working capital position." Current assets to current liabilities was 9 to 1, cash in hand was $426,790, and its real estate value of hangars, airports and buildings was $170,317.42. It had fifty-one single-engine aircraft, one trimotor and eighty-seven aero engines. The directors hoped this healthy balance sheet would carry CAL through the gathering storm. All aircraft were repainted with orange wings and a dark blue fuselage and adorned with the WCA flying goose. The birth of the company flag occurred at this time as well, when CAL's Eastern Lines sported one on the fin of all their aircraft. Used in advertisements and flown at the company offices at St. Hubert, the flag was identical to the checkerboard design used by Canadian Pacific steamships. In 1931, it was replaced by the flying goose insignia.

Radio receivers were installed on the mail aircraft, with mixed results:

When the operators started eavesdropping on each other's frequencies and directing their pilots to swoop in and purloin traffic intended for someone else, the benefits of radio were greatly diminished. When one operator gave false information over the air in order to lure eavesdroppers to a far distant point, there to pick up non-existent passengers or goods, the benefits became burdens. Yet this sort of thing happened, and is fairly typical of the destructive bitterness that competition had fostered.[15]

CAL's first international route, the 66-mile (106 km) flight between Winnipeg and Pembina, North Dakota, was inaugurated that summer. Only when the Canadians landed at the American airport and marvelled at the spotless Ford Trimotors of Northwest Airlines with their uniformed crew (who served their passengers hot coffee, chewing gum and biscuits), did they realize how backward their own airline was. But not for long, if James Richardson had his way.

The Mackenzie Air Services Bellanca
66-70 Aircruiser, shown here on Great
Bear Lake in later CAL insignia, gave
James Richardson stiff competition for
freighting contracts. The Bellanca's
distinctive airfoil-shaped stub wings
and outer struts provided extra lift,
allowing the aircraft to carry up to
fifteen passengers or two tons of
cargo. CAA (UNN)

CHAPTER THREE

"I Froze My Toes
In Barkley-Grows . . ."

Like some medieval plague, by the winter of 1931 the Great Depression was remorselessly destroying the fabric of the country, especially on the drought-stricken, windswept prairies. An unnatural winter, which brought no snow to cover the soil, was followed by a hot spring and summer, foretelling an environmental catastrophe of biblical proportions. After the Wall Street crash industrial eastern Canada was in no better shape, as production declined to a third of what it had been in 1928. Thousands of proud and independent Canadians came to know the humiliation of applying for relief.

They put their hopes of getting out from under the blackness in the man they elected as prime minister. Richard Bedford Bennett came to office only because—as historians have since observed—no one knew much about him. Unlike his opponent, William Lyon Mackenzie King, he had no past to criticize, but his inexperience in high office was made worse by the unprecedented economic depression.

Bennett had campaigned on the promise of cutting all unnecessary government expenses, and the air mail services, which in his opinion benefited very few voters, were popular targets. He cared little for national defence and less for civil aviation, so his lumping the two under one budget came as no surprise. He vindicated this move by pointing out that Canada, being situated between two oceans, would never come under air attack and the RCAF would justify its existence by taking over all government contracts and flying the mail as well. James Richardson, seeing that his hard-won prairie mail contracts were in danger of withering away, cabled Colonel Mulock to use his influence as a war hero and exert all possible pressure in Ottawa to prevent this from happening.

On March 4, 1931, a disastrous fire occurred at Canadian Airways' Winnipeg hangar at Stevenson field, destroying seven aircraft and fourteen aero engines and causing $170,000 damage. Although the company was insured, the temporary loss of the aircraft reduced the number of mail runs and was reason enough for the Post Office to call for a re-examination of CAL's contract. Quick to criticize, the assistant postmaster, P.T. Coolican, wrote to J.A. Wilson in April concerned, he said, about what he had heard of CAL from others in the aviation industry that justified cancellation of all its contracts. Wilson came to the airline's defence, replying on May 6 that he had "known many of the staff for years and that they were men who do not lie down on the job . . . and that by listening to hangers-on around the airports, I quite understand, as such men would gladly see the air mail service split up into little contracts which they might have a chance to get at."[1]

The Maritimes air mail services were the first to feel Bennett's economies: on April 14 the Moncton, Montreal and Charlottetown air mail contracts were not renewed. Pan American's Juan Trippe spotted this breach and immediately secured a Post Office contract to begin daily mail runs between Boston, Saint John and Halifax. To deflect the anticipated criticism, Ottawa did not allow Pan American to pick up passengers en route, to the consternation of Saint John and Halifax, which had built airports for that purpose. For Trippe (who rather strangely used a Sikorsky amphibian on the route) the loss of passengers was hardly a problem because he now had a launching point for future transatlantic flights. Once again, Richardson wondered how—after not permitting Sherman Fairchild to bid for air mail contracts because his airline was American-owned—the Post Office could justify Trippe's Canadian venture.

The cancellation of the Maritimes business was partially compensated for in May, when CAL acquired Wop May's Commercial Airways, its competitor on the Mackenzie River route. It also won a Vancouver–Victoria mail contract, which was to begin October 1, 1935. The Post Office specified that CAL buy a de Havilland Dragon Rapide, CF-AVJ, for the purpose.

Wop May was made CAL superintendent in Edmonton, with Con Farrell as the chief pilot and Dave Stirton the office manager. Rooms were rented in the Magee Building in the city centre, and a maintenance base was set up at the municipal airport. Mail was taken to Fort McMurray and then from there on a weekly or monthly schedule to isolated posts on the Athabasca, Slave and Mackenzie rivers.

The mail contracts elsewhere were becoming a bad memory, so Richardson looked to increase his company's cargo-carrying capacity instead, especially as the Hudson's Bay Company promised him a huge contract to fly 480 tons of freight that October. Impressed by the versatility of the W34s, on July 31 Richardson purchased a Junkers giant single-engine Ju 52, of which only five models were built. With its corrugated duralumin skin (unable to obtain aluminum, Junkers developed his own "Elektron" metal) and three main wing spars, the Ju 52 was ideal for bush conditions; it was even being licence-built in the Soviet Union for Siberian operations.

The choice of engine was optional, and in keeping with Richardson's policy of buying the very best he sent Tommy Siers, his engineering manager, to tour the Junkers plant in Dessau, Germany. Siers influenced the decision to replace the standard L88 engine with a twelve-cylinder, liquid-cooled, 685-horsepower BMW engine.

The Junkers was shipped to Montreal, and when assembled at the Fairchild factory it caused a sensation. Its cargo hold was 21 feet long, 5¼ feet wide and 6¼ feet high (6.3 x 1.6 x 1.9 m) and it could take off within 17 seconds with a 7,490-pound (3370 kg) load. Nicknamed "the Flying Boxcar," the Junkers CF-ARM was the largest aircraft in Canada until World War II. Richardson expected that it would substantially increase revenue by being able to fly more freight at lower prices and that all heavy hauling contracts would automatically come to CAL from now on.

Memories of the Junkers' size and idiosyncrasies lived on long after its use. Albert Hutt, a CAL engineer, never forgot that the engine was cooled with ethylene glycol:

It was a hell of a job draining that thing, heating the glycol and putting it back. Now I had just bought a house . . . and replaced the gas-heated water heater with an electric one, so I took the old gas heater down to the shops and we installed two valves on the bottom of the radiator and hooked up the heater to them. Then we

CF-ARM was the last of Hugo Junkers's five single-engine Ju 52s. CAL's decision to power it with a twelve-cylinder BMW engine in 1931 proved an expensive disaster. The BMW was replaced with a Rolls-Royce Buzzard five years later. CAA (UNN)

This photo provides some idea of the size of CF-ARM, beached here at Brandon Avenue, Winnipeg. Pilots thought "she was an awkward old cow on the water because you couldn't operate the throttle, watch where you were going and work the water rudders at the same time." It remained the largest aircraft in Canada until World War II. CAA (UNN)

heated the coils with a big Clayton-Lambert torch and that heated the glycol in no time at all. Boy that glycol made a helluva mess in the engine if a gasket leaked or a cylinder cracked. The coolant pipes had flanges on the ends and clamps around them, and that gave us some trouble until I figured a way to beat that . . . We could make a damn good connection out of a hockey puck. I used to go downtown and buy up to fifty pucks at a time, and they always wondered what I was doing with them, especially in summer.

Scotty Moir, a pilot, remembered:

She was an awkward old cow on the water because the pilot couldn't operate the throttle, watch where he was going, and operate the water rudders at the same time. She wasn't bad in the air, except she was slow. She had hinged flaps and floats as big as all outdoors! Standing beside those floats, the step was above you! Some of the places we went into didn't have docks that could handle her and in fact, on a big freighting job into Summit Lake, they had to build a barge, put two outboard motors on it and unload the Junkers in the middle of the lake, then bring the freight on shore in the barge. That gives you an idea how big she was![2]

CAL was now able to sell its original workhorses, the Fokker Universals, to small operators. G-CAGD *The City of Toronto* was sold to Independent Airways in 1931, and the young president of the company immediately used it to circle Edmonton's Blatchford airport and complete the necessary hours for his commercial licence. Grant McConachie always was audacious.

George William Grant McConachie was born in Hamilton, Ontario on April 24, 1909. His father was the chief mechanic for the CNR in Calder, Saskatchewan, and Grant could have had a job in the railway for life. As a teenager, Grant did toil as a CNR stoker during the summer, long enough to know he wanted no part of it. The middle-class respectability of the McConachies was tarnished, however, by the visits of a relative whose scruples were somewhat suspect. Uncle Harry sold "miracle hair tonic" at carnivals, always making sure he kept his own hat on to hide the fact that he was completely bald. Uncle Harry's showmanship and outlook on life profoundly influenced Grant and, through him, commercial aviation in the next three decades.

Jimmy Bell, the Edmonton aero club's director, said that McConachie was one of the best students he had ever taught. Even when he "grounded" Grant for charging for rides without having a private pilot's licence (let alone a commercial one), Bell was forced to rescind the order so that McConachie could participate in the club's upcoming aerobatic display. On graduation in 1931 Grant discovered that flying jobs were impossible to get during the Depression; he made up his mind to emigrate to China and join its air force as a mercenary. To prevent him from going to what would have been certain death, Uncle Harry mysteriously found the money for his nephew to buy CAL's Fokker Universal and start his own air company.

Flushed with pride that day, the young man loaded his passengers on board the Fokker—in cages. Professor William Rowan of the University of Alberta had a theory that birds flew south in the fall and north in the spring because of the effect of the number of hours of daylight on light-sensitive cells behind their eyes. To test this theory, the professor had used artificial lighting to re-create for the crows the daylight hours of summer, right up to December. Grant was hired to fly him and his crows as far south as fuel would allow, where they would be released. Anyone spotting them and notifying the university would receive five dollars for each legband sent in. If more crows were found to have flown north of the drop-off point, Rowan's theory would be proved.

McConachie flew Rowan and his crows to a farmer's field in Drumheller. On landing, the farmer and his sons helped unload the cages and release the birds. The pilot and professor then gratefully accepted the farmer's invitation to come into the house and have a cup of tea. Seated in the kitchen, Rowan was delighted to see out of the window that the crows were hopping about in a confused state. Eventually curiosity got the better of the farmer, who asked what it was all about. The professor explained the experiment, mentioning the five-dollar reward for every yellow legband sent in. When they took off, to thank the farmer for his hospitality Grant circled the farmhouse to wave goodbye. The farmer and his sons didn't wave back, because they were too busy blasting away with their shotguns. A week later Rowan got the legbands of fifty dead birds and a bill for $250. It was, after all, the Depression. McConachie must have wished he had joined them, because it was months before he got another customer.[3]

In Winnipeg that winter of 1931–32, James Richardson seemed to

A confident Grant McConachie (in flying suit), with five crates of crows. Professor William Rowan, standing next to them, looks doubtful about the whole endeavour. The other well-wishers are Jimmy Bell, the Edmonton airport manager, McConachie's Uncle Harry (who always wore a hat because he sold "miracle hair tonic" and was bald) and Bert Haddow, the Edmonton city engineer, in a fur coat. CAA (UNN)

attract bad news. Leigh Brintnell resigned over a difference in aircraft choice, disagreeing with CAL's increasing reliance on Junkers aircraft. To no one's surprise, on December 28, 1931, Brintnell opened his own air company, Mackenzie Air Services (MAS), taking John Hunter with him. He had been able to do this with financial backing from Anthony Fokker, to whom Richardson had given so much business. Brintnell signed a contract with the Eldorado Mining and Exploration Company to transport fresh vegetables to its mine at Great Bear Lake and to return with ore—business that should have gone to Canadian Airways. The MAS–Eldorado relationship worked so well that within a few years Brintnell's distinctive green Bellanca Aircruiser became known as the "Eldorado Radium Silver Express."

That winter also saw the winding down of almost all air mail routes. By March 1932 the Post Office had reneged on most of its air mail contracts, causing CAL's revenues to decline to one quarter of their previous levels. It was of no use for Richardson to point out that the contract had been for four years and that because of it he had invested large sums of money in F.14s and hangars. Although the contract still had two more years to run, it contained a clause that allowed the post-

The tiny, four-passenger de Havilland 83 Fox Moth was flown by Canadian Airways from 1933 to 1937. Note the nose hangar, invented by H.A. "Doc" Oaks. The model was reborn after World War II, when de Havilland brought out the more powerful 83C. WCAM 29

master general to annul the agreement "if it was in the public's interest." Years later, R.B. Bennett would explain: "With 300,000 of the population receiving some form of relief, there was very little gratification in seeing an aeroplane passing by day after day when the unfortunate owner of the soil could hardly see the aeroplane because his own crop had gone up in dust."

Meanwhile the RCAF, having accepted the role of "bush pilots in uniform" to save itself at the expense of commercial aviation, heard on March 31 that its own appropriations were to be reduced by two thirds. The Department of National Defence was forced to dismiss 78 officers, 100 airmen and 110 civilians—almost 20 percent of its total force. The following month, to its continuing shock, the still-reeling air force was told to take over CAL's Quebec City–Anticosti and Moncton–Magdalen Islands air mail routes. In addition, it had to initiate a ship-to-shore service for the Imperial Conference being held in Ottawa that summer.

In hindsight, Richardson was getting off lightly. When the CNR's annual deficit topped $60 million, Sir Henry Thornton, its president, was hounded out of office and a vindictive campaign was waged to ruin his reputation. Less than a year later he died suddenly of a previously unknown cancer. He was replaced on the CAL board by the new CNR president, S.J. Hungerford, a man who knew well enough to keep his head down.

Without the mail runs, CAL was destined to lose money. In his annual report to the shareholders in April 1932, Richardson said that he "felt a moral responsibility to those who depended on them to carry on the bush services" and hoped the government would recognize its moral responsibility toward the company. The immediate problem CAL faced was that without the steady income from the government, there was no ready source of cash. Those who patronized CAL in the bush had no use for banks. "Punch" Dickins, who was then in charge of the Mackenzie District, observed that because they were not part of a cash economy, the traders, trappers and lumberjacks settled their accounts as they themselves were paid: credits to fur trading companies, bartered furs or "wolf bounties," all forms of currency that didn't impress aviation fuel companies and aero-engine manufacturers.

Then there were the peculiarities of prospectors.[4] Because of the secrecy involved in mineral discoveries, miners disliked using what they considered "large air carriers" like CAL to take them to their claims, preferring the anonymity of one-man outfits. Since most prospectors could afford only a single flight to stock up for the winter, they were tempted to overload the aircraft with supplies. Inevitably this made for longer takeoff runs, which caused the aircraft engine to overheat when it was most needed.

Richardson knew that if the DND air inspectors discovered any overloading in CAL's planes, they would shut him down. His pilots and engineers were ordered to ensure that the letter of the law was scrupulously followed. This hardly endeared CAL to potential customers, and encouraged rival operators to quote lower rates, flout weight regulations and steal much of the company's business away. When Richardson complained to J.A. Wilson, he was told that DND had too few inspectors to enforce the loading regulations everywhere in the bush.[5] It was little consolation to Richardson that most of the "fly-by-nighters" eventually went bankrupt.

As airline owners in trouble had done before (and have since), the accountants at the Winnipeg office reduced costs by cutting staff and salaries. Of the 234 employees in CAL on January 1, 1932, only 161 remained by May. The pilots had already taken a 30 percent reduction

Most prospectors could afford only a single flight and were tempted to overload the CAL Junkers. Aware that if the government inspectors discovered overloading, they would provide the excuse Ottawa needed to shut his airline down, Richardson ensured that the regulations were scrupulously followed. CAA 988-138-8

in salary the previous September and were saved the 10 percent salary cut that everyone else suffered. Richardson ensured that the reduction was applied with discretion for employees earning under $100 a month. The Montreal office was closed, and all operations were moved to Winnipeg. By the winter of 1932–33 passenger rates had declined to half of what they had been in 1931, and Richardson was forced to borrow $400,000 to keep his company operating.

The airline's poverty seems to be in contrast to the apparent state of commercial aviation then. On February 20, J.A. Wilson, the controller of civil aviation, wrote to "Billy" Bishop (now vice-president of McCall-Frontenac Oil) that although the government had cut appropriations by 70 percent, the amount of mail carried by air had only been reduced from 470,000 to 412,000 pounds (211,500 to 185,400 kg) and

The fortunate residents of Pickle Crow, Ontario, still had Canadian Airways Ltd. air mail service on February 22, 1935. Elsewhere, Prime Minister R.B. Bennett's cost-cutting measures had reduced mail contracts to a memory. CAA (UNN)

that commercial aviation in Canada was finally self-sustaining. The optimism of this statement may have caused Bishop and others to consider entering the industry themselves.

Richardson would not have agreed with Wilson, at least from the state of his own company's books, and in March 1933 he arrived at a decision. If the government allowed CAL to go bankrupt, it should be aware that the void would be filled by airlines from the United States and that the country's commercial aviation industry would in future be run from banking houses in New York or committee rooms in Washington. Richardson wrote to Prime Minister Bennett personally and phoned A.G.L. McNaughton, informing them of the seriousness of the situation. The construction of the Trans-Canada Airway was proceeding on schedule, thanks to the Depression and the ever-expanding labour pool of unemployed men that it had provided, and McNaughton and Wilson both thought it time the federal government designated an airline as the nation's "chosen instrument."

Bennett had no sympathy for Richardson. He replied that if CAL could not become profitable, it might be best to close it down. However, Bennett took the precaution of asking Wilson for his recommendations. The controller of civil aviation confirmed Richardson's view that if CAL were not helped immediately, it would be the end of an independent airline industry in Canada. In true Ottawa fashion, this precipitated the

Bush pilot Ginger Coote operated Vancouver–Zeballos and Vancouver–Port Alice flights, his business helped by a local gold rush. His company was bought by Canadian Pacific for $63,000 in 1941.
Photo courtesy Margaret Fane Rutledge

naming of a committee to decide whether the air carrier for the Trans-Canada Airway should be private or government-owned. The committee began by recommending that if a private airline was to be used, it could either be an existing one like CAL or one to be created by the government. When it reported in January 1934, ominously McNaughton (who was retiring from the DND to become president at the National Research Council) was leaning toward eliminating CAL from the running. Even Wilson, it seemed, was now having doubts, thinking it unwise for any one private company to have the monopoly on the whole Airway.

What allowed CAL to survive the Depression was the business it received from the mining and mineral-exploration industries. "To every budding airman on the west coast all avenues led to an office on Seymour Street in Vancouver," wrote George Lothian, "the windows of which sported an insignia encircling a Canada goose. This was Canadian Airways, presided over by Don MacLaren with Wally Courtney as his number one man."[6]

Besides carrying eager miners, CAL transported anything needed to maintain and provision a "strike" or a mining camp—from gasoline engines, lengths of drill rods and bags of cement to horses and cows. In 1934, the company carried 60 percent more freight than all US airlines combined and double the total carried by Imperial Airways. Richardson continued to consolidate his position by purchasing Spence McDonough Air Transport, which had been giving Punch Dickins problems on the Mackenzie run.

CAL's closest rival in British Columbia was Ginger Coote, who owned a small bush flying business at Vancouver and expanded it to Zeballos, employing Margaret Fane Rutledge as a co-pilot, radio operator and dispatcher. Rutledge would become a charter member of the famous "Flying Seven," a group of female pilots from the Vancouver area.

Primarily because of Quebec's heavy political representation in Parliament, the province's mail routes on the lower St. Lawrence had been designated as essential services and were kept operational through the Depression. Beginning in 1927 two Canadian Transcontinental Airways pilots, Romeo Vachon and "Duke" Schiller, using Fairchilds equipped with skis, flew on a mail run between Quebec City and Seven Islands. No navigational aids and few airfields existed en route, so the mailbags were thrown from the aircraft to the isolated postal stations along the way. This was hardly satisfactory, and on one occasion,

> a complaint was received by the postal authorities at Quebec City that pilots were dropping the mail too far from the Godbout post office. To oblige them the stalwart on the next trip successfully bounced a mail bag off a snow bank in the Godbout Postmistress' back garden through her kitchen window. There were no more complaints.[7]

Rather than attempting to obtain the Quebec routes under a CAL tender, and to prevent a local entrepreneur, Desmond Clarke and his Quebec steamship company, from starting his own air carrier, Richardson created Quebec Airways in December 1934 and asked Clarke to take 49 percent of the shares in it. Clarke didn't take up this option, and CAL's share eventually became 75 percent. Romeo Vachon, who had become famous when he flew medical aid and the media to the crew of the crashed *Bremen* in 1928,[8] was made assistant manager of Quebec Airways.

Ironically, there was now little need to fear an American invasion in eastern Canada. On February 19, 1934, the White House cancelled all civil air mail contracts and transferred the work to the US Army Air Corps. Northwest Airlines lost its Pembina–Winnipeg run and CAL took it over, using a Laird mailplane. The cancellation was a major blow for United, Transcontinental & Western, Eastern and American Airlines, all of whom had invested heavily in the development of the Boeing 247 and Douglas DC-2. The US Air Mail Act of 1934 required the separation of airlines from aircraft manufacturers and forced them to begin wooing passengers rather than chasing mail contracts. The act's social implications transformed US commercial aviation drastically. Now there were advertising campaigns to convince families to fly, lobbying for more comfortable airports, airline cuisine, automated ticketing. The switch in markets from mail to the middle class would have profound implications for all airlines.[9] In Canada, more than a decade would pass before commercial aviation evolved to this stage.

In the run-up to the federal election of 1935, Mackenzie King's Liberals attracted other contenders for the mail monopoly on the Trans-Canada Airway. The most prominent was the British North America Airways Co., sponsored by the "Toronto Group," whose dyed-in-the-wool Liberal members included, at one time or another, James Henry Gundy, President of Wood, Gundy & Co.; John C. Elliott, soon to become postmaster general; C.G. McCullough, of *The Globe and Mail*; Roy Maxwell, chief pilot of the OPAS; Percy Parker, and later E.P. Taylor and the Hon. J.L. Ralston. Billy Bishop was also associated with the company, lending considerable prestige to what was a "paper airline." Certain that Bennett would lose the election, the Toronto Group set its sights on cultivating the neophyte Liberal candidate from Port Arthur, Ontario, Clarence Decatur Howe, correctly guessing that he would be in Mackenzie King's cabinet.

They had every reason to be optimistic about their chances of starting up a new air carrier. CAL was suffering its worst year. From a high of 15,889 passengers and 5,763,600 pounds (2,593,620 kg) of air freight in 1934 it had slipped to 10,609 passengers and 5,286,794 pounds (2,379,057 kg) of freight a year later. Added to Richardson's aviation woes was Bennett's election strategy concerning the Winnipeg Grain Exchange. As a last-ditch effort to appeal to western agricultural voters in the upcoming election, on June 10, 1935, the prime minister established the Canadian Wheat Board, undercutting the free-market power of the Exchange forever. Richardson rightly feared that Canadian Airways Limited could expect no better.

The Junkers Ju 52 also was proving to be a disappointment, and fortunately for Richardson the Hudson's Bay contract was not as large as it had originally promised to be. The Junkers' problems were legion: the conventional ski undercarriage fell apart after repeated landings on hard snowdrifts, the BMW engine's vibrations seemed to be tearing it apart and the "Elektron" metal came unbolted under stress. Although these problems were ingeniously and repeatedly dealt with by CAL

A sister aircraft of the five de Havilland 89 Quebec Airways Rapides, CF-BBG was bought on June 10, 1937, by George McCullough, the air-minded publisher of *The Globe and Mail*. When it was destroyed by fire on August 2, the undaunted McCullough led the "Toronto Group" to bid for the air mail monopoly against James Richardson.
Bombardier Archives 47040

mechanics, the constant breakdowns of the BMW engine were beyond even their skills. In its first year the "Flying Boxcar" was grounded for 297 days, and efforts to replace the engine began as early as 1932. Pratt & Whitney Canada, whose "Wasp" engine powered CAL's other foreign-made aircraft, did not want to get involved in what it considered a white elephant. It was only when Rolls-Royce suggested that its twelve-cylinder "Buzzard" engine might work that Richardson hoped to recoup some of his investment. The Buzzard engine was installed into the airframe, and after a nail-biting flight over Winnipeg on January 17, 1937, the Junkers began to earn its keep, becoming a familiar sight at every large construction project in the bush. Richardson tried to buy a second 52 from Junkers, but the single-engine models were no longer made, and now that the Nazis were in power, trade with Germany became politically undesirable.[10] The Flying Boxcar later gave good value,

Repairs on the pontoons of Junkers Ju 52 "Flying Boxcar" being carried out at low tide; this time, flattened stove piping is used to sheet them.
Schade family photo

transporting the money-carrying bulky loads at a slow 97 miles per hour (155 kph) and, as the pilots discovered, crabbing along with the left wing slightly more forward than the right.

As for the Fokker F.14s, of which CAL expected so much—three had crashed by now, one had been destroyed in the Stevenson Airfield fire and, with the cancellation of the mail contracts, the survivors were put in storage, never to be used again. Too specialized for the rough flying conditions, the F.14s were gradually replaced by a true bush plane that was Canadian-designed and built. On November 14, 1935, at Pointe aux Trembles airfield, Montreal, test pilot W.J. McDonough put the Norseman CF-AYO through its paces for twenty minutes and in doing so changed bush flying in Canada forever. The aircraft was strong, had flaps and could lift up to 3,000 pounds (1350 kg) of freight. Its designer, Robert B.C. Noorduyn, had worked with Anthony Fokker on the Universal and later for the Bellanca and Pitcairn companies before moving to Montreal in 1934 to form Noorduyn Aircraft Ltd. The first aircraft designed expressly for the Canadian bush, for decades afterward the Norseman's rugged durability appealed to aviation enthusiasts and airline accountants alike. The prototype itself, CF-AYO, would be sold to Dominion Skyways and end up with Canadian Pacific Airlines years later. CAL bought its first Norseman, CF-BAU, in December 1936 for $23,500, complete with floats and skis. Twenty-seven of the aircraft

The dogmatic C.D. Howe distrusted the private sector's commitment and capability to run transcontinental aviation. He created Trans-Canada Airlines for the purpose, ensuring that in his lifetime it had no competition.
WCAM 2976

would serve with Canadian Pacific, the last being disposed of only in 1955.

On October 14, 1935, the Depression having broken Bennett in body and spirit, Canada had chosen "King not Chaos" and the Liberals came to power. C.D. Howe was made minister of railways and canals and of marine, the astute King recognizing his business acumen. After four years of Conservative budget cuts, an airline network had ceased to exist in most of Canada; the only air connection that had been renewed was between Charlottetown and Moncton. This was not done for any commercial purpose but to maintain a token presence against incursions by Pan American into the Maritimes.

Mackenzie King had been vaguely committed to an aviation policy as prime minister in the 1920s, and his initiative to build St. Hubert Airport (albeit for the Empire Airship route) and begin the Trans-Canada Airway were commendable. But it is safe to say that had Bennett not cut the mail contracts, King as leader of the Opposition would have fought them as financial extravagances. Now in power, and having inherited a nearly completed Trans-Canada Airway, King was not adverse to reaping the rewards of Bennett's intolerance of commercial aviation.[11]

Hardly had the prime minister moved into Laurier House when Richardson tried to get an audience with him—not, as he said, in the role as the owner of CAL but in his capacity as chancellor of Queen's University. He offered to appoint King's friend, O.D. Skelton of External Affairs (whom Bennett had disliked), to the university's board of governors. King was experienced enough as a politician to pass Richardson on to C.D. Howe, now grappling with decades of mismanagement and corruption in the federal transportation system. On February 12, 1936, Howe assured Richardson that every consideration would be given to making CAL the country's "chosen instrument" on the Airway. He also told Tommy Thompson, CAL's assistant general manager, that he "intended to form one large company with Canadian Airways as the backbone." Although Richardson had his doubts about how much influence the "Toronto Group" wielded, he countered Billy Bishop, their war hero, with one of his own, sending Donald MacLaren, CAL's superintendent of the Pacific region, to Ottawa. Having won fifty-four victories in the Great War, MacLaren was the third-highest-ranking Canadian ace, after Bishop and Raymond Collishaw.

Howe attempted to revamp the national transportation system within a year and set about uniting the railways and marine ministries into the Department of Transport, creating the National Harbour Board (and thereby eliminating many political patronage jobs, causing members of Parliament from the Maritimes to rant about the American immigrant who was destroying "local traditions"). He threw himself into aviation wholeheartedly, stating that he would have "a definite program in place to fly mail and passengers from Winnipeg to Vancouver on July 1, 1937, extending this to Halifax by July 1, 1938." In the first parliamentary session, Howe drafted legislation for a civil aviation branch in order to take it away from the military. Through the 1936 summer recess, he studied the condition of commercial aviation in the United States, catching a Ford Trimotor from New York to Los Angeles and meeting with Eddie Rickenbacker of Eastern Airlines and Juan Trippe of Pan American.

The CAL air mail service between Vancouver and Seattle was inaugurated on October 1, 1935, with E.P. "Billy" Wells as the pilot. Until the Vancouver airport opened on Sea Island on July 22, 1931, aircraft had landed at Minoru Park on Lulu Island. The sleepy, out-of-the-way airfield was being used by only the Aero Club of British Columbia and a few small bush outfits, and the airport manager, Bill Templeton, realized that he had to attract scheduled carriers for it to prosper. The single CAL flight was not enough, so Templeton went to Seattle and persuaded United Airlines to service Vancouver in 1936, thus giving CAL some incentive to compete. The American pilots were nervous about crashing into the mountains surrounding Vancouver, so a radio was installed at the airport to tell them what the local visibility was.[12]

The site of the Vancouver airport was chosen by Don MacLaren. When this administration building opened in 1931, it sat on the south side of the field in lonely splendour until 1935, when Canadian Airways began its Vancouver–Seattle service. CAA 989-51-3

The CAL Dragon Rapide was thus equipped with one of the company's first two-way radios and a low-frequency radio range receiver. Radio range facilities (which transmitted the two Morse code letters A and N) had not yet been installed at Sea Island airport, but they were at Seattle because American pilots had come to rely on flying across the continent from radio beam to radio beam. Fortunately for CAL, Billy Wells was an ex-Royal Flying Corps fighter pilot who had flown reconnaissance missions in the Middle East for the fabled Lawrence of Arabia. Making blind approaches into both Seattle and Vancouver was child's play to him.

Early in 1936, having the ability to fly by using instruments became mandatory for anyone wanting a Canadian pilot's licence. The RCAF had been offering an instrument flight rating (IFR) course at Camp Borden for several years, and a few of CAL's pilots had taken it. Most bush pilots were convinced that flying by instruments alone was impossible and that radio beams could never replace "gut" instinct. South of the border, all commercial pilots had been required to hold an instrument licence since 1933. CAL, hoping that its Seattle flights would help it break into the transborder market, had tried to get its pilots places on the Camp Borden IFR courses. When the pilots were consistently told that the next session was full, the company sent Z.L. Leigh to the Boeing School of Aeronautics in Oakland, California, to learn the latest IFR methods.

One CAL pilot who must have wished he could have flown by instruments was Art Schade. For Schade, as for the other pilots, the mail runs could be anything but routine. In 1936, he was returning in the Fairchild 71 CF-AAT from Red Lake, north of Sioux Lookout, with the mail and three passengers when he ran into low clouds. Because he couldn't fly with instruments, Art reduced his altitude to stay below the clouds, almost tree-hopping within sight of the ground.

The gas gauges on the Fairchild 71 were infamous for their inaccuracy, and to avoid running out of fuel during the final approach to landing, experienced bush pilots always switched to the fullest tank on descent. Five miles out of Sioux Lookout, Schade moved the lever to the full reserve tank . . . and the engine quit. Unknown to him, the fuel tank was blocked. With a crash landing imminent, Schade scanned the trees below, warning his passengers to brace themselves. Without the luxury of altitude, he aimed for a small clearing in the forest. As it came down, the Fairchild's right wing hit a tree and was ripped off, causing the aircraft to tumble over in mid-air at least twice before coming down right side up. One passenger was injured and the others shaken up. After lighting a fire, Schade set out on snowshoes for Sioux Lookout. He fought his way through the snow and forest to Lac Seul and then followed the lakeshore into town. It took two hours in bone-chilling cold, but Schade staggered into the CAL office to report the crash. Then, borrowing another aircraft, he flew to the edge of the lake, as near to the crash site as he could. He trekked back and helped load the injured passenger onto a toboggan and dragged him to the aircraft, flying to the Sioux Lookout hospital.[13]

In the summer of 1936 Richardson planned to holiday in England, but before he left—remembering that the Post Office had levelled complaints about the unsuitability of his Fokker F.14s—he bought two new Lockheed 10 Electras, CF-AZY and CF-BAF, at $60,000 each, strengthening his hand for the Airway mail run. Lockheed, not having the resources of Boeing or Douglas, concentrated on smaller, faster aircraft. Its Electra came in two models, the "10" and its smaller version, the "12." With a crew of two and the capacity to carry ten pas-

Schade's Fokker Universal after crashing into a tree at Lac Seul on March 12, 1936, shearing off the wing and tumbling over twice before coming down right side up. Schade family photo

The toboggan that Schade would use to drag the injured passenger to hospital is in the foreground. Schade family photo

sengers, the "10" had a range of 1,000 miles (1600 km). It was chosen by Amelia Earhart[14] and Howard Hughes for their pioneering flights.

The competition from the United Airlines' Boeing 247 (which allowed for stewardess service) on the Seattle–Vancouver run prompted the immediate replacement of the little Dragon Rapide with one of the Electras. There was no doubt that this would draw customers away from the Boeing (especially when they discovered that the main wing spar passed through the cabin area). But, best of all, the Lockheed would allow the CAL crew to wear uniforms for the first time. Like the "flying goose" insignia, uniforms demonstrated professionalism. At a time when airmen were still seen as barnstorming gypsies—until 1925 those who flew aircraft were classified by the US government as "entertainers"—uniforms reassured the public and symbolized the maturing of the industry.

When Donald MacLaren and Billy Wells flew CF-AZY from the Lockheed plant at Burbank, California, to Vancouver, it was the first modern airliner to be put on the Canadian registry. Wells and Maurice McGregor were checked out on the Electra 10A by MacLaren, becoming "captains," as the pilots were now called, and were put on the Vancouver–Seattle run with it. The purpose, McGregor noted, was to gain experience on the Electra. They lived in Seattle, and at night, during the layover, would do instrument and nightflying, using local radio beam facilities. Both hoped that their experience would put CAL in a better position to get the Trans-Canada Airway contract.

Knowing that Richardson was in Europe that summer, C.D. Howe and J.A. Wilson arrived at the company hangar in Winnipeg one morning and demanded from a surprised CAL staff to charter the second Electra, CF-BAF, which was imminently expected. Hinting that the Airway contract depended on getting the aircraft, Howe negotiated for its hire—complete with a CAL crew—for a period of three months. As one of the two most modern aircraft in the country, CF-BAF could be used by the government to test the Airway's radio ranges. Howe also ensured that one of the conditions of the contract was that CAL was not allowed to publicize its co-operation in this venture.

Howe was entranced by the features of the Electra, with its retractable landing gear, flaps and variable pitch propellers (and perhaps by the company's slogan, "It takes a Lockheed to beat a Lockheed"), and all his life the minister remained a diehard Lockheed

CF-BAF, temporarily fitted with skis, was one of the two Lockheed 10 Electras that James Richardson bought for the Trans-Canada Airway mail contract. The first modern aircraft in Canada, it was later sold to Trans-Canada Air Lines for training its crews. WCAM (UNN)

fan. When his own Department of Transport came into being on November 2, he ensured that its first aircraft was also a Lockheed—the smaller 12, CF-CCT.

On his return to Winnipeg in mid-August, Richardson, hearing that the Electra had been commandeered, must have hoped that this would improve his chances. There were rumours that other members of King's cabinet openly disagreed with Howe's transport policies, adding fuel to the lobbying for the Airway contract, which by the end of the summer was intense. Richardson informed civil servant Norman Lambert that he suspected Howe "was in league with the Post Office Department against Canadian Airways." When Lambert got to Ottawa, on one of their customary walks around Rockcliffe, Howe told him that Richardson's word was no good and "that he doubted whether Richardson wanted to work with him at all."[15] But Howe was buying time—prevaricating between the Toronto Group and CAL until he could formulate a policy of his own and gain enough support in King's inner circle.

Richardson sent Edward Beatty to Ottawa to plead his cause—a choice of messenger that ultimately proved disastrous. Mackenzie King didn't want the CPR's tentacles anywhere near the new field of air transport; there were still many Canadians who held that the country had been created to justify the CPR's existence and not the other way around. To King, the CPR's immense power (one in twelve Canadians were employed by it) and wealth were a reminder of the recent colonial past. As a young politician, King had witnessed CPR president Thomas Shaughnessy's attempt to force the amalgamation of all the railways in Canada under his control.

If the railway's tainted past was not bad enough, King knew too that most of the CPR's shareholders were British and that in 1935 Beatty had been knighted by King George V "for services rendered to the British Empire." A knighthood for a Canadian was not uncommon then, but it was reported that the CPR president had played the open-mouthed, colonial boy around Buckingham Palace at the investiture. If the CPR came on the government's airline board, London's influence behind the scenes would always be suspected. In addition, King was just then preparing to leave for the Imperial Conference in London, where he wanted to discourage the British from counting on massive Canadian co-operation in the growing European turmoil.

There was also the influence of the Department of External Affairs. Entrusted to negotiate bilateral agreements for international air routes, the department knew that other governments, particularly the British, were reassured by the presence of a "chosen instrument" that had the authority of the prime minister behind it. A "fly-by-night" outfit would be unacceptable. A government-owned airline would assure both London and Washington that they were dealing with a reputable firm. Keeping these factors in mind and using Howe as his mouthpiece, King manoeuvred Beatty and Richardson (counting on their pride) into taking positions from which they could not publicly stand down.

By late August Howe had sent word to Richardson that it was no longer necessary for them to meet because he "was not yet clothed with the proper authority" and "everything was working out along the lines of our previous discussions." Richardson suspected that he was falling out of favour, but then Howe asked that CAL submit to him a detailed proposal about its trans-Canada service. Reassured, Richardson rallied his staff to draw up a comprehensive fifty-page document, complete with diagrams, photographs of aircraft and financial charts of CAL's plans, and delivered ten bound copies to Ottawa. CAL staff had been working on this plan since the Liberal victory; the study laid out the schedules of services, the four stages in which the airway was to be inaugurated and a complete, detailed breakdown of all the costs of each segment of the airway. Beatty was to be the plan's salesman because he frequently visited Ottawa in his private railroad car and Howe seemed quite amenable to meeting with the CPR president.

Richardson proposed that a new company, "Canadian Airways," be formed to fly between Vancouver and Halifax, with a branch line between Lethbridge and Edmonton. The existing Canadian Airways Ltd. would be renamed "Airways Limited" and continue with the bush flying that it did so well. The two railways, the CPR and CNR, would each hold 37½ percent of the new company's shares, and Canadian Airways Ltd. would control the remaining 25 percent. Canadian Airways would need a fleet of 12 Lockheed Electras, 21 pilots, 21 co-pilots and 166 staff across the country.

The minister seemed very receptive and even suggested a few changes: that the name be the "Trans-Canada Air Line Company," that the railways and their aviation interests take 50 percent each of the shares and that the federal government pay the company a subsidy for its actual operating loss, for its first two years. Richardson did not like the idea of other air carriers being involved, but this turned out to be irrelevant, since cabinet turned the proposal down—something that he believed Howe held him responsible for.[16]

On November 2, Howe hinted at what was to come when he spoke about a transcontinental air service at a Canadian Club meeting in Montreal: air transportation would never be self-sustaining, at least not in sparsely settled Canada; it would always require the railways to bankroll it, just as they had been supported by the government when starting out. Canadian Airways Ltd.'s fate was finally decided on November 26, when four senior cabinet ministers—Ernest Lapointe, Charles Dunning, Tom Crerar and James Ilsley—met with Howe and King to hammer out what the country would know in the next session of Parliament: the monopoly to operate on the Trans-Canada Airway would not be given to either the Toronto Group or CAL; neither would it be run entirely as a government-owned enterprise. Instead, the two railway companies would be asked to provide $2.5 million each to the monopoly's capital stock of $5 million and would nominate most of its directors; some representation would come from the government. Howe was now "clothed in the proper authority."

After Christmas 1936 a new bill was drawn up, stipulating that the board would be composed of four directors from each of the railway companies, and the fifth would be appointed by the minister of transport. Beatty thought this unacceptable because he felt the majority vote would always be with CNR (which, as part of the government, he suspected would always vote with the minister of transport). Howe had the bill redrafted, changing the composition of the board to nine directors—three from each railway company and three from the ministry of transport. On March 6, 1937, Howe forwarded a copy to Beatty, informing him that this time the bill had passed through cabinet "after much opposition." On March 12, Beatty told Howe that he had discussed it with the CPR's executive committee, and their view was that if the railway was to provide half the capital, it should have an equal voice in determining the airline's policies. His own board of directors, he said, could not allow CPR shareholders to invest in what was a "politically directed corporation."

Howe explained that the cabinet felt that having provided for the Airway's airports, radio and meteorological services, Ottawa had a greater responsibility for the effective operation of the company than either the CPR or the CNR, and this justified the government having some representation on the board. Furthermore, he could not agree with Beatty that the government had any control over the management policies of the CNR. Beatty remained unconvinced and on March 16, 1937, withdrew from further participation. His last letter to Howe stated, "It is not, in the circumstances, necessary to continue our correspondence and I should appreciate it if, in the Bill as introduced, you will omit any reference to this Company . . . I hope that your company will realize its highest hopes."

The Trans-Canada Air Lines Act (Bill 74) was introduced in the House on March 22 and passed on April 6. The amendments were that the number of directors be reduced from nine to seven, that the government would assume the airline's deficits for a period of only two years and that the CNR would subscribe to 51 percent of the shares; the remaining 49 percent were to be offered to other aviation interests. Howe said that he hoped these would be CAL and British North America Airways, and on April 26, in return for feeder services into the transcontinental airway, he invited Richardson onto the board. Like Beatty, Richardson refused, echoing the CPR president's fear that against such odds he would never be in any sort of managerial position. Howe must have been confident of this reply because three days before, at a CNR board meeting, he had the railway company subscribe to all the shares in Trans-Canada Air Lines (TCAL) and nominate four of its members to the TCAL board; the three remaining board positions went to federal civil servants.

The news that CAL was not going to operate on the Trans-Canada Airway caused several of its employees to reconsider their future. On April 28, Donald MacLaren asked for his release from the company. Howe had been very impressed by MacLaren and invited him to join Trans-Canada Air Lines as assistant to its vice-president, P.G. Johnson. Richardson was reportedly furious about this as MacLaren had been privy to the company's plans at the highest level. Like most of TCAL's technical advisers, Johnson was an American recruited by Howe to get the airline running. In response to criticism about the dearth of Canadian advisers, the minister blithely pointed out that the first two presidents of the CPR, William Van Horne and Thomas Shaughnessy, had also been American. By the end of 1937 thirteen other pilots had also transferred over from CAL, as did eight other staff.

Of British North American Airways, little was ever heard again. Disappointed when the Airway monopoly didn't come its way, the Toronto Group then aimed its sights at Queen's Park, hoping to pick up all Ontario Provincial Air Service contracts and even buying two de Havilland Dragon Rapides for the purpose. Here, too, it failed, and by year's end the company no longer existed.

To keep his promise of a transcontinental service by July 1, 1937, C.D. Howe arranged for a "dawn to dusk" flight from St. Hubert to Vancouver in the Department of Transport's new Lockheed 12A, CF-CCT. It would be July 30 before most of the Trans-Canada Airway was even remotely ready, and even then many of its radio stations were not working. As a result, the party got lost looking for Sioux Lookout and almost ran out of fuel, but it touched down at Vancouver's Sea Island airport seventeen hours and thirty-four minutes after leaving St. Hubert. Richardson and Beatty were both shocked at Howe's effrontery and luck.

There was little need now for CAL to keep the fast Electras, and although CF-BAF was temporarily fitted with skis it was hardly a bush aircraft. To recoup some of the cash, Richardson offered the pair of them, as well as the Vancouver–Seattle route, to the new airline. Trans-Canada Air Lines purchased both Electras as its first aircraft, along with a CAL Stearman biplane and some equipment at Vancouver airport, for

$169,176. Thus, the first TCAL flight, from Vancouver to Seattle, was piloted by Billy Wells and Maurice McGregor—both still wearing their Canadian Airways Ltd. uniforms and using that airline's tickets.

For the remainder of his life, James Richardson blamed "certain officials in the Post Office" for turning C.D. Howe against him. He always held that the Post Office made a handsome profit on the backs of the struggling air carriers. This was true even after the cancellation of most of the intercity air mail services during the Depression. In 1936–37, for example, the Post Office paid $284,000 for its air mail operation, but its revenues were $605,000.[17] Richardson wrote:

> These profits are made at the expense of a struggling industry of prime national importance . . . [A] healthy air transport industry . . . should not be left in the hands of civil servants who after all, have the limited outlook of their own particular department and no particular knowledge of the air transport industry . . . [T]heir zeal has resulted in the lowering of rates for carrying mail which has inflicted serious hardship on the struggling industry of vital importance to the welfare of the north . . .
>
> The Department of Transport was powerless to take care of the existing situation as long as the Post Office had controlled the awarding of the airmail contracts. But the Trans-Canada Air Lines Act that authorized the formation of a publicly owned company to carry the air mail at a remuneration which would cover the actual cost plus a fair return on the capital invested meant that the Post Office would no longer decide. If this was a wise policy for a public corporation supported by the taxpayers' money, why was it not a wise policy for the privately owned corporations who had developed commercial aviation in Canada to the benefit of the north country at large?[18]

For all the anxiety that they entailed, the mail contracts had not been paying for CAL's operating expenses since 1935. What saved the company from total bankruptcy were the mineral strikes of the mid-1930s. The mining industry, led by Gilbert Labine's Great Bear Lake mine, was responsible for the volume of air freight in Canada exceeding 24 million pounds (10.8 million kg) in 1937. Freed from chasing after the Trans-Canada Airway contract, Richardson now concentrated on heavy freighting, especially to construction projects at gold fields or mines at

A rare photograph of a rarer aircraft: the biplane Fleet Freighter, with United Air Transport pilots in uniform. Left to right: Grant McConachie, Charlie Tweed and Sheldon Luck, talking to the postmaster at Prince George. CAA (UNN)

Yellowknife and Pickle Lake. He also bought controlling interests in competing air carriers Dominion Skyways Ltd., which had the Montreal–Rouyn–Val d'Or route, and Arrow Airways Ltd., which operated out of The Pas. Because of the air mail contracts they held, both companies kept their own names to prevent Canadian Airways Ltd. from being charged with monopolizing the air mail.

Richardson felt that for politically motivated reasons, CAL was increasingly being treated as a leper and deliberately starved by the Post Office of further air mail contracts. Leigh Brintnell's Mackenzie Air Services underbid it on the Mackenzie River route "by a few cents," as CAL's staff bitterly noted. In Ontario, CAL was suddenly notified that its Red Lake route was to be given to Starratt Airways. No tenders had been called, nor had there been any complaints about the CAL rate for the carriage of mail—10 cents a pound. Only later was it revealed that Starratt had offered 9½ cents a pound. In what must have seemed the ultimate insult to Richardson, an air carrier called United Air Transport (UAT) which hauled frozen fish—the lowest occupation in the air freight industry—was given the Edmonton–Whitehorse air mail contract without tenders being called at all.

The three Beechcraft 18s used for the Maritime routes were ideal as transitional bush airliners because they could be equipped with skis, floats and wheels. Note the Canadian Airways crew in uniform. CAA 988-31-4a

The owner of UAT was the same Grant McConachie who had purchased the original WCA Fokker Universal and now operated from Cooking Lake, a floatplane base near Edmonton. In March 1937, one step ahead of his creditors, McConachie bought a Ford Trimotor from the RCAF and on July 5 used it to inaugurate a weekly service between Edmonton and the Yukon. Later, he would secure the Peace River–Dawson Creek–Fort St. John mail route as well.

The freighting opportunities apart, CAL sought out feeder routes to connect passengers and mail with the Trans-Canada Airway. A rise in the price of gasoline in 1937 added to its hardships, as did the cost of maintaining an aging fleet. Many of its aircraft dated from the late 1920s and were in dire need of replacement. Of the Fokker Super Universals that had been CAL's mainstay for so many years, only one had survived to 1937—incredibly, it was G-CASQ, which had fallen through the ice during the MacAlpine rescue in 1929.

The airframe and engine revolution in the United States that had spawned aircraft with all-metal construction, automatic pilots, cowled engines and retractable landing gear had, except for the Lockheed Electras, passed CAL by. At a time when competitors like the up-and-coming Prairie Airways and Starratt Airways were buying twin-engine, all-metal airliners, Richardson could not afford to do so. From 1937 onward, economies like keeping the aircraft in their aluminum finish (except for orange bands painted on the wing tips for identification in case of emergencies) became more evident. Unsure of its future and unable to buy, CAL took to leasing Stinson SR-9 Reliants from Canadian Car & Foundry and renting a Fleet 50 from Fleet Aircraft. Only with the acquisition of three Beechcraft A-18s (two "D"s and one "A") at $33,000 each for its Maritime routes in 1940 would CAL maintain any pretence at being more than a bush operator. First flown in 1938, the nine-passenger Beech 18 was ideal for a bush airline because it could fly effectively on wheels, skis and floats.

Like many of Richardson's aircraft, Grant McConachie's fleet was still on pontoons, switching to skis in the winter and grounded for eight weeks each year during fall freeze-ups and spring thaws. The former fish-hauler too had pretensions of graduating to airline status. As the first step toward this, he made his UAT crew wear blue serge uniforms—at least on landings and takeoffs. More seriously, McConachie looked to replace his floatplanes with radio-equipped wheeled aircraft, which would free their pilots from the tyrannies of the weather and climate. Always an optimist, he set about the expensive task of constructing air-fields (unpaved and unlighted) along the routes at Fort St. John and Fort Nelson for these aircraft long before acquiring any.

More and more, Canadian air carrier operators were discovering that economic pressures, combined with increased passenger expectations and the availability of heavier, faster aircraft, were forcing them away from the traditional lakes and rivers to lighted airports with paved runways. Waiting to hear if the ice had melted on a certain lake in the spring or frozen over in the fall cut into profits. The increased use of radio compasses that homed in on airports freed pilots from relying on visual references and flying only in good weather. But apart from those airports or wireless stations on the Trans-Canada Airway or near urban centres, there were few such locations elsewhere in the country; the first fully equipped control tower opened at Montreal's St. Hubert Airport on April 13, 1939.

Using his famous charm, McConachie talked the directors at Canadian Car & Foundry into leasing him three Barkley-Grow wheeled aircraft for a token dollar each. The metal, twin-engine, low-wing monoplanes (which could be converted to pontoons), had been designed in 1937 by two Americans, Archibald Barkley and Harold

Only eleven Barkley-Grows were built, but they served with Yukon Southern Air Transport, Quebec Airways, Canadian Pacific Airlines and Pacific Western. They enabled bush companies to progress from floats to "all-year-round" wheels. CAA 988-31-8

The same Barkley-Grow, now on wheels. Mr. Barkley and Mr. Grow built rugged aircraft that "didn't want to do nothin' but fly." Unfortunately, both the USAF and the RCAF preferred the Beech 18, relegating a good aircraft to the status of an aeronautical curiosity. CAA 988-31-18a

Grow. To fulfill director Victor Drury's ambition of entering the aviation field, Canadian Car & Foundry had unwisely acquired the Canadian distribution rights for the aircraft. It was initially seen as a good decision because the RCAF and the US armed forces were re-equipping for the coming war, and the Barkley-Grows were cheap and perfect for use as trainers or communications aircraft. However, although the RCAF bought one, both armed forces preferred the Beech 18. The Barkley-Grow Company lost heavily, before being bought by Consolidated-Vultee. By then, eleven of these aeronautical curiosities were built and sold at bargain prices to bush operators who would not otherwise have been able to afford modern aircraft. Canadian Car must have been equally pleased to get rid of three Barkley-Grows to McConachie.

The aircraft, which would be flown by Yukon Southern Air Transport and, later, Quebec Airways, Canadian Pacific Airlines and Pacific Western Airlines, had both detractors and enthusiasts. A bush pilot once said of them, "Them Barkleys was murder to work on. Every one was different and nothin' would fit another but it was hell of a good airplane though—didn't want to do nothin' but fly." Alf Caywood, who would later fly with Wop May, remarked:

> I think that the Barkley-Grow was about the ultimate plane for bush flying. It may not have had good single-engine performance, but one engine could take you a hell of a long way. I took off one day when both engines quit and we were heading for the bloody trees! Up in the cockpit, we were switching everything—I don't know what went wrong, but the engines suddenly caught again and we got out. But that Barkley was a lovely aircraft![19]

As McConachie's first modern aircraft, his three Barkley-Grows, with their uniformed crew and radios, allowed him to incorporate UAT into a rudimentary airline, Yukon Southern Air Transport (YSAT) on January 16, 1938. It flew according to published schedules from Edmonton and Vancouver to Yellowknife, and best of all, its passengers enjoyed lunches of sandwiches and cookies, served by the flight engineer.

By the end of the 1930s, Canadian Airways Ltd. was still the largest air carrier in the country. Despite its woes, it accounted for most of Canada's commercial flying with its aircraft scattered across five

Homemade advertising by Grant McConachie. Two ladies with luggage and Ted Field's convertible glamorized his shiny new Yukon Southern Air Transport Lockheed 14. Ahead of his time, McConachie understood the value of sex appeal to sell commercial aviation. CAA (UNN)

provinces. Its competition consisted of 127 small operators, chief of which were five:

- **Leigh Brintnell's Mackenzie Air Services.** Flying between its headquarters in Edmonton and Aklavik, Coppermine and Stoney Rapids, this company used Bellanca Aircruisers and Noorduyen Norsemen. Brintnell also had financial interests in YSAT and Prairie Airways. Seeing what the cutthroat competition was doing to all of them, in July 1939 he recommended to the Department of Transport that Canada be divided into a number of areas and that the various companies in each area be amalgamated into a single organization. For obvious reasons, CAL didn't support Brintnell.

- **Grant McConachie's Yukon Southern Air Transport.** YSAT had its own network of airfields and fifteen aircraft, chief of which were the three Barkley-Grows equipped with radio compasses and two-way radios. On paper this might have looked good, but the problem was that the airfields had bankrupted McConachie—he blamed the

The Custom Waco of Wings Limited. Started up by former CAL pilots to compete with James Richardson around Sioux Lookout, Red Lake, Kenora and Winnipeg, Wings was typical of the bush outfits purchased by Canadian Pacific in 1941. CAA 988-31-12

horses used in their construction for eating their weight in oats (which had to be flown to the sites).

Worse, only two of YSAT's aircraft were insured at any one time. The insurance coverage was moved around from aircraft to aircraft, in the hope that if a crash occurred, it would be the one currently insured. Grant was lucky in this regard when in 1939, a novice RCAF pilot smashed his Hurricane fighter into the old Ford Trimotor at Vancouver airport; McConachie was awarded $52,000 in loss of air freight contracts and another $5,200 from the insurance company.

The cost of attempting to convert a bush outfit to a year-round "all wheels" airline was beyond even McConachie's creative bookkeeping, and in 1938 YSAT lost $70,000. Although this was reduced to $40,000 in 1939 and $20,000 in 1940, when the first Barkley-Grow touched down at Fort Nelson, YSAT was in serious financial trouble and close to bankruptcy.

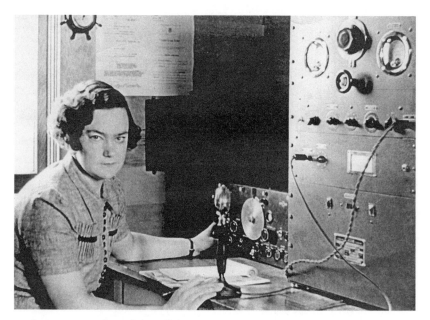

A charter member of the "Flying Sevens," Margaret Fane Rutledge
worked as a radio operator, dispatcher and co-pilot for Ginger Coote.
Margaret Fane Rutledge photo

■ **Prairie Airways.** Begun in Moose Jaw on March 15, 1934, Prairie
flew between Regina and North Battleford.

■ **Wings Limited.** Incorporated in 1934, this company was created by
former CAL pilots Francis Roy Brown, E.W. Stull, M.E. Ashton and
Jack Moar, who bought Waco and Fairchild aircraft and had routes
around Sioux Lookout, Red Lake, Ilford, Gods Lake, Kenora,
Winnipeg, Little Grand Rapids and Flin Flon. Wings Limited and
Starratt Airways were potentially CAL's major competition.

■ **Ginger Coote Airways.** The Vancouver-based company flew
between Zeballos and Port Alice. During the 1930s Vancouver
Island gold rush, prospectors flocked to Coote's airline to be
dropped off and picked up along the coast. Margaret Fane
Rutledge, who worked for Coote, remembered the reservations
system: "You take a shipping tag and write the person's name on
it. We had cup hooks all along the wall from one to thirty-one. If
Coote was going on the fifth of June or the fifth of July, about the

third of the month you'd pick the ones off number five . . . then you'd call all these people and tell them to be ready to go at such and such a time in the morning. If you got three of them you'd call out the Waco and if you got ten you'd call out the Norseman. If you got ten for the Norseman you'd have to take all the seats out and they'd all have to sit on their baggage. That was the reservations system."[20]

All the companies teetered on the edge of a financial abyss, a situation comparable to the one air carriers in eastern Canada had faced a decade before, but now no American airline was interested in salvaging them. This time, these airlines looked to the two Canadian institutions that had the wherewithal to bankroll them: the federal government and the CPR. But, having launched Trans-Canada Air Lines, Mackenzie King and C.D. Howe were now preoccupied with events in Europe. The CPR's Edward Beatty, however, seemed amenable.

That summer, perhaps to console his colleagues, Fred Barron, the CAL radio operator at Edmonton, penned an unsolicited poem, "Ode to Canadian Airways":

> Bellanca, Norseman, Junkers, Rapide,
> Mighty messengers of speed,
> Ply the airlanes to the North,
> Then South again, back and forth.
>
> Fairchild, Beechcraft, Vegas too.
> Fly through the azure skies of blue,
> Like winged Pegasus, as of old,
> Through summer's heat and winter's cold.
>
> Dragon, Stinson, Waco, Moth,
> To the Airways plight their troth.
> Of mercy flights from coast to coast,
> Proudly, bravely do they boast.
>
> The blue goose emblem of C.A.L.
> Throughout the North is known well,
> And of't at night brave deeds are told
> By searchers of elusive gold;

95

Of engineer's and pilot's skill,
When there's a way there is a will;
As in the past, their work done well,
We drink a toast to C.A.L.

On June 26, 1939, James Richardson suffered a fatal heart attack at his Winnipeg home. His sudden death symbolized the passing of a time when the dreams of an individual owning an airline had been possible. He introduced his countrymen to commercial aviation, against the mealy-mouthed and envious in the Post Office, and persevered to keep it under Canadian control. So identified was he with the fortunes of CAL that no one could adequately take his place, and the company that he had guided seemed about to collapse under the weight of competition and lack of capital. To forestall this, in August its comptroller, J.W. Tackaberry, led a team to reorganize the carrier and look for a godfather. The company newsletter notes:

It is with deep regret that we remember that Mr. James A. Richardson is no longer with us, to follow, as he did, with great personal interest, the actions and well-being of the Company as a whole, and the labours of our individual hands. We believe that we can show no greater love and respect for his memory than by individually and collectively doing all that is humanly possible to maintain and improve the enviable reputation our Company has gained in the Air Transport field.[21]

The role of the "Father of Commercial Aviation" was tersely summed up at Richardson's induction into Canada's Aviation Hall of Fame many years later:

In the annals of this nation's flying history, no businessman gave more of himself for less reward to the everlasting benefit of Canadian Aviation.

As a postscript, in 1945, when Trans-Canada Air Lines wanted to double the number of its transcontinental flights, the Post Office's P.T. Coolican threatened to make trouble. The conflict revolved around scheduling: for the convenience of passengers, TCA wanted the flights to leave during daylight hours, when the Post Office would have preferred midnight departures—after the mail had been collected.

Prime Minister Mackenzie King had had enough of the Post Office interfering with commercial aviation and sent his private secretary, the intractable Walter J. Turnbull, to "sort out" the department. On hearing that the government of Peru had requested a foreign adviser for its mail services, the first action Turnbull took was to ship Coolican to Lima.

In August 1940, as desperately needed Hudsons, Liberators and Mitchells collected at Dorval airport, Montreal, the British government turned to the Canadian Pacific Railway to ferry them across the Atlantic. National Archives PA 114759

CHAPTER FOUR

Ideas Above Its Station: The Canadian Pacific Railway Buys In

On the night of September 3, 1939, within hours of Britain's declaration of war, the liner *Athenia*, outward bound from England, was torpedoed by a German submarine. Of the 112 lives lost, a dozen Canadians drowned. One was Margaret Hayworth, a ten-year-old from Hamilton, Ontario—the first Canadian victim of the war to be brought home for burial. Her funeral, at St. Andrew's Presbyterian Church, Hamilton, drew thousands of mourners and received nationwide coverage. A propagandist's dream, her death inflamed the country against the German submariners as never before.

Echoing the sentiments of many Canadians was an editorial in CAL's newsletter:

The World's Most Valuable Pelt. This is not the pelt of the Russian sable. Neither is it the most superb silver fox, platinum fox or ringneck fox. At the moment of writing, the most valuable pelt in the world has not as yet been taken off. It is a pelt of coarse human hair, coming down in a cowlick towards one eye. The human hyena who grew it has been a disaster to his own people and a severe pain in the neck to the world in general. The pelt will be available in due course. The pelting season approaches.[1]

The declaration of war against Germany affected commercial aviation throughout the Dominion, no matter how distant the enemy. Margaret Rutledge, now a co-pilot for Ginger Coote Airways, was with Sheldon Luck on the day that war was declared. They landed for the night in Kamloops and heard on the radio in the lobby of the Plaza Hotel about the sinking of the *Athenia*. She wondered what would happen to the pilots and airlines as a result of the hostilities. "Some of the pilots," she recalled, "did go into the air force, but the airlines managed to carry on. There was very little civilian flying, because the seats and cargo space were used on a priority basis by the military."[2]

CAL pilot Gordon Ballentine was on the Vancouver–Victoria run with the de Havilland Dragon Rapide in 1939–40. Because he carried government personnel and flew over the dry docks at Esquimalt, Ballentine was ordered to carry a side arm and have the passenger windows of the Rapide blacked out. He recalled that pilots also had to "guard all the airplanes during that part of the war. You couldn't leave an airplane unguarded. If you were overnight someplace, you either had to sleep under it yourself or hire somebody, which was a cursed nuisance."[3]

It was especially galling to North Americans that the supply convoys bound for England were coming under attack from German submarines, frequently within Canadian or American coastal waters. The sinkings of supply boats brought the war home to those Canadians who, like the prime minister, were wavering between involvement and neutrality. Feeling that Canada was protected by the Atlantic Ocean and the might of the United States, Mackenzie King was reluctant for it to be embroiled yet again in what he viewed as another mess that the Old World had gotten itself into. The sacrifices of the Great War were still fresh to many of King's generation, and if the British cause could be aided without Canada providing troops, it would be enough. Accordingly, as early as 1938, when London requested air crew training facilities in Canada like those the Royal Flying Corps had enjoyed in 1917–18, King was more than happy to comply. Canada had unlimited air space and clear skies. It was close to American aircraft factories, and its own tiny aeronautical industry could only prosper from the program. Best of all, there would be no casualties as at Vimy Ridge or a conscription crisis within French Canada.

Representatives from the United Kingdom, New Zealand, Australia and Canada met in Ottawa in October 1939, to discuss a joint air training scheme to be held in Canada. The British Commonwealth Air Training Plan (BCATP) was signed on December 16 (the day before King's birthday); each of the four countries was to contribute men and money based on its "trainee population."

The BCATP was exactly what Canada's faltering bush air companies needed. All had lost their younger men to the armed forces. Many of those men beyond military age, like Tommy Siers, CAL's maintenance manager, volunteered their hard-earned expertise in aviation. Siers went to Ottawa temporarily to take charge of the overhaul of all RCAF aircraft engines. Two CAL pilots, Walter "Babe" Woolett and Peter Troup, tried to rejoin the Royal Air Force but were persuaded to remain in Canada to help organize the civilian air observer flying schools for the BCATP.

As the country struggled to get organized, CAL picked up some unexpected but lucrative contracts. On September 23, 1939, Walter Gilbert, the company's superintendent for the BC region, sent pilot Russ Baker a confidential memo: he was "one of a small selection of picked men . . . being retained by the company throughout Canada under instructions of the Minister of Transport," and he would not be allowed to enlist in the armed forces without permission from the minister of defence.[4] Baker was told to report to the Consolidated Mining & Smelting Company at Pinchi Lake. The mine produced mercury, which during the war would be used in fungicides for tropical clothing, gunsights, scientific equipment and explosives. Until a road was built to Pinchi Lake, CAL was charged with flying in all supplies and taking out thousands of flasks of mercury. Like many airline pilots, Baker became part of a "behind the scenes" operation far from the European battlefields that, though unglamorous, played a key role in the war.

The airline also began carrying the mail between Moncton, Halifax, St. John, Summerside, Charlottetown and, in the winter months, the Magdalen Islands. Because this involved overwater flying in harsh weather, the airline purchased three Beechcraft 18s. Not surprisingly, after having paid for the Beechcrafts, CAL lost the contract, on April 15, 1941, when Ottawa gave it to Trans-Canada Air Lines (TCA).

During the war the aviation industry in particular required a local source of aluminum, for which an unlimited supply of electricity was necessary. When the federal government planned an aluminum plant at Arvida, Quebec, CAL was asked to provide an aerial survey of the watersheds in the region so that hydroelectric dams could be built. Two

In 1940, Capt. Wilfrid Reid "Wop" May was general manager of No. 2 AOS Edmonton, operated by Canadian Airways Ltd. as part of the British Commonwealth Air Training Plan. CAA M2699

dams were erected at Lake Manuan and Passe Dangereuse on the Peribonca River, and CAL air-lifted all men and material from Lake Onatchiway to the sites. Bulldozers, diesel shovels, trac-tors, bags of cement, a complete sawmill, all the necessary fuel, a twenty-foot boat, baggage, bed-ding, narrow-gauge dump cars, six horses (the first of which was not doped), four bulls, a cow to supply the camp with fresh milk—all were flown to the site. The *Montreal Star* noted that the airlift not only flew the contractors their complete housing needs "and juicy T-bone steaks as well," but "the romantically minded company even accommodated the love-lorn by flying them out to keep dates with their girl friends."

From August 12 to the freeze-up in November 1941, 660 tons of freight were packed on board nine CAL aircraft, especially the Junkers "Flying Boxcar." Lake Onatchiway, the supply site, was named "Canadian Airways City," and the operation was supervised by Harold Smith. The onset of winter halted the flights until the lake froze suffi-ciently, and then skis were fitted on all aircraft. The operation was restarted in December, and when it was completed in October 1941, CAL had flown 3,000 tons of freight.

The airlift was greatly assisted because some of the aircraft, notably the Junkers, were equipped with an oil-dilution system developed by Tommy Siers. No longer did engineers have to suffer the lengthy and dangerous early morning start-up procedures of blow-pot heating to warm the engine and circulate the oil. Invented by Weldon Worth, a US Army Air Corps engineer, the oil-dilution system had come to Siers' attention in 1937 when he attended a maintenance course at Wright Field, Ohio. He experimented with it on his own car in a Winnipeg win-ter and installed it initially on a CAL Norseman in 1938. After two years of trials, the system was ready for large-scale use. For his work on the

Worth oil-dilution system, Siers was awarded the McKee Trophy in 1940—the first time it had been given to anyone other than a pilot.

The logical places to teach basic instruction for the Elementary Flying Training Schools (EFTS) were the municipal flying clubs, but the operation of the Air Observer Schools (AOS) to train air crews was put out for tender to commercial carriers. Each carrier was made a limited liability company that had to raise $35,000 in local capital and assure the RCAF that it could provide adequate instructional, administrative and technical staff, and all were to be run on the basis of making a small profit. All other revenues were to be returned to the government. The government supplied all aircraft, Link trainers and navigation instructors, and the company was responsible for the facilities and the pilots, mechanics, radio operators, clerks, storekeepers and parachute packers. Initially each AOS facility had 120 pupils; later, this was standardized to 189. Nine AOS were opened; of these, seven belonged to carriers that were or would soon be connected with CAL. Supervised by C.R. Troup from the Dominion Square offices in Montreal, they were:

■ **No. 1 AOS Toronto Island** (later Malton), operated by Dominion Skyways, opened in April 1940; W.A. McLeod, general manager. This facility had been opened by Woolett and Troup and served as the model for all other schools in Canada. Both men insisted that their salaries be frozen at prewar levels, enabling the AOS program to be run at a profit throughout the war.

■ **No. 2 AOS Edmonton**, operated by Canadian Airways Ltd., opened on August 5, 1940; W.R. "Wop" May, general manager.

■ **No. 3 AOS Regina**, operated by Prairie Airways; Dick Ryan, general manager.

■ **No. 5 AOS Winnipeg**, operated by Starratt Airways and Wings Limited, opened on January 6, 1941; D.S. Ormond, general manager.

■ **No. 7 AOS Portage la Prairie**, operated by Yukon Southern Air Transport, opened on August 28, 1941; W.L. Parr, general manager.

■ **No. 8 AOS Ancienne Lorette, Quebec**, operated by Quebec Airways, opened on September 29, 1941; G.R. Beck, general manager.

■ **No. 9 AOS St. John's, Quebec**, operated by Dominion Skyways, opened on July 7, 1941; W. Woolett, general manager.

Besides Wop May, several experienced bush pilots taught in the AOS. Mike Finland assisted May at No. 2 AOS and, from 1943, at No. 8 AOS. Matt Berry was second in command at No. 7 AOS, and Dick Ryan, whose Prairie Airways was soon to be bought by the CPR, served at No. 3 AOS. Ryan remembered the air carrier's role in the BCATP: "In the spring of 1940, Prairie Airways was contacted by a Mr. Apedale of the DND and advised that No. 3 Air Observer School was being located at Regina airport and would be operated under civilian management. Mr. Apedale asked if Prairie Airways wanted the contract."[5]

Ryan called a meeting of the directors and, on receiving their approval, formed a company he called Prairie Flying School Ltd. to operate No. 3 AOS. The government was already completing the necessary hangars and classrooms at Regina airport by the time it was in operation. The thirty aircraft supplied on loan to No. 3 AOS were Avro Ansons, which, Ryan wrote, had an unusual feature: "The undercarriage was retractable, but it had to be cranked up manually. It was rather hard work because to do so the crank had to be turned about one hundred times, and it wasn't easy to turn. It was much easier to lower the undercarriage since its weight was working for you. The civilian pilots often prevailed upon the student observers to do this job for them."

Each morning and afternoon, thirteen or fourteen Ansons would lumber out on a three-hour flight. The air carrier's pilots flew the courses given them by the students being trained in navigation under the eye of the RCAF instructor. Before Pearl Harbor, many of the pilots Ryan hired were American. Older and more experienced than the Canadians, they were also, he admitted, less trouble. Ryan recalled that the Canadian pilots frequently broke regulations on and off base:

> In order to put a stop to this, we put into effect a system of fines. The fines started at $25 and increased with the seriousness of the offence. When a pilot was fined, the money was sent to the Red

The ubiquitous Avro Anson began in 1935 as a fast, twin-engine airliner for Imperial Airways. In service with the RAF's Coastal Command in 1938, its deficiencies were obvious in comparison with the Lockheed Hudson. The Anson really came into its own as a BCATP trainer for bomber navigators. After the war the "Bamboo Bomber" was widely used by Canadian Pacific, PWA and several bush companies. CAA (UNN)

Cross as a donation in the name of the offender. The pilot in due course received a nice thank you letter from the Red Cross, who were oblivious of his motivation or lack thereof.[6]

This early in the war, life at the schools was cheerfully chaotic, and someone at No. 2 AOS penned the following poem:

I remember, I remember when this place called Number Two,
Was nothing but a bunch of scrub, with here and there a slough.
Just now we're short of several things, like sugar, tea and butter,
But when we hear of other folks, we really should not mutter.
Keep smiling friends, the time will come when we shall all be
    merry,
With Hitler and his crowd no more, while we drink beer and sherry.

In addition to operating its Air Observer School, because there was no flying club nearby, Quebec Airways also ran No. 11 EFTS at Cap de la Madeleine, which opened on October 14, 1940. Aircraft at the base were overhauled by CAL's St. Maurice Aircraft Repair.

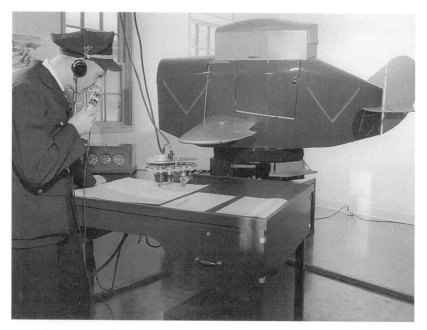

CAL personnel using the Link trainer at No. 2 AOS Edmonton to teach blind flying to BCATP trainees. CAA 989-53-27

Commercial carriers also managed the five repair depots built by the Department of Munitions and Supply to overhaul BCATP aircraft. CAL and Wings Limited had two depots in Winnipeg under M.E. Ashton, and Prairie Airways' H.H. Burke was in charge of the Moose Jaw depot. All CAL offices, no matter how small, set up blood donor clinics or raised money for the Red Cross. Sir Edward Beatty's Christmas Message for 1940 summed up the company's approach: "It should be a source of great pride to all of us that Canadian Airways has been called upon in an ever-increasing measure to play its part in the Commonwealth Air Training Plan. It has been stated that the Battle of Waterloo was won on the 'playing fields of Eton,' and this war may very well be won on the 'flying fields of Canada.'"

The airline was soon to be involved in an adventure as daring as any out of Hollywood. Egged on by Canadian actor Raymond Massey (who was a distant relative of Beatty's), Warner Brothers would make a movie about a group of bush pilots who enlist in the air training plan but, in keeping with their rough-diamond characters, run afoul of air force bureaucracy. *Captains of the Clouds*—the title was from a speech given by Billy Bishop—starred James Cagney and was partially filmed at the BCATP school at Uplands airport, Ottawa; the stars stayed at the Chateau Laurier hotel. To dramatize the Clayton Knight committee's recruiting drives for American pilots, it was shown in the United States, which in the fall of 1941 was still neutral but becoming apprehensive about the war. In the last reel of the movie, the heroes redeem themselves by ferrying Hudson bombers to England.[7] That this was an initiative of the CPR would have amazed American moviegoers.

The CPR's part in ferrying aircraft across the Atlantic Ocean was a direct result of Sir Edward Beatty writing to Mackenzie King on June 14, 1940. The CPR president had been appalled by the loss of shipping— some of those ships going to the ocean bottom were CPR liners—and suggested to King that because of the U-boat menace, a fast air mail service be set up across the Atlantic for the duration of the war. As the representative in Canada for the British Ministry of Shipping, Beatty feared for Canada's growing isolation from Britain. Pan American's Atlantic air services had been stopped by the United States Neutrality Act, and Imperial Airways (now renamed British Overseas Airways Corporation, or BOAC) had lost two of its flying boats to enemy action in Norway. Having just succeeded James Richardson to the presidency of CAL, Beatty volunteered the company's aircraft facilities to help run a proposed Atlantic air mail service from St. Hubert Airport. It was typical of him: irascible to his CPR employees, he gave away thousands of dollars from his own salary to children's charities, and the death of Margaret Hayworth no doubt affected him deeply.

To appreciate what was involved, it must be realized that to the public in the early 1940s, the Atlantic was an ocean of fear and challenge. Few people had flown over it; more than a decade after Charles Lindbergh's flight in 1927, fewer than a dozen adventurers had lifted off from the airstrip at Harbour Grace, Newfoundland, to attempt crossing the void between North America and Europe. Aircraft then didn't have the range to complete the crossing; neither could they reach the heights required to fly above the ocean weather, which especially in the winter months could be fatal. For a while, nations put their faith in airships and flying boats. When the United States captured the lead in flying boat technology, Britain's Imperial Airways concentrated on overland air routes to Australia. The company then put its hopes of conquering the Atlantic in a de Havilland–built landplane called the

Albatross and constructed a large airport at Hattie's Camp, Newfoundland, to receive it. Initially built at a railway stop 113 miles (181 km) from St. John's, the airport was renamed Gander after a lake nearby. The wooden construction of the Albatross proved unsuitable for flying the Atlantic, however, and it was transferred to European routes, leaving Gander deserted.

Even before the Germans invaded Poland, the British government had been impressed by Imperial Airways' Lockheed Electras—one of them was immortalized when it flew Prime Minister Neville Chamberlain to Munich to meet Hitler—and it placed orders for 250 Electras of the military version. Anticipating military contracts from the hard-pressed British and French, Lockheed disguised its airliner in wolf's clothing, calling it the "Hudson," a name that had English and Canadian connections. Although unarmed and untried in maritime conditions, the Hudson was desperately needed for coastal reconnaissance to replace the even more innocuous Avro Anson aircraft.[8]

Initially the Hudsons were to be disassembled and loaded onto ships for the transatlantic voyage, but due to the war other cargoes assumed greater priority. Soon they gathered at dispersal points in North America because of the shipping logjams—to the embarrassment of President Roosevelt and his neutrality policies. The German invasion of France worsened the situation as orders for American aircraft escalated to a scale undreamed of in peacetime: a staggering total of 26,000 aircraft of all types. But before any of the Hudsons, and now Liberators, Mitchells and B-17s, could be delivered to the British, they were flown to Canada for testing and repainting in Royal Air Force markings. Some were towed by horses across the US–Canadian border at Pembina, North Dakota, just south of Winnipeg, to pay lip service to American neutrality.

The idea that these aircraft could be flown on to England was not revolutionary, but the aeronautical establishment of the day—especially the experts in the Royal Air Force Coastal Command—thought this would be suicidal during the coming winter months, when icing would be a problem. Engines, they warned, would cut out with ice in their carburetors, propeller blades would lose their efficiency and the ice that piled up on the wings would make them lose their lift over the Atlantic, far from emergency airstrips. More of the precious aircraft would be lost in this way, they warned, than by the U-boat sinkings. The truth was that much of the weather over the ocean was still a mystery.

The Lockheed Hudson maritime patrol bomber was the military version of the Lodestar. Its gun turret was later removed for installation in England.
Lockheed PC016-003

Mackenzie King's cabinet discussed Beatty's Atlantic air mail idea and turned it down. C.D. Howe, now the minister of munitions and supply (and despite having TCA in his portfolio), replied that he preferred that CAL concentrate completely on the domestic market during the hostilities. Beatty's proposal, however, circulated on to more receptive channels, and it was read at the British embassy in Washington and by the former Imperial Airways managing director, George Woods-Humphrey, now retired in New York. It is impossible to know precisely when the idea of an emergency air mail service was transformed into a plan to deliver aircraft, but since no one knew more about flying the Atlantic than Capt. Don Bennett of Imperial Airways, he was brought into the picture.

It took another Canadian to get the ferry scheme under way. Max Aitken, Bennett's mentor, a newspaper magnate and now Lord Beaverbrook, had been chosen by Prime Minister Winston Churchill to be his minister of aircraft production. If the British were to survive against the Luftwaffe—and in the summer of 1940 that was uncertain—they needed an immediate, continuous supply of Spitfires directly from the factories to the squadron tarmacs. With so much at stake,

Churchill assigned "The Beaver" to expedite fighter production, knowing that only he could bully his way through industrial labour disputes and prewar red tape. The idea of doing the same on a larger scale—flying aircraft from American factories across the Atlantic to the front line—caught Beaverbrook's imagination and no doubt renewed his connection with Canada.

His agent in North America was Morris W. Wilson, president of the Royal Bank of Canada and a board member of CAL. On July 8, 1940, Wilson informed Beatty that the offer to place the CPR's air resources at the disposal of the war effort had come to his attention. Could this be done to help ferry the gathering Hudsons across the Atlantic? Beatty was already overworked with shipping difficulties and planning to amalgamate the various bush air companies, but the opportunity to help the war effort in so tangible a way was beyond anything he had imagined. The CPR president threw himself into the project wholeheartedly, working as a colleague remembered, "twenty-four hours a day," at great cost to his health.

Amazingly, in the midst of a war, this plan was from the start a private arrangement between "old boys"—"The Beaver," Beatty, Wilson and a few powerful Canadian anglophiles. Having been rebuffed by Ottawa, Beatty did not approach Mackenzie King or C.D. Howe, or ask Trans-Canada Air Lines for help. The treatment accorded James Richardson and the subsequent fortunes of CAL might have had something to do with this arrangement. But to avoid ruffling too many feathers in Ottawa, Beatty ensured that the proposed CPR air ferry be known as ATFERO, the Atlantic Ferry Organization. Howe and C.G. "Chubby" Power, the minister of defence for the air, were well aware of this private initiative, but because its patron was a man of Beaverbrook's stature—a man who had Churchill's ear—they were reticent about condemning it. They sanctioned the use of the airports at Montreal and Gander (the British had asked Canada to assume protection of Newfoundland during the war) and of the transport department's radio operators and meteorologists.

The agreement to fly the Hudsons (and now Catalina flying boats) over to England was signed between the British government and the CPR on August 16, 1940, and the Canadian Pacific Air Services Department (CPAS) came into being. Beatty was named the chairman, Woods-Humphrey the vice-chairman, and Lieut. Col. H. Burchall (also ex–Imperial Airways) as general manager. With his stature in the

Montreal business world, the CPR president was able to persuade several "dollar-a-year" executives also to work for the ferry organization. They came from publishing, merchandising and banking circles, and like Beatty they wanted "to do their bit for the war effort." Their influence and connections smoothed the way through wartime red tape to obtain the necessary fuel, equipment and personnel.

The CPR was made responsible for hiring and paying the air crew, servicing the aircraft and arranging for fuel. Windsor Station would be ATFERO's administrative centre, and until the new airport at Dorval was ready, the CAL hangar at St. Hubert Airport was to be the base for all the engineering. Salaries for the air crew and bills would be paid by the CPR accounting office and later reimbursed by the British Ministry of Aircraft Production. Because of the times, the CPR could find only automobile mechanics to prepare the Hudsons' Wright Cyclone engines for the Atlantic run; fortunately, when the later models arrived with Pratt & Whitney Twin Wasps, Pratt & Whitney Canada at nearby Longueuil could be called in to help.

It must have been difficult for a privately run endeavour such as this to find fuel and aircraft spares during a war, even with its high-powered connections; but hiring experienced pilots and navigators as ferry crews was impossible. Any Commonwealth citizen with the slightest aptitude for flying was already in his national air force or at that moment "square bashing" in the BCATP. The neutral United States, thanks to its extensive airline network, was a source of trained pilots, and there was no difficulty in recruiting American civilians by offering them attractive salaries of up to US $1,000 per month to fly to England. When Beaverbrook saw a photograph of the first American volunteers, he is reputed to have said, "They don't look like angels, but perhaps they can fly."

From his office at Windsor Station, Don Bennett involved himself in every detail of the operation. As one of the few pilots who had actually flown over the ocean, he knew at first hand the hazards to be faced. For the meteorologist stationed at Gander, Andrew McTaggert-Cowan, to be able to predict the weather in the mid-Atlantic with any certainty, an arrangement had been made with the Anglo-Newfoundland Development Corp. to have radio operators on two of their ships report on the weather on their way to and from England. The Hudsons, George Lothian from TCA discovered, were Lockheed 14s with extra fuel tanks. Their gun turrets had been removed for reinstallation in

England, and in place of the upper turret a hatch had been let into the roof of the cabin for "star shots." In the cabin itself a large fuel tank was secured in the space where the passengers would have sat, and another was located in the bomb bay. The engines could be run off the cabin tank, which was refilled by the navigator, who wobble-pumped fuel from the bomb bay at regular intervals. Navigation and oxygen equipment were also fitted in while the hired crew were checked out to verify their exaggerated claims of airline experience. One boasted that he had been thrown out of every airline in the United States. Those selected—the "Foreign Legion," as the recruits were nicknamed—familiarized themselves with operating the Hudsons from St. Hubert, signalling with the Aldis lamp and flying in formation within sight of each other.

George Lothian and Jock Barclay were two of the pilots who reported to Punch Dickins at ATFERO. Wally Siple was in charge of the civilian pilots, and a young Al Lilly tested the Hudsons once they had been prepared for the Atlantic crossing. Lothian later wrote that "Punch must have had his hands full, as crews consisted of as wide a mixture of backgrounds, experiences and cultures as had been gathered together . . . [T]here was Clyde Pangborn and Hugh Herndon, who had made the first crossing of the North Pacific in a single-engine Bellanca, ex-TCA captains Jack Bradley and Bud Scouten, Archie McMahon a bush pilot and part-time undertaker from Sault Ste. Marie, Louis Bisson a well-known Arctic pilot and Don MacVicar who would pioneer the Crystal Route . . ."

When Lothian and Barclay asked what it was like to fly the Atlantic, Lew Parmenter, an old bush engineer, gave them some sage advice drawn from his experience of sitting behind a variety of pilots over the years: "I don't say anything. I sit behind the pilots and when the going gets tough I watch the neck of the one who is flying the airplane. When it starts to get red and the colour starts going up to his neck, I poke him and point to the other pilot and say, 'Give it to him.'"[9]

Hoping that some commercial good would come out of a transatlantic ferry service, on October 2, 1940, Ed Beatty and George Woods-Humphrey wrote to C.D. Howe, outlining a plan whereby CPAS and TCA could "jointly accept responsibility for initiating the development of this vital enterprise. We have searched the American market for suitable planes and . . . there are two types of military aircraft, a few of which could probably be obtained quickly . . . and that through a separate company jointly controlled and bearing a distinctive name such, for example, as 'Canadian Atlantic Air Lines,' Canada would be making a very important contribution to the prosecution of the war . . ." Howe discussed this proposal with Captain Balfour, the United Kingdom's secretary of the air and Sir John Reith, chairman of Imperial Airways, and concluded that neither was willing to help. No matter how desperate Britain's plight, an Atlantic air service run by Canadians was, in the long run, still considered a commercial threat to British aviation. If anything, the British were consolidating their hold on oceanic flying as BOAC purchased six long-range Liberator bombers to use as transports. As a result Howe replied to Beatty that although he "saw advantages to his proposal, the move was one that should be considered after the war."

The first ferry flight was to be a seven-plane formation with Bennett in the lead Hudson. The aircraft were to fly within sight of each other from Gander to Aldergrove, Northern Ireland, the closest British airbase. The flight was planned to occur at night, when the stars could be used for navigation. If trained pilots were scarce, however, experienced navigators were nonexistent. Navigation had gone out of use in the United States when commercial pilots had converted to flying from one radio contact to another. Nevertheless, a single navigator was found and assigned to Bennett in the lead aircraft. Beaverbrook's plan was that the pilots could lock on to it and "play follow the leader" across the Atlantic. The pilots knew that they could not afford to become lost or deviate from the course in any way, because there was just enough fuel to get them to Northern Ireland. So desperate were the British for the scheme to succeed that the air ministry held that even if three of the seven aircraft got through, the ferry service could be repeated ad infinitum.

As the recruiting and training continued, all in ATFERO knew that fall was drawing near and that no one flew the ocean so late in the year. But not until late October did Bennett consider the first seven crews ready and allow them to fly their Hudsons to Gander. Navigation en route was by dead reckoning, as a final practice for the crossing.

The once-quiet Newfoundland airport had been transformed by the war after Ottawa stationed an RCAF squadron of Douglas Digby maritime patrol aircraft and a regiment of the Queen's Own Rifles there to guard the perimeter. This influx of the military severely strained what little housing there was, and the CPAS were hard pressed to find any accommodation for its ferry crews. But since Gander also remained a

Crews of the first ferry flight at Gander airport, November 1940, with the Newfoundland Railway sleeping cars behind them. Composed of Americans, Britons, Canadians and the single Australian D.C.T. Bennett (standing at far right), the ferry flights were about to usher in mass Atlantic air travel. DND PMR 85-475

railway community, two sleeping cars and a diner were requisitioned from the Newfoundland railway system and parked on a siding for CPAS use.

On November 9, 1940, the first ferry flight was prepared to leave Gander, but a snowfall and an overnight freezing mist delayed its departure to the next day as ground crews chipped ice off the wings. The next evening was clear and Bennett made the decision to go. At this stage no one knew what weather to expect in the mid-Atlantic because U-boats had sunk both the Anglo-Newfoundland ships that provided meteorological information. The British had replaced them with two more ships with meteorologists on board, but by October they too had disappeared without a trace—presumably they were sunk as well. Without their reports, McTaggert-Cowan (known as "Mr. McFogg" to the pilots) had to estimate where he thought the crews would encounter the icing storms and headwinds.

In the gathering darkness of November 10, the seven Hudsons were fuelled and positioned line abreast on the wide runway so that the lead plane would not have to waste precious gasoline circling around for the others. Bennett had prepared the flight plan and presented each crew with a card, showing courses and times for alterations of each leg of the course. The radio was not to be used because Luftwaffe aircraft were known to range out into the Atlantic.[10] If a crew ditched, there was a slim chance of being rescued by the convoys, who were themselves dodging the U-boats. When the nine Americans, six Britons, six Canadians and a single Australian (Bennett) climbed into their Hudsons, the drama of this historic flight was not lost on the spectators. Not only was the dreaded "Water Jump" being challenged, but it might be said that the outcome of the war was at stake.

It was the eve of Remembrance Day; poppies were distributed to the crews by the airport manager's wife, and to lend an air of drama to the occasion, the pipe band from the Queen's Own Rifles formed up to play an appropriate piece. Because of the winter darkness, however, the musicians could not read their sheet music, so they struck up a tune they knew by heart from church parade. Thus, as the Hudsons moved down the runway into the unknown, the mournful strains of "Nearer My God to Thee"—the hymn reputedly played as the *Titanic* went down—were heard over Gander Airport. When the Hudsons took off, the band moved into "Lead, Kindly Light" and ended with the rousing "There'll Always Be an England."

Once they were airborne, it was up to Bennett and his pilots to complete what Beatty and Beaverbrook had begun. The aeronautical designers at Lockheed, the CPAS personnel, and McTaggert-Cowan all played their part in the epic Atlantic flight—the first of thousands to come. Over the ocean, the compasses failed on some Hudsons, oil leaked on others; the Bendix radios short-circuited, and one of the co-pilots passed out when the oxygen tube slipped from his mouth. The seven aircraft hit a storm front (exactly where McTaggert-Cowan had predicted) and almost collided with each other as the formation bucked its way through a blackness of heavy rain and turbulence. A strong southwesterly wind scattered them into a wide arc that stretched out into the grey dawn. Eleven hours and twelve minutes after they had left Gander, Bennett's and two other Hudsons touched down at Aldergrove Airport. A little over an hour later, the remaining four had safely landed.

The Atlantic Ocean had been conquered—en masse, in winter and without the loss of a single aircraft. That the Hudsons were Lodestar 18 airliners with roundels painted on them was not lost on Juan Trippe,

Eleven hours and twelve minutes after leaving Gander, the first three Hudsons touched down at Belfast's Aldergrove airport. Thanks to Beatty, Beaverbrook, McTaggert-Cowan and the Canadian Pacific Air Services, they had proved that aircraft could be flown from the factory to the front. DND (UNN)

C.D. Howe and George Woods-Humphrey, who now knew that commercial transatlantic travel was within their grasp. The fact that the historic flight had been co-ordinated by the CPR was soon forgotten.

The second crossing took place on November 29 and the third on December 17. Only on the fourth attempt, on December 28, did a Hudson crash on takeoff and another turn back with engine failure. The worst accident occurred on February 20, 1941, when a ferry aircraft crashed near Gander, killing all the crew and a famous passenger, Sir Frederick Banting, the Nobel Prize winner and co-discoverer of insulin.[11]

The war had the beneficial effect of putting both national railways in the black for the first time in their histories. Moving munitions and men between the coasts, at a time when gas rationing had eliminated competition from trucking, required the logistical organization that the railways thrived on. Three years into the war, the CPR and the CNR made profits of $40 million each. The executives at Windsor Station decided they could now afford to get into the flying business. Even before James Richardson's death, Beatty had made Laurence Unwin, the CPR's vice-president of finances, survey the sorry condition of all air carriers in Canada. Unwin discovered that as the country moved to a wartime footing, with customers, replacement aircraft, spare parts and gasoline scarce, the remaining members of the bush-flying industry were close to bankruptcy.

Taking advantage of this, Beatty had bought Mackenzie Air Services for $658,000 on November 9, 1940 and made Unwin the president and W.M. Neal vice-president of the CPR's new aviation branch. None of the expected sound and fury emanated from Ottawa over the purchase, indicating (Beatty hoped) that perhaps Howe was indifferent to it. He did not know that at that moment the minister was sailing to England (to negotiate the building of de Havilland Mosquito aircraft in Canada) when his ship was torpedoed in the mid-Atlantic. Encouraged by the official silence, Unwin set out across Canada to negotiate the purchase of all other major air carriers.

Unwin had emigrated from Kent, England, as a nineteen-year-old and began his railway career as a clerk at Chapleau, Ontario, in 1908. He served in the Canadian army during the Great War, was commissioned as a major and decorated with the Military Cross. After the Armistice, he rejoined the CPR, rose through the accounts department and became comptroller in 1932 and vice-president in 1935. Bald, bespectacled and always photographed in a pinstripe suit with a starched collar, Unwin looked every inch the "National Wartime Rationing Administrator" (which he was), rather than the saviour of bush pilots. No one could have known, but the former accountant was the first of the new breed of airline managers, with more experience of the "bottom line" than the cockpit.

The prize among the collection of near-bankrupt air carriers was undoubtedly Grant McConachie's Yukon Southern Air Transport (YSAT), which Unwin bought on January 13, 1941, for $1,057,000. Ginger Coote Airways went on February 1 for $62,743, and at the annual shareholders meeting it was announced that the CPR had also bought Starratt Airways for $425,000. Wings Limited followed for $190,000, Roy Brown's Arrow Airways for $76,000, Dominion Skyways for $142,000 and Prairie Airways for $528,000.[12]

To the surprise of many, on November 28, 1941, Grant McConachie was made an assistant to vice-president W.M. Neal. In hindsight, it was an inspired choice: Unwin and Neal were "railwaymen," and someone with expert knowledge of the aviation industry was needed. Putting Ginger

Laurence "Larry" B. Unwin was made president of the CPR's aviation branch in November 1940 as he set out across Canada to buy up the faltering bush companies. CAA 989-62-28

Coote, Roy Brown or Leigh Brintnell, the owners of the other companies, into this position would have torn the fragile alliance apart. It would have been difficult, if not impossible, for them or their employees to take orders from former competitors. McConachie, on the other hand, was universally liked: people couldn't help being caught up in his enthusiasm.

Besides, he brought with him a rich dowry of modern landplanes. His YSAT had three Barkley-Grows and a Boeing 247D, and soon it would own Lockheed 14s. In February 1940 McConachie had announced that YSAT would greatly extend its Edmonton–Yukon flights to Fairbanks, Vladivostok and Shanghai. He had become friend-ly with Pan American's Juan Trippe, who had just seen his Atlantic flights to Europe cut short because of the war and now turned his attention to the Pacific. Before Pan American could fly over Canadian territory to Alaska, Trippe needed a chain of radio-beam stations to be set up along the route. As an American he couldn't get a licence to operate a radio station in Canada, so he was very pleased when McConachie offered his own rudimentary radio cabins and airfields instead. Trippe's gratitude extended to giving McConachie flights in his Boeing Clippers and promising Pan American's co-operation in the Far East.

Disregarding the darkening war clouds in China, McConachie announced that the first stage of the YSAT flights to the Orient— through Fairbanks—was to begin that summer. At any other time this might have been Grant's bombast, but that June the Canada–United States Joint Defence Board considered the northwest route to Alaska and the Soviet Union of strategic importance if a war with Japan were to begin. "We had just laid down an airport at Fort Nelson in 1939 and were in the middle of putting one in at Watson Lake," McConachie later

Because of McConachie's familiarity with the staging route to Fort Nelson, Watson Lake and Whitehorse, the government pried two Lockheed 14s away from TCA for his company, YSAT. Here, one of them, CF-CPC, is shown before the newly opened Dorval airport. CAA 988-31-19

recalled, "when the US government asked the Canadian government to help them by finishing these airports so they could put land-based fighters through this good-weather route from Edmonton through Grande Prairie, Fort St. John, Fort Nelson, Watson Lake, Whitehorse, through to Fairbanks and eventually on to Nome and Russia. So these airports really were the starting of the Northwest Staging Route which supplied Russia with aircraft. And because these air bases required supplies and fuel for the fighters, they had to be connected by road. This in turn became the Alaska Highway."

Some of the airfields were also built for another wartime emergency. To safeguard a wholly internal oil supply, a pipeline 4 inches (10 cm) in diameter called CANOL (Canadian Oil) was constructed from the Norman Wells oilfield on the Mackenzie River, through the Richardson Mountains, to a refinery at Whitehorse for distribution to Alaska. One of the pilots chosen to organize and manage the flying operations was Bob Randall, the former operations manager of

Mackenzie Air Services, who was recognized for his flying in the aerial search for Russian pilot Sigmund Levanevsky in 1937. CPAL loaned Randall to Bechtel, Price and Callahan, an American construction contractor. The CANOL project required all equipment and personnel to be flown in during the short summers, and airfields were built at Fort McMurray, Hay River, Fort Simpson and Yellowknife. Reflecting the need for modern aircraft in so important a project, the US joint chiefs of staff themselves allocated seven Lodestars from the US Army Air Force to CAL.

The first US military engineers and construction battalions were followed by thousands of civilian workers, who built a two-way graded, gravel highway of 1,470 miles (2,451 km) from Dawson Creek, BC to Fairbanks, Alaska. Because McConachie's pilots were familiar with the terrain and the first airfields in the region were his, YSAT was given permission to buy whatever aircraft and gasoline it needed. The Barkley-Grows were too small for the increased traffic, and since McConachie had no money, the government prised two Lockheed 14s from TCA for him. In February 1941, the impossible happened: McConachie was given permission by C.D. Howe to buy nine more 14s, directly from the Lockheed plant at Burbank, California. They came in one colour only—military green, which was a drawback for planes flying (and crashing) in the bush—but this became an asset when they landed at the US Army Air Corps base at Great Falls, Montana, where their gas tanks were filled up without charge because no one would believe they were civilian aircraft.

When Unwin came to McConachie's door, YSAT had a total of five aircraft and a work force of seventy, including nine pilots, twelve ground crew and four stewardesses. After TCA, it was the most modern airline in Canada. But Unwin's accounting instinct told him that all of these facts were a facade. The airline's debts totalled $181,000, leaving the remainder of the purchase price to be divided up among the shareholders. More important than the aircraft and routes, as a condition of the sale McConachie was hired by CPR at a monthly salary of $500.

The Lockheed 14s were the first commercial aircraft in Canada in which the passengers had the luxury of flight attendants, and TCA began recruiting registered nurses as stewardesses in June 1938. Like the airlines in the United States, TCA and YSAT discovered that it was no longer acceptable merely to transport people from one airport to

The Lockheed 14s were the first commercial aircraft in Canada with flight attendants. These "well-bred ladies" fed the passengers, distracted them from being airsick, administered oxygen and were not allowed to accept "tips." Here Jewel Butler, the first CPA stewardess, practises serving colleague Jessie MacLean. CAA 1356

another. Once on board, passengers had to be distracted from the perils of flight by being fed (and no longer by the disgruntled engineer); they had to have oxygen administered to them as the aircraft strained over the mountains, be cleaned when airsick and shepherded to and from the transit rooms at refuelling stops—all this by what American Airlines (who had hired their first stewardess in 1930) called "well-bred" young ladies with nursing qualifications.

In his quest for airline respectability, and aware of what would entice male customers on board—a reccurring theme in Canadian Pacific's future—as soon as he got the Lockheeds in June, McConachie started hiring stewardesses. The rigours of serving food (still cold) and coffee from a thermos while giving comfort on a bucking Lodestar were nothing compared with what the first YSAT stewardesses had to endure on the many stopovers. One recalled being stranded as the only female in a remote construction camp of eight hundred men. That night she was awakened in her cabin by the howling of wolves, but with relief she

remembered having been locked in by airline staff to prevent "the attention of wolves—the two-legged kind."

In Montreal, Canadian Pacific's Atlantic ferry service was now better organized. It was preparing to move to the new airport at Dorval,
which was scheduled to open in September 1941, and as more navigators were made available from BOAC, one was assigned to each aircraft,
allowing Bennett to stop the dangerous formation flying. Aldergrove
Airport was replaced by Prestwick in Scotland, and a thirty-three-room
hotel called the Eastbound Inn was opened for the crews at Gander
Airport. Best of all, McTaggert-Cowan got his weather ships. George
Lothian explained how it came about:

> Just before Pearl Harbor, a United States admiral enroute to the
> U.K. on a mission he felt would brook no delay was held up at
> Gander for several days because of McTaggert-Cowan's "no-go"
> forecast. This was due to complete lack of any information on what
> was happening in mid-Atlantic. In the course of continually bug-
> ging "Mr. McFogg" (as McTaggert-Cowan was called) for permis-
> sion to proceed, the admiral learnt a good deal about getting
> accurate data in mid-Atlantic . . . and vowed to do something about
> it on his return to Washington. Within a month, four U.S. naval
> weather ships were positioned in the Atlantic, fully armed to com-
> bat and destroy submarines.[13]

The deliveries a success, ATFERO's fame spread, at least in official circles. Now everyone wanted to be part of it, especially Howe, who saw
an opportunity for his TCA pilots to gain from the experience. The
ferry crews were initially returned to Montreal by ship, but in May
Bennett took some back in the bomb bay of a Liberator, unknowingly
beginning the return ferry service that BOAC (with three TCA air
crews) would take over, ensuring a continuous rotation of ferry crews.
Despite the CPR's efficiency, the system was becoming too complex for
the railway to run by itself. By June 1941, an estimated eight hundred
aircraft of all sizes and types were expected—well beyond the abili-
ties of the forty pilots that Bennett had personally checked out. The
implications of American civilians flying British bombers into war
zones prompted President Roosevelt to ask the British government to
take over the ferrying and end the arrangement with the CPR. On
May 27, 1941, Beatty received a letter from the British Ministry of

Aircraft Production asking for the immediate termination of the contract.

> I regret to inform you that owing to conditions over which we have no control it is necessary to give you formal notice of termination of the agreement dated August 16, 1940 . . . under which you performed certain services in connection with the ferrying of airplanes across the Atlantic. The exigencies of the war have made it necessary that this service be operated in future direct by the British Ministry in close collaboration with the Governments of the United States and Canada. The services performed by your company have been a notable contribution to the war effort.[14]

This letter might have been a shock, but the CPR president was beyond caring: in the fall of 1940 his pride and joy, the Canadian Pacific liner *Empress of Britain*, was torpedoed and sunk off the coast of Ireland. Beatty went into a slow mental and physical decline and never recovered. On March 17, 1941, he suffered a massive stroke that froze his handsome face into a deathlike leer.

Despite this, on August 11 Beatty presented Howe with a plan that became the first of many proposals to merge the two companies. He reminded the minister that the CPR held a minority interest in CAL and that with the war and Richardson's death it could "secure a controlling interest . . . on favourable terms . . . as well as other minority shares . . . allowing the government and the CPR to own, control and . . . easily acquire all air services for a completely co-ordinated system of air transport which could link with Trans-Pacific and Trans-Atlantic services." Beatty then suggested that:

■ the CPR should be permitted to acquire 49.8 percent of the stock of TCA;

■ the government should purchase 50 percent of Mackenzie Air Services, Yukon Southern Air Transport, Ginger Coote Airways and Starratt Airways;

■ the Richardson estate holdings in CAL should be purchased jointly and equally by Ottawa and the CPR, with the government taking over 50 percent of the shares now held by the CPR;

■ both the government and the CPR should subscribe in equal shares the capital necessary for Canadian participation in transatlantic and/or transpacific ocean air services. Not only would this eliminate costly competition and duplication of services, but Canadian travellers would have the benefit of the CPR's experience and the Canadian government's power in dealing with foreign governments.

Howe replied that all this was a good idea, but until the war was won, "no steps would be taken to work out a permanent solution to Canada's air transportation." Believing that he had official blessing for the CPR's entry into aviation, on December 3, 1941, Beatty told the minister that the Richardson family had approached him to buy CAL and that as his directors felt it could be secured at a good price, he intended to do so. Howe answered on December 6 that he had no objections to this. In trying to keep all the balls in the air simultaneously, Beatty was literally working himself to death; but he had been the supreme ruler of the Canadian Pacific empire for too long to delegate authority. Laurence Unwin was away negotiating, and in his absence, W.M. Neal took over more of the airline organization, becoming chairman of United Air Services in Edmonton, a holding company for CAL and Mackenzie Air Services.

Eventually paralyzed and unable to speak, humiliated by requiring help both in his personal and corporate functions, Beatty died on March 23, 1943. Thousands of Montrealers lined the streets to watch his funeral cortege pass by. Mackenzie King gave a moving tribute at his funeral: "Few men in our country have given so much of their time and energies to serious and useful work for the benefit of the community. His patriotism was of the highest order, his example of quiet and efficient public service will long be remembered." J.A. Wilson had long since taken over the running of the CPAS, and when Woods-Humphrey resigned, Punch Dickins replaced him as vice-chairman and manager.

D'Alton Corry Coleman had been elected CPR's new chairman and president in December 1942 and was committed to ensuring that Beatty's wish to extend the air network be pursued. The British government's letter had called for the termination of ATFERO by June 30, 1941, but not until August 1 did the Royal Air Force Ferry Command assume all the CPAS's duties, and even then many CPR staff remained on loan to it. Now that the British were in charge, Ottawa could encourage ATFERO openly and ensure that there would be no shortage of air

D'Alton Corry Coleman was appointed chairman and president of the CPR in December 1942. A former newspaper editor, he crossed swords with C.D. Howe on TCA's monopoly, forcing Ottawa into clarifying its aviation policy. CAA 989-62-25

crews. It had the RAF, the RCAF and the graduates of the BCATP to choose from. Lord Beaverbrook, who had made few friends by his bullying, had also left, having resigned from Churchill's cabinet over a dispute concerning aircraft production.

The CPR's momentum to acquire a domestic airline network continued unabated with the purchase on November 14 of Wings Limited for $190,000 and Prairie Airways for $527,743. Not unexpectedly, Unwin and Neal then met with James Richardson's widow, Muriel, to buy CAL and its subsidiaries, Quebec Airways, Dominion Skyways and Arrow Airways. The result was that CAL's general manager, G.A. Thompson, received the following letter on December 17 from Muriel Richardson, informing him of the sale of CAL to the CPR.

Dear Mr. Thompson:

It has been apparent for some time that more constructive measures should be taken to eliminate wasteful competition in the field of commercial air transport in Canada, and the move towards this desirable end has been underway for many years . . . the Canadian Pacific Railway indicated to me their desire to purchase our interests and bring all commercial air services under one management. The advantages of this arrangement are so obvious and are such a fulfillment of all that my husband aimed to accomplish that I felt no alternative but to accept their offer. My action also implements an understanding entered upon many years ago between my husband and Sir Edward Beatty, when they agreed that whenever the Canadian Pacific Railway decided to enter extensively into the field of air transport, Canadian Airways would be available as one medi-

With the acquisition of Canadian Airways, the CPR redesigned its logo, displacing the Flying Goose and its own beaver with a corporate shield featuring a Lockheed Electra. This is the CPAL captain's badge. CAA 989-40-4

The flight attendant's badge. During the war, stewardesses were required to sell Victory Bonds to a captive audience in flight. CAA 989-40-2

um whereby this entry might be made. Notwithstanding all this, the decision has caused me real regret. I shall never forget my husband's interest in the opening up of Canada's northland by the pioneer flights of Western Canada Airways' pilots nor later his pride in the achievements of the personnel of Canadian Airways . . . His disappointments were many—as you know—but they were never caused by the failure of anyone within the organization.

Yours Sincerely,
(Signed) M.S. Richardson
(Mrs. James Richardson)[15]

Mrs. Richardson's letter was published in CPA's newsletter, *The Honker*, which on Unwin's express orders would be retained by Canadian Pacific with A. Graeme Macdonald as its editor until 1944. The "Flying Goose" continued on its masthead and on the CAL aircraft, but was gradually displaced by the corporate Canadian Pacific shield. With its interests so diversified, Canadian Pacific took the beaver off the old railway logo and replaced it with the words "Canadian Pacific" on the crest, adding the appropriate branch symbol beneath—a hotel, a ship, a truck or a telegraph pole. In the airline's case, it was a Lockheed Electra. But the old beaver was only hibernating: to mark the postwar expansion of the company in 1946, it would be awakened to sit on top of the shield and the symbols would disappear—forever.

Melding a group of rivals together would have been difficult even if they hadn't been aerial gypsies who prided themselves on self-reliance and nonconformity. If all the managers were pleased at being taken over by a secure, wealthy company like the CPR, many of the bush pilots chafed under the company's bureaucratic rule and quit. Having been independent for so long—quoting their own rates for freight, assigning their own staff, flying wherever they wanted (and settling differences of opinion with a bottle or a fistfight)—the bush pilot fraternity resented the railway men. Directives poured out of Windsor Station ordering McConachie, Gilbert and Brintnell to check with head office before committing their aircraft to contracts. Walter Gilbert was one of many who didn't remember the CPR merger with pleasure: "The spirit was entirely different from what we had in Canadian Airways," he said. "It was just too darn much army. When they brought in that brownie system of merits and demerits like they use in the railroad, that just finished it. You sign here, and you keep this copy, and we keep this one . . ."[16]

Earl Gerow, who had begun his career with Donald MacLaren's Pacific Airways, spoke for many of the old-timers:

> When you were in the bush operation, you got up at 4 o'clock in the morning when it was early daylight, and you worked until 10 o'clock in the evening. If something came up and you wanted the day off, well, you took the day off. When you came into the airline industry and its schedule, you worked your five days a week and you had your two days off. You became schedule-conscious after a while. It took a while to get used to it and, for people who came and went as they wanted to, it was a big change. A lot of people don't like to be told, "You do this, you do that, you work from 8 o'clock to 5 o'clock." It became a "living-to-schedule" situation.[17]

The railway's purchase of CAL gave it a motley collection of seventy-seven aircraft of twenty-seven types, the most numerous being the ten ancient Fairchild 71s. In addition there were two Lockheed 14s loaned from TCA, and five Boeing 247s that Quebec Airways had borrowed from the RCAF for the strategically important North Shore routes—giving CAL a total of 84 aircraft of 29 types. The czars of Windsor Station then merged all the carriers and their forty-eight operating licences. The whole was incorporated as Canadian Pacific Airlines Ltd. (CPA) on May 16, 1942. Following the mergers, the organization of CPA stood as follows:

Head Office: Windsor Station, Montreal
D.C. Coleman: Chairman
L.B. Unwin: President
W.M. Neal: Vice-President
T.H. Moffit: Treasurer
W.R. Patterson: Comptroller
General Office: 620 Dominion Square, Montreal
C.H. Dickins: Vice-President and General Manager
T.W. Siers: General Maintenance Manager
D.B. Wallace: Assistant to Vice-President
E.L. Smith: General Traffic Manager
F.V. Stone: Personnel Manager
Western Lines: Municipal Airport, Edmonton
G.W. McConachie: General Manager
C. Becker: Assistant to General Manager
B. Phillips: Special Assistant to G. McConachie
H. Hollick-Kenyon: Supervisor of Operations and Training
D. Patry: Chief Pilot
N.E. Dennison: Superintendent of Maintenance
R. Terpenning: Assistant Superintendent of Maintenance
H.S. Jones: Superintendent, St. Lawrence District. 101
    Confederation Building, Montreal
L.J. Dalton: Superintendent, Northern Quebec District
W.H. Coates: Assistant Superintendent, Northern Quebec
    District
W. Kahre: Superintendent of Maintenance, Rimouski
Air Lines Agencies: Royal York Hotel, Toronto
Miss C. Connon: Assistant Manager
Central District: 164 Royal Alexandra Hotel, Winnipeg
U.S. Wagner: Superintendent
T.W. Brown: District Traffic Manager
W. Field: Superintendent of Maintenance
H.A. Schade: Assistant Superintendent, Sioux Lookout[18]
Saskatchewan District: Regina Airport
W.T. Bunn: Assistant Flying Superintendent
J.L. Green: District Traffic Manager
F. McNair: Crew Chief
Mackenzie District: Municipal Airport, Edmonton
W.J. Windrum: Superintendent

Lockheed Lodestar CF-CPB, in military brown and green, was one of two allocated by the Joint Chiefs of Staff, Washington, to Canadian Pacific. Many years after the war Unwin revealed that they were allocated because the aircraft were needed to carry the raw material from Uranium City, Saskatchewan for the making of the first atomic bomb. CAA 989-71-38

    D. Stirton: District Traffic Manager

    J.W. Busby: Supervisor of Maintenance

Yukon District: Whitehorse

    E.R. Field: Superintendent

    J.W. Barber: District Traffic Manager

Vancouver District: A.M.F. Vancouver

    W.E. Gilbert: Superintendent

    G.A. Scott: District Traffic Manager, Hotel Vancouver

    G.C. Wilson: Crew Chief

    F.R. Baker: Flying Assistant Superintendent, Fort St. James

James Richardson's dream had lasted fifteen years—four as Western Canada Airways and eleven as Canadian Airways Ltd. The *Financial Post* estimated that between 1927 and 1941, these companies had carried a total of 60 million pounds (27 million kg) of cargo, 8 million pounds (3.6 million kg) of air mail and 250,000 passengers.

CPA's first crash occurred on December 20, 1942, when Lockheed Lodestar CF-CPD, en route from Prince George to Vancouver, hit the side of Mt. Cheam, killing the pilot, co-pilot, stewardess and ten passengers. The cairn was erected by CPA employees in their memory. CAA 988-3-151C

Besides giving the railroad some glamour in the skies, the airline also made it money. Because the airlines were bought by the CPR and not CPA, it was the railway that claimed the income tax capital cost allowance; this arrangement continued into the jet era. For its first year of operation, CPA and its affiliated companies reported:

Operating Revenues

Passenger	$1,805,409
Mail	725,045
Other transportation revenues	1,467,581
Incidental revenues	299,317
Total	**$4,297,352**

Operating Expenses

Aircraft operation and maintenance	$2,999,731
Ground operation	747,554
Incidental services	108,916

Traffic and general administration	547,135
General taxes	11,425
Total	**$4,414,761**
Net Operating Income	**$117,409**

CPA's revenues and expenses for its first year constituted 30 percent and 40 percent, respectively, of the totals for all Canadian air companies operating over licensed routes.

The airline suffered its first crash on December 21, 1942, when one of the two ex-TCA (and ex-YSAT) Lockheeds, CF-CPD, en route from Prince George to Vancouver, hit the side of Mt. Cheam, BC, killing the pilot, co-pilot, stewardess and ten passengers. Acquiring replacement aircraft for the airline's increasing passenger load was almost impossible. It was only through the personal intercession of President Roosevelt that the airline was able to get ten Lodestars. Many years after the war, Laurence Unwin recalled that the deciding factor in this largesse was that CPA was chosen to carry the raw material from Uranium City, Saskatchewan, that would be used to manufacture the first two atomic bombs.

In August 1943, when the war was at its height, the notices in the airline's newsletters reflected the mood:

> It is with the deepest regret that we read of the death of Captain W.V. "Bill" Walker, whose service to his country was ended in December 1942 somewhere over the North Atlantic. Bill was with No. 1 AOS for over two years, leaving his post as Flight Commander in the Fall to continue his flying with RAF Ferry Command. We know his quiet confidence and unassuming friendliness made him as highly regarded in Ferry Command as he was here, and our most sincere sympathy goes out to Mrs. Walker and their little girl.[19]

Albert Mah, a twenty-three-year-old Chinese-Canadian pilot flying for Quebec Airways, became one of the many CPA air crew that the war siphoned off. In 1942, when Mah heard that his thirteen-year-old sister and aged mother were starving in the part of China then under Japanese occupation, he consulted with his brother Cedric—also a CPA pilot but based in Winnipeg[20]—and decided to do something about this. He obtained permission to leave the company and got a job flying in Kunming, China, for the Pan American subsidiary China National Aviation Corporation (CNAC). Although he couldn't speak the local

Vancouver airport during the war. The huge Boeing factory, built in 1939, was sold after the war to Canadian Pacific Airlines and later Pacific Western Airlines. CAA 989-51-4

dialect, on his "vacation" Mah made his way to the war zone. Still in his uniform, he floated down a coastal river in a junk for three days, then bargained with sedan chair coolies to carry him through Japanese lines. He walked for days over trails and rice paddies until, staggering into his mother's village, he was met by his sister, whom he had last seen in Prince Rupert, BC. His mother took his sudden appearance in stride and calmly invited him in for tea. Too old to make the journey out of China, she was hidden in Canton with friends. The former CPA pilot took his sister to safety and returned to his CNAC job, flying the "Hump." Looking back on the trip Mah said, "In many things, I did look stupid. I guess I'm just lucky."

C.D. Howe, now at the height of his power and flushed with his success in reorganizing the aircraft industry, summoned D'Alton Coleman to Ottawa and warned him that the government intended to look into the legal ramifications of the CPR's swallowing of the bush companies, especially their forty-eight licences to operate those routes. The Board of Transport Commissioners held that the all-important licences were

Air Commodore Herbert Hollick-Kenyon, supervisor of Canadian Pacific Airline operations at Edmonton, with H. Hardham, flight dispatcher. McKee Trophy–winner Hollick-Kenyon later became the airline's first chief pilot. CAA (UNN)

not transferable to CPA; new licences would have to be applied for, and the old ones were suspended. In January 1943, CPA applied for fifteen new licences that would have consolidated all the services now being operated under the existing forty-eight licences. Although no new airports would be served, the grouping of destinations was more logical. The Board of Transport Commissioners recommended to the governor-in-council that the grouping be renamed so that the new licences could be issued in a consolidated form, but the council refused.

On August 5 Howe told Laurence Unwin that Ottawa especially wished to look into the ramifications of CPA acquiring the licences of subsidiaries like Quebec Airways; the consolidation of its routes would place CPA in direct competition with TCA. Howe was losing patience with CPA and felt that he had every right to be indignant. From holding a single seat on the board of Canadian Airways Ltd. three years before, the railway's aviation interests had developed into the largest civil air company in Canada.

If ever an aircraft was cursed: CF-BNG Rapide at Esquimalt in 1943. Pilot Joe Fecteau and two passengers died at a trapper's cabin in Labrador on BNG's original flight from Rimouski. The aircraft ran out of fuel and was not found until spring. It was flown across Canada to be used on the West Coast and then, in spring 1946, returned to Quebec, caught fire and crash-landed on the St. Lawrence. CAA (UNN)

CPA celebrated its consolidation by issuing the first timetable in November 1943. By the year's end, it had carried 70,000 passengers, 2.2 million pounds (990,000 kg) of mail and 9.1 million pounds (4,095,000 kg) of cargo. Passenger traffic and mail had increased by 21 percent and 30 percent, respectively, from the previous year; air cargo, however, had declined by 6 percent because of a decrease in mining and construction project activities. CPA's 84 aircraft had flown 6,030,000 miles (9,648,000 km), the average flight of 315 miles (503 km) contrasting with 228 the previous year.

CPA's employees were distributed across Canada as follows:

Vice-President/General Manager's Office, Montreal	40
Western Lines, Winnipeg	991
St. Lawrence District	162
Northern Quebec District	20

The three most powerful men in commercial aviation in 1944 were (left to right): C.D. Howe ("Minister of Almost Everything"), H.J. Symington, president of TCA and J.A. Wilson, director of civil aviation. National Archives of Canada PA139755

Aerial Surveys Division	7
General Auditor's Office	52
St. Maurice Aircraft Repair, Cap de la Madeleine, Que.	158
Mid-West Aircraft, Winnipeg	478
CAL Engine Overhaul, Winnipeg	173
Prairie Airways Repair, Moose Jaw	767
CPA Repair Plant, New Westminster, BC	1383
No. 1 AOS Malton, Ont.	570
No. 2 AOS Edmonton	740
No. 5 AOS Winnipeg	618
No. 7 AOS Portage la Prairie	707
No. 8 AOS Ancienne Lorette, Que.	789
No. 9 AOS St. John's, Que.	725
11 EFTS, Cap de la Madeleine, Que.	267
Total	**8,647**

Captain George Jarvis, F/O Larry Dakin and stewardess Eleanor Block pose before starting the inaugural daily service between Vancouver and Whitehorse on May 24, 1944. The route was CPA's contribution to the war effort and the construction of the Alaska Highway and the CANOL pipeline. CAA (UNN)

In February 1944, Coleman was granted interviews with Mackenzie King and later C.D. Howe in Ottawa. Before joining the CPR, Coleman had been the editor of the Belleville *Daily Intelligencer* and later the Port Huron *Daily Times*. For a former newspaperman, the injustice of the TCA monopoly was too much to bear, and he pressed for a CPA share in the business. Confronting Howe, he asked outright that his airline be allowed to begin transcontinental and international flights immediately. When Howe predictably refused, Coleman wanted to know, what was the federal government's policy on private airlines?

Committed more than ever to the government's domination of commercial aviation, Howe had gone so far as to discuss with King the complete removal of the CPR from the airline sector. On March 9, the prime minister hinted at what was to happen by saying that Ottawa had to have "a perfectly free hand in the world organization of the air and in the decision upon routes, what was needed for defence, strategic reasons etc. We should not link up with private interests where there were these vaster national considerations." In reality, King feared making a firm commitment on aviation; the Tories might take issue with it and gang up with the Social Credit Party and the CCF in the House of Commons against the Liberals.

But Coleman continued to press him for an answer. With a Commonwealth Conference to begin in May (when the other Dominions would also be asking for a declaration of Canadian principles), King could stall no longer. On March 17, 1944, he had Howe table the country's postwar aviation policy in the House of Commons— the ramifications of which would guide (or straitjacket) civil aviation in Canada until 1966. The prime minister was in a foul mood because the

The CPA repair plant at New Westminster employed 1,383 personnel to rebuild locally used military Cansos, Ansons and Bolingbrookes. A female welder wrote, "We kept the aircraft flying for our boys in the hell of Dieppe and Bataan." CAA 989-53-39

newspapers and CPA already had copies of the speech before it was tabled, lessening its impact.[21]

Howe's speech, the most important policy statement in Canadian aviation to date, stated that:

1. Canada had proposed the creation of "an international air transport authority" to license and regulate air traffic between nations.

2. The authority would be based on the freedom of licensed aircraft operators to fly over, land in, and carry passengers and freight to or from, but not within, any country.

3. The Canadian government was not about to change its policy that Trans-Canada Air Lines was the sole Canadian agency that might operate international services.

4. Within Canada, Trans-Canada Air Lines would continue to operate all transcontinental systems.

The Canadian Pacific Airlines Social and Athletic Club Choir, seen here on May 19, 1944, directed by William Pilling. They entertained at the annual staff banquet with the following selections: "Stout-Hearted Men," "Dear Land of Home" (Finlandia) and "Tramp, tramp, tramp," and concluded with "The Lost Chord." CAA (UNN)

5. Competition between air services over the same route would not be permitted, whether between a publicly owned service and a privately owned service or between two privately owned services.

6. The two national railways would remove themselves from the ownership and operation of airlines within a year of the war ending.

7. An Air Transport Board (ATB) would be created to advise the government on the licensing, regulation and control of all commercial air services in Canada. It would have the power to assess all route applications and grant franchises from now on.

Howe elaborated that a second coast-to-coast service in Canada would be wasteful and unjustifiable, adding that "the newly formed Canadian Pacific Airlines had lost no time in challenging the non-competitive position of Trans-Canada Air Lines and reaching out for new franchises; this

Were you there the day Jack Benny and Gerry Foote (billed as "New Westminster's answer to Dinah Shore") came to entertain? CPAL PA010

at a time when it alone seemed to be able to buy new equipment." As to the small carriers, he said that the bush could continue to be developed by them and that the new airports along the Northwest Staging Route would provide employment for returning servicemen to start their own outfits. Finally the federal government had acquired the rights, on the recommendation of TCA, to manufacture the Douglas DC-4, a four-engine transport aircraft that would be suitable for domestic and international use.

The content of the speech was overshadowed only by its appeal to all political parties. No one could argue with its promise to safeguard the government's investment in Trans-Canada Air Lines. All would agree that the duplication of routes in Canada's two railways should not be repeated in the air. Best of all for the wartime workers, the Liberals pledged continued employment in the aircraft industry.

Only at the CPA offices, especially at Windsor Station, was there weeping and gnashing of teeth. D'Alton Coleman discussed the implications with his board of directors before writing to the prime minister.

ATTENDING SUPERINTENDENTS' MEETING HELD IN EDMONTON.

Back row: K.B. Phillips, Supt., Alberta District and Acting Supt., Yukon District; Russ Baker, Asst. Supt., Yukon District; O.H. Johnston, Regional Traffic Manager, Western Lines; W.J. Windrum, Superintendent, Saskatchewan District; W.T. Bunn, Asst. Supt., Mackenzie District. Front Row: E.R.R. Field, Superintendent, British Columbia District; G.W.G. McConachie, General Manager, Western Lines; C.H. Dickins, Vice-President and General Manager; R.W. Ryan, General Supt., Western Lines; W.E. Gilbert, Superintendent, Mackenzie District. February 12, 1944.

The superintendents' meeting, Edmonton, 1944. CAA (UNN)

He reminded King of the CPR's contribution to the war and national development in general. Then he let drop the legal clincher: "by the Act of 1919, the Company was specifically empowered to operate aircraft for the carriage of passengers and goods within or without the limits of the Dominion of Canada. Your statement of government policy would not only deny the exercise of these powers but limit the opportunities of expansion . . . which this Company has developed over the years." The CPR was not about to give up its airline without a fight.

To help the war effort, all CPA stewardesses were required to sell Victory Bonds to a captive audience in flight.[22] There were now enough stewardesses to warrant a chief—Miss M. Tamney, based at Edmonton—and some form of training program was seen as necessary. CPA asked American Airlines to train its stewardesses, and for three months Ella Mannix, Jean MacDonald and Jean Mills attended American's school in New York. The *Honker* reported that "Miss MacDonald seems to have stolen the show . . . Mr. F. de Wolfe Sanger, supervisor of training for American Airlines, commented that Miss MacDonald ranked first in every phase, securing 98% in 'Information Procedure' and 97% in the 'Reservations Procedure.' Her score in the Otis Mental Ability Test bore out her scholastic achievements, placing her at the top of her class. All the Canadian girls were conscientious, co-operative and wholesome young ladies." Jean Mills, the chief stewardess on CPA's Eastern Lines, would later present Laurence Unwin with the coveted "Three Star V" pennant, which was awarded to CPA by the National War Finance Committee that year: 95 percent of CPA's employees had subscribed an average of $117 each, a total of $975,850, to buy Victory Bonds.

Inevitably, the newsletter would also carry announcements like: "Stewardess Eileen Butler will be leaving her position on the Edmonton–Whitehorse run, trading her job of looking after endless passengers and of feeding petulant pilots for that of looking after one man, the one and only, who is indeed fortunate, as many can vouch for her unfailing good humour and cheerful and radiant smile. May we extend all good wishes for a successful flight through married life, and may the skies be clear and the air smooth." Stewardesses were required to resign if they married or were even engaged to marry—a universal rule among airlines then.

At about the same time, the CPA agent at Fort Selkirk, 200 miles (320 km) from Whitehorse, showed that the airline's traditions were still firmly rooted in the bush. Agent G.I. Cameron was alone at the office when she saw a "big animal skulking in the yard, but thought it was a stray malemute dog." When it crawled into one of the dog kennels behind the cabin, she picked up a broom and went outside. She was banging on the kennel with it when two Natives came by. The wolf (for that's what it was) charged out at them, "snarling viciously." Cameron ran into the office, grabbed a rifle and gave it to one of the Natives, who shot it. For many years after, the CPA airline office was adorned with a six-foot-long pelt of a black wolf.

In Ottawa, Mackenzie King and C.D. Howe hadn't finished with the CPA yet. When the act to create the Air Transport Board (ATB) was amended, the government's position was made very clear. All commercial air transport was classified and put under a licensing system in which the ATB issued regulations on schedules, rates, route protection and levels of competition. To prevent a recurrence of the chaotic bankruptcies of small air carriers, competition was to be strictly regulated in the public's interest. Finally, the ATB would always be in agreement with whatever government was in power. The ATB's first order was, "Steps will be taken to require our railways to divest themselves of ownership of airlines within the period of one year from the ending of the European War. Transport by air will be entirely separate from surface transportation."

The sturdy Noorduyn Norseman at Edmonton's Municipal airport in 1945. Until the appearance of Boeing 737s, the Norseman was the most popular aircraft in CPA's fleet; at one time twenty-seven were in use. The Mark V was a civilian version of the military Mark IV, and after the war was available in large numbers. CAA 989-71-12

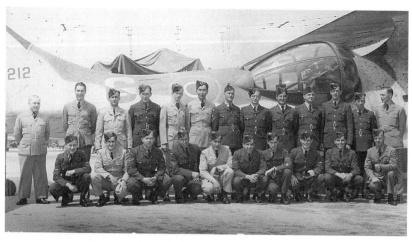

Refurbishing PBY Cansos for coastal defence was a major part of CPA's war effort. CPAL PE011

In 1945, after the Allied victory over the Axis powers, the airlines divested themselves of their wartime trappings, perhaps too quickly to be appreciated. The work performed by Canadian Pacific Air Services and Canadian Pacific Airlines may have been a footnote in the conflict, but the Atlantic Ferry and the Air Training Plan were one of history's greatest air operations, and the CPR had played a pivotal part in it. Edward Beatty's patriotic, unselfish gesture of starting an Atlantic air mail service ultimately led to the ferrying of nine thousand aircraft to England.

Better known are efforts of the five Air Observer Schools operated by the CPR, which trained 29,130 air crew and refunded $945,000 to the government, and the CPR repair depots, which, as a welder wrote, "kept the aircraft flying for our boys in the hell of Dieppe and Bataan." All of the company's staff helped make flying routine and laid the foundations for the postwar expansion of CPA. Unfortunately, government censorship and C.D. Howe—who had his own axe to grind—ensured that the CPR received no recognition for any of these accomplishments.

Overleaf: The DC-3 interior, with stewardess Mary Ann Bow. For early flight attendants the hours were long and the pay less than what their male colleagues made, but as trained nurses they thought flying was more glamorous than working in a hospital.
CAA M4337

CHAPTER FIVE

"Oh You Can't Go to Heaven on CPA 'Cause They Just Fly the Other Way!"[1]

After World War II, a worldwide explosion of interest in commercial aviation occurred as airlines took advantage of the abundance of aircraft, pilots and wartime innovations such as radar and pressurization. Carriers on both sides of the Atlantic hoped to profit from the public's hunger for flying. In recognition of Canada's new-found status as a power broker between the Americans and British, the International Civil Aviation Organization (ICAO) and the International Air Transport Association (IATA), set up to regulate commercial aviation, were both headquartered in Montreal. IATA's first home was at the Windsor Hotel, a few blocks from Canadian Pacific Airlines' offices. The future of CPA was in grim contrast to that of all other burgeoning airlines. Unless its fortunes changed, it was doomed to disappear.

With the deadline approaching for the CPR to orphan CPA, D'Alton Coleman pressured the Air Transport Board to allow the airline to withdraw from bush flying and concentrate on intercity routes. To prepare for this, he attempted to standardize the airline's fleet by having fewer planes, but those remaining would be larger, wheeled airliners.

Canadian Pacific Airlines ended the war with the following inventory:

6 Barkley-Grows
3 Beechcraft 18s
3 Bellanca Aircruisers
6 Boeing 247Ds
2 de Havilland Rapides
10 Fairchild 82s
4 Junkers 33/34s
12 Lockheed Lodestars
12 Noorduyn Norsemen
1 Travel Air

The age of the company's aircraft ranged from the Norseman Vs, factory-fresh from Cartierville, to the last Fairchild 71C, of 1930 vintage, which fell apart in 1945. Punch Dickins warned Coleman of the situation in January, pointing out that one of the most modern aircraft in the inventory, the six Barkley-Grows, were "not only limited in passenger carrying capacity but as they had not been manufactured for eight years . . . all parts and spares have to be made by hand which is extremely costly." The six Boeing 247Ds were soon to disappear from the inventory. They had been loaned to Quebec Airways by the RCAF and were scheduled to be returned in March 1946. The only twin-engine replacements available were the locally made Avro Anson Vs, of which the airline bought nine. Too small (and too cold) for passenger use, the Ansons were allocated to photographic work and freighting. Of the Beechcraft 18s, or "Shakey Jakes," that had been casualties of the short-lived Maritimes contract, the two "Ds" had been sold to a Venezuelan company and the remaining 18A would be bought by Northern Wings, of Quebec City, in 1946.

The indestructible Junkers remained, but since spares from Germany were unobtainable the airline relied more on its Norsemen to operate between the frontier settlements. The lack of larger aircraft placed an increasing strain on CPA by preventing it from turning a profit or bidding for more routes. The Lockheeds were kept on the Regina–Saskatoon–North Battleford and Winnipeg–Red Lake–Kenora runs. Although they carried fourteen passengers, they compared unfavourably with TCA's reconditioned DC-3s, which took thirty people in comfort. Without the urgency of the war, unless some accommodation was reached with the federal government regarding both aircraft and routes, the chances of a switch in market strategy were nil. C.D.

Two DC-3s replaced the Rapides in the Eastern region; this one was based in Montreal for the Quebec City–Saguenay route. Directly below it is the company headquarters at Windsor Station. CAA 988-31-29C

Howe now controlled the War Assets Corporation, and when Dickins asked for permission to purchase DC-3s from military stocks, he was curtly advised by E.J. Bennett, the minister's chief executive assistant, that all such aircraft had already been allocated, but "perhaps a few Norseman Vs would be declared surplus soon."

As in the war, the United States rode to the rescue. In early 1946, the US War Surplus Board sold Dickins two C-47s—the military version of the DC-3—for $20,000 (US) each plus $6,000 in tax, and two 1830 Pratt & Whitney Twin Wasp engine spares at $7,000 each. Each C-47 cost another $25,000 (Cdn.) to be converted by Canadair into a twenty-eight-passenger airliner. Considering that the Lodestars had cost $90,000 each and the Norsemen $33,000, the DC-3s were a bargain. Even so, since the government's order for the railways to sell off their airlines had not been lifted, the CPR's accountants questioned the wisdom of the investment.

More DC-3s were supplied by the aircraft contractor C.H. Babb Inc., of Glendale, California, which sold the airline twelve over the next two

147

years. Others came from Aircraft Industries, Peace River Northern Airlines and Doris Yellowknife Goldmines. The DC-3 was the first modern CPA airliner to serve many communities,[2] especially in the Eastern Region, where it replaced the ancient de Havilland Rapides.[3] One DC-3 was based in Montreal for the Quebec City–Saguenay run; the other was based in Mont Joli for the North Shore, where the passenger traffic had doubled since 1944. For these runs, Raymonde Ouellette, the first French-speaking female flight attendant, was hired.

For the stewardesses the hours were long and exhausting, but these women were trained nurses who "felt very privileged" to have such a glamorous job. Whatever the hazards of flying, they were preferable to pulling twelve-hour shifts in a hospital. "We were all so young we weren't too concerned about the long hours," said Evelyn (Donnelly) Curan who, flew for CPA from 1946 to 1960. "We found it rather amusing that we could do it."[4] Besides, the war had just ended and a new world beckoned, where everything was possible.

The public perception of their occupation was that while this was acceptable employment for a middle-class girl to engage in between homework and housework, it was definitely casual. Everyone knew that the "girls" were doing this as a lark and that within a couple of years they would "catch" a husband (usually a pilot) and retire from the airline. Unfortunately, there was some truth to this perception. Whether it was because of the low wages, long hours or marriage, the attrition rate among CPA stewardesses well into the 1950s was between a year and eighteen months. Any who remained more than three years were considered very senior. The airline exploited the high turnover for a constant supply of youthful looks and naiveté and actively discouraged women from flying at the end of their twenties. As late as 1954, CPA stewardesses had "to sign a paper stating they would quit at thirty."

The only male flight attendants in Canada at that time were employed by TCA for its overseas runs on the converted Lancaster bombers known as Lancastrians. Stewards were viewed very differently by the public and their employer, who considered them professionals and family breadwinners. Thus they were paid more than their female colleagues, who until 1949 were kept on domestic routes.

More numerous and experienced than their CPA sisters, the TCA stewardesses were the first to organize their profession. Muriel Peacock joined TCA as a stewardess after leaving the military in 1946. What sparked her to crusade for a flight attendants' association was a very mundane but essential issue: the women wanted the airline to pay for their transportation from the city centre to the airport when they caught their flights.[5] On November 11, 1947, the Canadian Air Line Stewardess Association (CALSA) came into being, not as a union—a word that still had unsavoury connotations for women who, as nurses, had taken an oath to serve mankind—but as an "association to bargain with the airline management for salaries, seniority, hours and holidays." A year later, when TCA's male flight attendants joined them, CALSA became the Canadian Air Line Flight Attendants' Association (CALFAA); however, the association's membership was limited to TCA attendants.

For veteran bush pilots like Ernie Boffa, World War II had been a strange experience. One of the original barnstormers of the 1920s,[6] Boffa, like many of his profession during the war, instructed for the BCATP. He enjoyed passing on his skills to recruits at No. 6 Elementary Flying School at Prince Albert until the winter of 1942–43, when the RCAF took over. As he remembers it, everything changed then: the instructors were issued sergeants' uniforms and were expected to drill the recruits. "Suddenly we really had to toe the mark." Having to conform to the King's Regulations was too much for an old bush pilot, and in February 1943 Boffa had walked off the parade ground. Bob Randall, a friend and former Mackenzie Air Services pilot, was now flying on the CANOL and advised Boffa to see Grant McConachie for a job.

Recognizing a natural bush pilot when he saw one, McConachie hired Boffa to fly the CPA Norseman on mail runs around Yellowknife. The region's mines were still closed because of the war, and Boffa discovered that his best customers were the local trappers.

> In late August all these trappers go back to the Barrens and spend their winters trapping. You had to get them there early enough so that they can put up their firewood, scrub spruce, stuff like that. It takes a lot of wood fuel to see you through the winter, so they went in early.
>
> We'd leave Fort Reliance for the Slave Lake bunch, but we also flew some in from Camsell Portage, near Goldfields on Lake Athabasca (in northern Saskatchewan), and from Stoney Rapids . . . There would be maybe 10 or more trappers with all their dogs; all their gear, everything they'd need for the winter. You

By 1946 the picturesque scene of the Bellanca Aircruiser and the dogsled only typified Canadian Pacific Airlines' problem—unprofitable Northern routes and obsolete aircraft. CAA 989-71-11

can imagine the barking and howling that went on. The dogs were all staked out, each family had its own tent in the spot across the bay from the police barracks at Fort Reliance. So Canadian Pacific would have a bunch of gas stored there, barrels and a wobble pump.

Away we'd go. Fill her up with gas and load all these dogs into the plane. First of all, we'd put in all the gear—flour, groceries and stuff—then the guy would get in with the dogs and he'd sit there with a club in his hand, ready to clobber them if they started fighting or anything. Well, we just kept gassing up and flying out a load at a time out to the trapping camps. Most of the time you just got one trip in during the day. If the guys were closer in, you could sometimes make two trips.[7]

When the war ended, the influx of men and money caused the North to experience another gold rush. A "staking" frenzy was in full swing around Yellowknife, its streets crowded with prospectors all wanting to make their fortunes, all wanting to be dropped off by CPA at some secret location with their supplies. Prospectors were known for their paranoia; Ernie remembers some folding a map into tiny squares and thrusting it at the pilot, pointing to the lake where they wanted to be left. Boffa soon developed an uncanny knack for finding exactly *that* lake, and the

CPA Norseman was constantly in the air. "It was easy, because all the staking went on not too far from Yellowknife," Boffa said. "It was mostly within a radius of twenty or thirty miles. You didn't have to be a wizard after flying around there for a year or so, taking prospectors all over the place. Surely, you'd recognize those lakes."

In December 1945, Ernie and his Norseman became famous. Business had been so good that he was sent to Montreal to pick up another Norseman—and to buy Christmas gifts for his family. In his absence, Jack Herriot and mechanic Gordie Brown took Boffa's CF-CPS to do the regular mail trip to Coppermine, carrying a passenger, the Hudson's Bay inspector, L.A. Learmouth. They left Yellowknife on the morning of December 18, landed at Eldorado along the way and then made for Coppermine, where everyone was looking forward to the Christmas mail. When the plane didn't arrive that evening, with the temperature at 30 below zero and visibility limited, the worst was feared. Because of the low cloud, the inhabitants of Coppermine lit flare pots along the airstrip. Late that night, when there was still no sign of the plane, the CPA office at Yellowknife was notified, and it called the airline's operational headquarters at Edmonton. Although this was the festive season, Alf Caywood[8] and CPA superintendent Wop May, both experienced at search and rescue, immediately took off from Edmonton's Blatchford Field for Eldorado.

Ernie Boffa was on his way home with the new Norseman CF-BHV when he was radioed about the search. He picked up a doctor at Eldorado and began searching the Coppermine River area, turning off at Bloody Falls and making the mouth of the Nipartoktuak River before darkness fell. They landed on the sea ice without having seen any traces of the downed aircraft. Because of a full moon that night Boffa continued the search in four-hour flights, taking off again and again from a makeshift runway lit by flares. Caywood and May scanned the area around the Kendall River north and east, staying in radio contact with Boffa.

By December 24, as the survival chances of Herriot, Brown and Learmouth diminished with every hour, all members of the search party were discouraged. Taking Dr. Baker with him, Boffa flew farther into the unmapped country south and east of Coppermine. Knowing that he was running out of areas to look, he carefully circled over each lake from a height of four hundred feet. At the farthest point of his search, as he was turning westward over one of the lakes, barely visible below

Posing for a photo after the search for the missing Norseman, Wop May is next to Ernie Boffa, who temporarily overcame his shyness to shake hands with superintendent Dick Ryan. CAA 1228

appeared the outline of a Norseman. Two figures staggered out of that plane on hearing his engine. Boffa set his own aircraft down about half a mile from CF-CPS. It was Christmas Day, and for the first time in four days he relaxed and "grinned his delight." Herriot had got lost in the low cloud and decided to land at a small lake south of Coppermine until help arrived. The three men were flown out; Learmouth was the only casualty, with a frozen foot. Boffa had flown a total of thirty hours in the search—twelve of them by moonlight.

Many veterans returned to the North to resume their prewar lives. In frontier communities across Canada, loud reunions took place on main streets, in beer parlours—and at CPA ticket offices. It was a good time for bush air carriers; wartime restrictions on gasoline had eased and there was a surplus of aircraft, ranging from J-3s to the ubiquitous Ansons. The Americans may have gone home, but their ferry routes to the Soviet Union and Western Europe (the Crimson Route) had dotted the North with hundreds of paved airports, all of which Prime Minister Mackenzie King had paid for. Once more, just as after World War I, "demobbed" pilots used their savings to go into business for themselves.

C.D. Howe kept the promise he had made in 1944 about encouraging local operators, and a spate of regional licences appeared. More and

In March 1945, Jim Spilsbury began Queen Charlotte Airlines with the unlikeliest of airliners, two Stranraer flying boats, taking sixteen passengers from Sullivan Bay to the Charlottes. CAA 3636

more, TCA was not the only competitor CPA had to contend with. In British Columbia, Russ Baker quit the company and, with the help of Karl Springer, a mining executive, began Central British Columbia Airways (CBCA) on July 1, 1945. Together they sought a contract from the provincial government to fly forest patrols. During the war, the Japanese had sent ingeniously designed incendiary balloons that drifted across the Pacific Ocean. Taking advantage of the winds, these balloons were intended to sabotage Canadian industry by starting fires in the vast British Columbia forests. The RCAF flew fire patrols until the war ended, when the BC Forest Service hired private contractors to continue the aerial surveillance. When CPA bid for the contract, Grant McConachie reminded the bureaucrats in Victoria that his pilots, whether YSAT or CPA, had always "spotted" when flying over the forests, without charge. This argument must have struck home, because the first civil fire-patrol contract was awarded to CPA. For the critical period—June 15 to October 15, 1945—the airline flew 276 hours of patrols, earning $12,362.

But Baker wasn't done yet. The contract specified twin-engine aircraft, so he bought two war-surplus Cessna Cranes and teamed up with Walter Gilbert (who at that time was still working for CPA) to get the contract for the summer of 1946. Using a ploy that would serve him well in the future, Baker posed as a local boy familiar with the territory. The Air Transport Board granted him a charter licence on March 21, in time for the worst summer of forest fires in British Columbia history.

The most colourful rival Baker and McConachie had in the region was undoubtedly Queen Charlotte Airlines Ltd. (QCAL), an air carrier typical of the time and place. Ashton James "Jim" Spilsbury was an avid radio ham who had been talked into purchasing a couple of the war-surplus Stranraer flying boats, which were lying idle at Jericho Beach. He had heard from a cousin that the logging companies wanted a scheduled air service in the Charlottes. His friend Wally Siple, who had been involved with ATFERO, had just bought the Stranraers "as is/where is" in order to sell them off in South America. The Stranraers' Bristol Pegasus engines made them powerful enough to take off with a two-ton load in twelve seconds from a standing start, and Siple was selling them for $25,000 each. Spilsbury raised the money to buy a pair of the ungainly looking contraptions and start up his own airline. On March 5, 1946, the first QCAL Stranraer was ready for business, taking sixteen passengers from Sullivan Bay to the Charlottes.

In Yellowknife, Ernie Boffa and the CPA Norseman rescued an ambitious former BCATP instructor named Max Ward from a youthful misadventure. Flying for former WCA pilot Jack Moar and his company, Northern Flights, Ward had used enough of his "demob" savings to buy a new de Havilland 83 Fox Moth CF-DJC and form his own company, Polaris Charters, in 1946.

In Ontario, the Atlantic provinces and Quebec, small airlines like Ward's were taking away the type of contracts that had gone to CPA. In northwestern Ontario in 1946, two WCA veterans, Milt Ashton and Francis Roy Brown, who had sold their carrier, Wings Limited, to Canadian Pacific in 1941, re-entered the airline business to begin Central Northern Airways Ltd. With ex-CPA Norsemen, they began flying from Sioux Lookout to Pickle Lake, Lac du Bonnet, Norway House, Ilford, Gods Lake, Flin Flon and Sheridan, Manitoba, taking over the bush routes that CPA gratefully wanted to be rid of. In Charlottetown, Maritime Central Airways (MCA), which had barely kept a Boeing 247D in the air during the hostilities, now prospered with government contracts, lobster hauling and ice patrols. In 1945 it had bought two

The youthful Max Ward, proud owner of Polaris Charter Co. Ltd. With his savings from instructing in the BCATP and flying for Jack Moar's Northern Flights, Ward bought Fox Moth 83C CF-DJC directly from the de Havilland plant in Downsview. Bombardier Archives 10951

ex-RCAF Cessna Cranes and later three DC-3s for a scheduled service to St. Pierre and Miquelon. In Newfoundland, Eastern Provincial Airlines (EPA) was begun to service Labrador by Chesley Crosbie, the father of John Crosbie, who would one day become minister of finance, justice and transport. In the Gaspé peninsula, Le Syndicat d'Aviation de Rimouski began providing charter flights and in 1948 took from CPA its old Rimouski–Sept-Iles route and changed its name to Rimouski Airlines.

Even CPA's eastern base at Dorval airport, Montreal, was no longer safe from competition. Boreal Airways was begun at Dorval on May 26, 1947, by François "Frank" Ross, who had bought a Norseman to take fishermen, hunters, trappers and surveyors around the surrounding area.[9]

This increased competition in the bush forced the executives at Windsor Station to re-evaluate CPA's market strategy and aim it toward scheduled routes between urban centres instead. Increasingly CPA turned over its "bush" operations to smaller carriers. In June 1946, for example, Russ Baker's CBCA was given the Prince George–Fort Ware route and was sold one of the old Canadian Airways Junkers W-34s to service it. But the strategic reorientation was easier said than done. Any plans the company had for expansion were at the mercy of the Air Transport Board and (it must have seemed) the whims of the government in power—and at the time Mackenzie King's Liberals looked like they might remain in office forever. Of immediate concern was the federal government's August 15 deadline for both railways to jettison their air companies.

If many of CPA's problems were beyond its control, in contrast, headquartered at the old Canadian Airways base in Winnipeg was TCA, the "nation's chosen instrument." As the country's flag carrier, TCA had an assured future. It enjoyed a monopoly on all transcontinental and international routes and the freedom to experiment with aircraft and markets, knowing that taxpayers were supporting its caprices. A snapshot comparison of the two airlines in 1945 shows their relative positions:

Total aircraft miles flown:	TCA 7,102,062	CPA 3,694,978
Total revenues:	TCA $7,105,793	CPA $3,581,627
Total expenses:	TCA $6,812,494	CPA $3,908,472
Number of employees:	TCA 2,892	CPA 1,018
Operating Ratio:	TCA 95.73%	CPA 109.14%

Although the aircraft that CPA had inherited from the bush carriers might be outworn and obsolete, the calibre of senior personnel that came with them was anything but that. Here were the legends of Canadian aviation—many of them the McKee Trophy winners, whose names were synonymous with aerial pioneering and bush flying. Relying on their experience and authority in the aviation fraternity, Laurence Unwin had treasured men like Punch Dickins, W.E. Gilbert and Wop May, and put them into positions of responsibility to overcome problems in morale that accompany any merger. The company's assets lay more in its employees' heads (and hearts) than in its aged, polyglot fleet.

But two events in early 1946 changed the airline's future forever. Laurence Unwin was elevated to the CPR's board of directors and William Neal succeeded him, becoming the second president of the airline. The son of Northern Irish immigrants who settled in Toronto (and anglicized the family name from "O'Neill"), Neal had left school after grade five to work in a factory, escaping it to join the CPR as an office boy. Within seven years he had been promoted to the position of a travelling railway car "counter," constantly studying to better himself. Lacking the polish, money and education of Edward Beatty and D'Alton Coleman, Neal laboriously climbed each rung of the corporate ladder by cultivating a ruthlessness that by 1934 made him vice-president of the railway's western lines.[10]

Neal was called "Old Slaughterhouse" behind his back by subordinates who thought him a malicious despot, ruling with all the pettiness that a huge bureaucracy is inherent to. Completely bald, he resembled the Fascist dictator, Benito Mussolini. With Unwin away negotiating for bush companies and Beatty ill, in 1942 the CPR directors made Neal vice-president of the struggling Canadian Pacific Airlines. Perhaps they recognized that beneath the tyranny lay a brilliant mind. "Buck" Crump, a future chairman and president of the CPR, thought Neal the best "transportationist" he ever knew. Neal's doctrine of "an integrated transportation system," in which rail, water, highway and air each best served the needs of the North, was far ahead of its time. His wartime work on the Allied Transportation Equipment Committee earned him the decoration of Commander of the British Empire in 1944.

Consumed by ambition to become the railway's chairman and president on Coleman's departure, Neal channelled this drive into CPA's bid to compete against TCA for the burgeoning postwar market. Many in the CPR tolerated Coleman as a caretaker until the war's end, knowing that Beatty's crown would then rightly go to Neal. Once he had moved into the presidential suite, the relationship between Windsor Station and Ottawa perceptibly altered. Whereas Coleman's directness had antagonized C.D. Howe, Neal's eloquence impressed him. Many years later, Unwin would acknowledge that while he had drafted the proposal to the government for the CPA's continued existence, Neal was the one who took it around Ottawa, meeting with both Howe and Mackenzie King.

Whoever its creator, the proposal caused Howe to do an about-face: it would be a mistake, the minister now said, for the two railways to orphan their air branches. He explained his reasoning when the legislation had been introduced in 1944: "Canadian Pacific Airlines were very ambitious to extend . . . over other parts of Canada. Today they find they have a sufficient task to develop the parts of Canada for which they were then and are now responsible, with the result that any new airlines that care to enter the business and are able to finance their operation can find a scope for their activities in parts of Canada not at present served."

Howe accepted his error in good style; as he frequently said, "If I'm right 75 percent of the time, its a pretty good average." He knew that if the CPR jettisoned CPA, the airline would suffer the same fate as Laurentide Air Services, Canadian Airways Ltd., Yukon Southern Air Transport, Starratt and Mackenzie Air Services—it would inevitably go bankrupt. Thus, on April 6, 1946, Howe noted: "We have found it necessary to extend the time for separating the railways from their airline operations for one year, due wholly to the fact that there seems no likelihood that anyone else can give the service without being heavily subsidized . . . C.P.A. have taken quite a beating during the war years in maintaining the many services for which they are responsible, and it seems only fair to give that company a chance at postwar business."

The social implications of a decision made in the excitement of the war affected thousands of lives and much of Canada's economy. Without the essential services that CPA performed in the North, the frontier communities would expect TCA and the federal government to take up the slack. Finally, the perennial spectre of American interests stepping into the void and buying up the Northern air services still stalked the corridors of Parliament.

Significantly, TCA was not in any condition to leave its parent, CNR, either. The much-publicized Lancastrian flights to London were draining its bank balance; the last year it had made any profit was 1944, when it barely cleared $7,400. Besides providing administrative support, the CNR would be carrying TCA financially through the foreseeable future. As a result the order for the railways to divest themselves of their airlines was quietly rescinded in August 1946, the wily Howe allowing Neal to think that he had persuaded him to stay the airlines' execution.

On December 23, 1946, at about 11 a.m. the CPA operations room at Mont Joli was jolted with the radio message: "May Day! May Day! We're going down in the river!" It was from Captain Jim Hartley, in the de Havilland Rapide 89A CF-BBC on the Godbout–Matane flight, with six passengers on board. The main passenger traffic so close to Christmas was the "woodsmen shuttle"—hordes of what Babe Woollett, the local CPA superintendent, called "bearded, smelly lumberjacks" employed by the pulp and paper mills—all eager to spend the season at home. The twin-engined Rapide was ski-equipped, but it did overfly the St. Lawrence River between Godbout and Matane for about 75 miles (120 km). In the sub-zero weather of late December the cold, salt water was overlaid with broken ice floes. The ground crew at the Rapide's last stop hadn't filled the aircraft's gas tanks up as usual; worse, the pilot did not "dip" (measure) them. While over the river one engine cut, and Hartley hoped to make the shore before the second one did. When it promptly died, he glided the Rapide toward the nearest ice floe, about two miles from shore. Fortunately CF-BBC had been equipped with radio that year. Hartley later said that the aircraft skimmed over the surface and that it was the best landing he ever made; but the Rapide was too heavy and sank within twenty minutes, forcing the seven men to scramble onto nearby ice floes. It was at this point that the party broke up: two stayed with the pilot, three drifted away and one made for the shore.

Woollett radioed Grant McConachie at Windsor Station, who directed one of the airline's DC-3s from Quebec City to the airfield. Because the weather was closing in, he and Captain North Sawle could not fly out of Dorval to take charge, so both caught the night train to Mont Joli. As soon as the DC-3 landed, the freight doors were removed and an inflatable life raft, borrowed from a visiting US Air Force aircraft, was thrown in. Woollett and J.G. "Pat" Twist took off into the dusk and deteriorating weather to search the area, but the white sheets of ice—sometimes dotted with seals—made spotting humans impossible. That night CPA staff phoned the villages along the shore to light fires and look out for the survivors. Woollett called on the nearest USAF base in Maine for help, asking for their helicopter.

McConachie and Sawle arrived the next morning, Christmas Eve, with news that a Grumman Goose amphibian belonging to one of the K.C. Irving companies was on its way from Montreal. It was being flown by James Wade, a former RCAF pilot who during Christmas 1942 had spent three months stranded in the Arctic, attempting to rescue three men on the Greenland ice cap.

Then there was a phone call from Les Mechins, a village on the river's shore, that three men could be seen on an ice floe two miles off-shore. Twist took off in the DC-3 and radioed back that he had spotted the men and was dropping survival gear to them. He returned to Mont Joli and briefed McConachie, Woollett and Sawle of the location. Wade landed at Mont Joli in the Goose, and with its engines still running,

Twist threw in the life raft, ropes and poles and accompanied him to the site. At about 5:30 p.m. Wade spotted them and brought the Goose down, landing "in the open space of a dime," as Hartley later reported, and the three scrambled on board. Three others were still drifting (the fourth having disappeared the night before); two USAF Flying Fortresses spotted them and dropped lifeboats as close to them as possible. In Quebec City, Walter Clarke, the president of Clarke Steamships, directed his ships to the scene. Boats from the SS *Colabee* picked up the three opposite Petite Vallee, while Twist circled frantically, looking for the last man. He was never found.

The six were suffering from hypothermia, and on December 27 McConachie commandeered a CPA Lockheed 14 and had North Sawle fly them to hospital in Montreal. The requirement for stewardesses to be registered nurses made their flight a lot easier. From start to finish, the courage of all involved, especially of Twist, Woollett and Hartley, was in keeping with the best traditions of the airline.

Though the bush communities were still its domain, CPA's eyes were increasingly turned elsewhere. Under Neal, the airline made a strong push to identify a competitive niche for itself outside the TCA monopoly—that of feeding into the government airline's transcontinental service. The raised commercial-aviation consciousness of Canadians after the war (when every town expected to be connected by air with the nearest city), were instrumental in the Air Transit Board's approval of several applications—but only after the same routes had been offered to and turned down by TCA. As a result, routes that came CPA's way in 1947 included Winnipeg–Dauphin–The Pas–Flin Flon, Vancouver–Prince Rupert, Seven Islands–Knob Lake and Vancouver–Penticton–Calgary.

Windsor Station's happier relationship with the federal government did result in the purchase of more and varied aircraft. Four PBY Cansos, which had been built at the Vancouver Boeing plant (soon to be CPA's base) were bought from Crown Assets Disposal. Since there was no runway at Sandspit, these amphibious panes were to be used for the Vancouver–Prince Rupert–Sandspit route. Fitted out in civilian comforts and dubbed "Lanseair," the Cansos picked up passengers on the tarmac at Vancouver airport, loading up to eighteen through the gun blisters.

The Cansos were flown seven days a week on a one-hour shuttle, rarely higher than 150 feet (45 m) above the water, on Visual Flight

In 1946, because there was no airport at Sandspit, four PBY Canso As were bought for the Vancouver–Prince Rupert–Sandspit run. One of them, CF-CRR, had, while in RCAF service on April 17, 1944, sunk the German submarine U-342. CAA 988-235-8

Rules. If the weather was poor, the pilot would "sit" it out at the last airport. Five decades later, Bill Green, one of its pilots, recalled that their cruising speed was 145 miles per hour (232 kph) and stalling speed 67 miles per hour (107 kph). The Canso was a large aircraft, 140 feet (42 m) long and 20 feet high (6 m) on landing gear, and it was ungainly out of its element—water.

Nina Morrison was a bank clerk working toward her nursing qualifications then and was, as she remembers, "desperate" to fly. As soon as she graduated, she applied at CPA's offices. Because she was from Prince Rupert, she could stay over there at no cost to the airline if the aircraft was grounded by bad weather—a factor in being hired for the Canso run. Besides ensuring that the life jackets were under the seats and serving coffee in paper cups and Peak Frean cookies to the passengers, Nina had to hose the seaplane down. Since the Canso's flight engineer was seated amidships, his feet hung down into her tiny galley, which was the size of a phone booth.

CPA overhauled its Cansos with a smart livery, calling them "Landseairs," adding an electrically operated stairway in the stern and installing blisters over the entry ports. CF-CRV would hit a log while landing at Prince Rupert on May 11, 1953, and flip over, killing the stewardess and a passenger. After that, a boat was always sent out to look for logs before takeoff. CAA 988-31-22

Fresh from his triumph with C.D. Howe, William Neal now looked toward the chairman's suite. Ever meticulous, he took steps to groom his own replacement, and the CPR's personnel department submitted a detailed list of candidates to him.[11] (page 163)

Punch Dickins, as vice-president and general manager, was the most senior candidate, and in the mythology that most Canadians associated with bush flying, Herbert Hollick-Kenyon and Wop May were icons. But in the infectious smile of Grant McConachie, Neal saw the company's future. Born in 1909, McConachie was in manner and spirit of another generation from the others. Too young to have fought in World War I or served his flying apprenticeship on Curtiss HS-2Ls in the OPAS, he had always been his own boss, running an air carrier at age twenty-two, barnstorming, selling rides and hauling fish to keep his company in the air. When McConachie entered the University of Alberta in 1927, Wop May and Punch Dickins were already famous. Between

Candidate	Age	Salary per annum	Remarks, Record and Personality
C.H. Dickins, vice-president and general manager. Ex–RFC, CAF, RCAF. McKee Trophy 1929, OBE 1934	46	$12,000	Good memory for detail, effective public speaker, shows initiative and organizational ability.
Richard W. Ryan, manager of Moose Jaw Flying Club 1930–40, Prairie Flying School 1940–43, general superintendent West. Ex–RNAS, RFC, RCAF	48	$7,700	Sound judgement, meticulous, refined manner, has Master of Arts degree.
T.W. Siers, general superintendent of maintenance, Winnipeg. Can. Army 1917–18, McKee Trophy 1940	49	$6,600	Very conscientious, hardworking, experienced in engine repair.
W.R. May, superintendent Mackenzie District, Edmonton. REC, RAF, McKee Trophy 1929, OBE 1934	49	$6,600	Management experience with OAS, well thought of by public.
W.E. Gilbert, flying superintendent. Prince George, BC. RFC, RCAF, McKee Trophy 1933	46	$6,000	Capable pilot, conscientious, hard worker, quick thinker, inclined to worry.
H. Hollick-Kenyon (from TCA 1937 to March, 1942), chief check pilot, Western Region, Edmonton. RFC, RAF, Honorary Air Commodore RCAF, Ellsworth Antarctic Expedition	48	$6,000	Pilot with very broad experience. Constantly developing new procedures. Rather reserved, inclined to be uncompromising. Respected by staff for technical knowledge.
G.W.G. McConachie, general manager, Western Region, Edmonton. McKee Trophy 1945	36	$9,600	Energetic, pleasant, aggressive, enthusiastic. A salesman, popular with staff and public.

The blisters made for unforgettable viewing, particularly when the aircraft was on "the step." Beyond the oval door was a telephone booth–size galley, suitable only for a snack of coffee and cookies. The last CPA Canso was disposed of in 1960. CAA (UNN)

1931 and 1938, McConachie had formed and lost three companies—Independent Airways, United Air Transport and Yukon Southern Air Transport—but if other bush outfit owners worried about bank overdrafts, he faced the world with a jaunty smile. His "stand-up comic routine patter" endeared him to bank managers and oil company executives, as did his preposterous dreams of flying to the Orient and Europe along a "Great Circle" route. Who else would boast that his air companies had never flown in the black—because they were pioneering in the development of the North—and get away with it? Perhaps seeing something of his own humble beginnings, Neal (like C.D. Howe) was genuinely fond of McConachie.

No matter how hard he worked for himself, McConachie barely paid his operating expenses. His aircraft were repossessed so often that in 1933 he had put them in the names of his pilots' mothers. Even when he was awarded the contract to fly mail between Edmonton and Whitehorse in 1937, he couldn't afford to look back—there were too many creditors gaining on him, led by the Royal Bank.

None of this stopped him from ordering new aircraft; McConachie

was fascinated by anything new.[12] In 1939 he had bought three bul-
bous-nosed Fleet Freighters, but because he was a bad risk, no insurance company in Canada would do business with him. Not so easily
thwarted, McConachie flew the first of the Freighters to Chicago, where
the mighty Aero Insurance Company had its head office. After charm-
ing the American directors and getting them to insure the aircraft, he
took them out to the airport to demonstrate the Freighter. When he
attempted a takeoff, it promptly caught fire and—before the fascinated
audience—burned to a heap of twisted metal on the runway.
McConachie did, however, collect enough from Aero Insurance to keep
his company flying a little longer.

What the cowed staff at Windsor Station recalled was that
McConachie lacked all fear of Neal. They couldn't know that Grant was
his Uncle Harry reincarnated. After bicycling to work from Morrison
Avenue in the Town of Mount Royal, he would barge past the secre-
taries in the presidential anteroom and burst into Neal's lair with a
cheery "And how are ya today, Mr. Neal?" Many times he came close to
being thrown out of Windsor Station; he was known to have said, "My
contract says they can't fire me, but I guess they could make me a mail
boy around here!"

Unlike Unwin or Neal, both CPR men, McConachie understood
what was wrong with the airline. It had outgrown the bush flying market and, denied the transcontinental routes, had to fly overseas. To the
two CPR directors, McConachie symbolized the optimism and prosper-
ity of postwar Canada that was sure to lead to a boom in air travel.
Pressurized airliners with tricycle landing gear, glamorous stewardesses, advertising campaigns to persuade Canadians to fly to the Far East—
it was what Grant had been working up to all his life.

D'Alton Coleman retired on February 1, 1947, and Neal, the former
office boy with a grade five education, became chairman and president
of the Canadian Pacific Railway. The transportation conglomerate was
still the largest company in Canada, but many of its locomotives predated the Depression, and its once-stately "Empress" liners, repainted
white to hide wartime scars, were feeling the competition from commercial aviation.[13] To escape the monumental task of rebuilding everything, Neal seized on CPA, hoping McConachie's excitement would rub
off on him. On February 7, as one of his first executive actions, he made
Grant McConachie its president.

For the ebullient thirty-seven-year-old, it was a long way from the

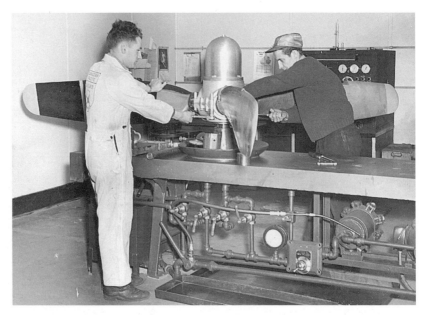

Lloyd Oatway, left, and Lou Green doing a feathering check on a DC-3 propeller overhaul. CAA 988-10-4h

Captains Wally Smith and Don McLean listen to the dispatcher describe the weather en route. CAA 988-10-4c

John Brush supervises a DC-3 R-1830 powerplant overhaul. The airline's seventeen DC-3s kept CPA as a good customer at Pratt & Whitney in Montreal, particularly when TCA began using Rolls-Royce Merlins for its North Stars. CAA 988-10-4d

CPA baggage handlers in the Winnipeg winter. CAA 988-10-4g

The ruthless, brilliant William Neal, second president of the CPA, persuaded the federal government to grant the airline a stay of execution and allow it to operate outside the TCA monopoly. CAA 989-62-27

blue Fokker and evading creditors. As if to celebrate the partnership, Neal and McConachie embarked on a tour of CPA offices that summer, culminating in Yellowknife. On August 28, they officiated in the opening of its new airport, begun in 1944 by Walter Gilbert on a strip parallel to Long Lake.[14]

Vice-president Punch Dickins was given the title of director of northern development—a job in which he soon lost interest. When Phil Garratt sought opinions from bush pilots about a replacement for the Norseman, Dickins left to join his company, de Havilland Canada, and the collaboration resulted in the first Beaver taking off from Downsview airport on August 16, 1947. The Beaver was the ultimate bush plane and became father to a dynasty of short take-off and landing (STOL) aircraft.

McConachie could empathize with the Beaver as he considered CPA's senior team. It consisted of:

R.W. Ryan: Assistant to the president.

W.G. Townley: General manager of operations. Former wing commander in the RCAF; later worked for Boeing Aircraft and Canadian Car & Foundry.

H.B. Main: Director of sales. Former RCN, joined the CPR in 1935, becoming secretary to the general manager (Western Region) in 1942.

C.H. Pentland: Manager of Pacific Operations. Engineering graduate; banker and life insurance clerk before turning to flying. Trained by Imperial Airways and BOAC.

R.B. Phillips: Manager of domestic operations. Friend of McConachie's from YSAT.

H. Fletcher: Superintendent of maintenance (overhaul). Served with Imperial Airways before coming to Canada in 1929 to join Canadian Airways Ltd. at Montreal. Thirty-one years of maintenance experience.

I.A. Gray: General superintendent of engineering. Bachelor of aeronautical engineering, joined CPA in 1943; worked on Canadair North Stars. Responsible for all modifications on C-4-1s and for expediting the supply of equipment from Montreal to Vancouver.

H. Hollick-Kenyon: Manager of flight, technical department. Associated with TCA and Western Canada Airways.

E.R.R. "Ted" Field: Assistant manager, Edmonton. Entered aviation in 1933 as a pilot for United Air Transport. Served as superintendent of CPA's Whitehorse depot until 1943, when he transferred to Vancouver.

Leo James Dalton: Superintendent of Quebec district. Began his career with Fairchild Aerial Surveys and joined CPA in 1941.

Backed by the deep pockets of the CPR, McConachie could now indulge his dream of flying to the Far East. To do this he needed two things: the permission of the Air Transport Board to operate outside Canada and suitable long-range aircraft to do so. Each depended on the other.

Transcontinental air travel had come a long way since Don Bennett's seven Hudsons ventured across the Atlantic Ocean in November 1940. As Juan Trippe and George Woods-Humphrey had known, the Atlantic crossing was a potential profit-maker. Except for the United States, all countries ensured that only their national carriers were permitted to fly the route. Since July 1943, TCA had been flying government VIPs in Lancastrians between Dorval Airport and Prestwick, Scotland. With the war's end, the airline allowed the general public on board, charging $572.12 for a one-way ticket.[15]

On the other side of Canada, the ocean was still virgin territory. Wider than the Atlantic and needing many more refuelling stops, the Pacific bordered on the war-ravaged countries of Japan and China, with only Australia and New Zealand affording any promise for air traffic. Unwilling to burden a deficit-ridden TCA with what was obviously a money loser, Howe evidenced no interest in flying the Pacific. After World War II, the Pacific was American "territory." Pan American Airways, with its network of bases from the flying boat days, flew DC-4s from San Francisco via Honolulu to Sydney and Wellington. The northern Pacific was covered by Northwest Airlines from Seattle through Alaska to Tokyo and Shanghai.

But the British empire hadn't quite died out. In February 1946, a Commonwealth civil aviation conference convened in New Zealand to discuss a possible Wellington–Sydney–Vancouver link: Australia and New Zealand had pressed for air services to their capitals from their North American sister in order to foster Commonwealth solidarity. Canada's reluctance to participate was evidenced by the sending of not a cabinet minister but Canadian High Commissioner Walter Riddell to the conference, and then only in the role of observer. (Ironically, in 1932, Riddell had been posted to the League of Nations and warned Ottawa of entering into compacts with other nations concerning commercial aviation.) The conference's main purpose, it turned out, was to organize a joint Commonwealth airline that would compete with Pan American Airways on the Pacific. On June 11, Canada and Australia concluded bilateral agreements that gave each other reciprocal landing rights. London, Canberra and Wellington then created the jointly owned British Commonwealth Pacific Airlines (BCPA), and there was some talk of equipping it with Canadair North Stars to entice the Canadians to join. But when that aircraft's debut was delayed, BCPA chose DC-4s, and in May 1947 its first flight touched down from Sydney in Vancouver.[16]

In the House of Commons, Stanley Knowles of the Co-operative Commonwealth Federation was badgering C.D. Howe to name a date for the TCA service to Australia to begin. Knowles represented Winnipeg North and had the interests of his constituents at heart; many of them worked at the main TCA base. The Australians engaged in some diplomatic arm-twisting as they too awaited the arrival of the first Canadian flight across the Pacific. Howe was at his evasive best and hinted that Canada was about to begin a Pacific routing as soon as possible.

The minister, for once, was powerless. This time it was TCA's new president, Gordon McGregor, who controlled the airline's future, and he refused to even contemplate a Pacific service. Shocked by the airline's deficit—caused in part by political interference and mismanagement—McGregor calculated that any service to Australia would require years of government subsidy. Besides, as a pilot himself, he had no intention of using the temperamental North Stars foisted on him by Ottawa to ply the lonely Pacific.

Christened "North Star" by Mrs. C.D. Howe, "Argonaut" by BOAC and "Canadair Four" by CPA, Canada's first airliner was a marriage of British and American technology and flew, in one guise or another, from 1946 to 1976. CAA 988-31-25

Into this rushed CPA's new president. Lesser men would have been content with starting the airline's international program with a minimal service, possibly Vancouver to Los Angeles or San Francisco. But Grant McConachie had sung his tune before too many bank managers and insurance executives to be put off now. He gave Howe the full treatment. Using his string and inflatable globe, he explained the Great Circle route from Vancouver to China and Australia. He requested not one but two Pacific services, north and south—to Japan and China and to Australia and New Zealand—neither of which, he stressed, would need a government subsidy. McConachie's audacity (or was it Howe's embarrassment with the Commonwealth cousins?) impressed the minister sufficiently for him to reconsider the official policy of allowing only the "chosen instrument" to fly overseas.

On May 11, 1948, in what was an unaccustomed role for him, transport minister Lionel Chevrier presented CPA's case before cabinet.[17] Although the airline was not the government's "chosen instrument," it was definitely in Canada's interest to have CPA fly the Pacific. His reasoning was impeccable: the CPA's maritime organization was already in place to provide support. If Canada was to participate in the opening up of Oriental markets, an air service was necessary. All Sydney–Vancouver flights could feed into TCA's transcontinental service. Finally, this plan was not going to cost the taxpayer a penny.

As might be expected, the strongest opposition came from the Department of External Affairs. Escott Reid and Lester B. Pearson chorused that the federal government was caving in to exactly the sort of pressure from commercial lobbies that the setting up of TCA in 1937 had been designed to prevent. Also, the Commonwealth allies had been led to believe that they would be dealing with TCA, Canada's official commercial air arm, not a jumped-up bush outfit.

Lionel Chevrier and C.D. Howe were, however, politicians, and expedient ones at that. On July 13 they granted CPA permission to operate over the ocean after which its railway parent had been named. Reid had been correct in his assumptions: the Australians and New Zealanders felt slighted. But Howe had no qualms; he had promised them and Knowles that a Canadian airline would serve the Pacific. It just wasn't going to be TCA. He knew that the route would never be profitable, and besides, he could make some capital out of this from CPA.

The pound of flesh that the minister demanded was soon made public. McConachie had to fly to Australia before entering the Far East, "to demonstrate Commonwealth solidarity." The second requirement was that CPA order four North Stars from Canadair, at $800,000 each. The first Canadian-built airliner had its origins in Howe's speech on March 17, 1944. The shipbuilders Canadian Vickers at Cartierville were awarded the contract to build fifty DC-4s, but when they backed out, Howe was persuaded by Ben Franklin, one of Vickers' managers, to set up Canadair in November 1944 and keep the momentum going.

With the war over, Donald Douglas was impatient to begin turning out the larger DC-6s and unloaded onto Franklin whole factories of C-54 (the military version of the DC-4) parts at scrap-metal prices. With these parts, Canadair employees were able to cobble together Canada's first airliner, the "North Star." The name was chosen by Franklin and used by Mrs. C.D. Howe to christen the prototype, CF-TEN-X, when it was rolled out at Cartierville on July 20, 1946.

The Douglas licensing agreement prevented Howe from badgering anyone other than the British, the RCAF and TCA to buy the North Star—except CPA. The four CPA aircraft, Nos. 47 to 50 on the production line, came off the line between April and June 1949. Nos. 45 and 46, which had been the first two BOAC aircraft, were called Argonauts, and the CPA models were identical to them. Their maximum weight, 82,300 pounds (37,035 kg), made them heavier than the TCA machines at 79,600 pounds (35,820 kg). The only difference was in the seating: CPA put in 36 seats, whereas BOAC had 40. Calling its North Stars C-4-1s, or "Canadair Fours," CPA christened them CF-CPI, *Empress of Sydney*; CF-CPR, *Empress of Vancouver*; CF-CPJ, *Empress of Auckland*; and CF-CPP, *Empress of Hong Kong*.

Now that he had the ATB's permission and the aircraft that Howe wanted him to buy, McConachie looked to secure the landing rights in Australia and the Far East. The United States permitted the Canadians to land in Honolulu in exchange for rights to the recently acquired Newfoundland airport at Gander, and to get to Japan, CPA was also allowed a refuelling stop at Shemya, a US air force base in the Aleutians. Exactly a month after signing the contract to buy the North Stars, a CPA team caught a Pan American DC-4 in San Francisco for Sydney and Canberra. With McConachie was Hugh Main, his sales manager; Walter Townley, general manager of operations; and Ian Warren, the CPR's traffic manager.

At first, it looked as if the four had flown a long way for nothing. The Australian government in Canberra refused to meet with anyone other than H.J. Symington, president of TCA; it was, they believed, what they had been promised at Wellington. As part of the Canadian Pacific empire, CPA stank of private enterprise—anathema to the Australian prime minister, Ben Chiffley. To no avail did C.M. "Crofty" Croft, the Canadian High Commission's commercial counsellor, make all the representations he could on CPA's behalf. Lionel Chevrier even telegrammed his opposite number in the Australian Ministry of Transport, confirming that CPA was indeed Ottawa's "chosen instrument" in the Pacific.

It was only the day before he returned home, at the farewell luncheon hosted by the Australian prime minister, that McConachie was able to work his magic and turn things around. Croft had casually mentioned to McConachie that Chiffley was proud of his origins as a railway engine driver and union organizer. In the course of the meal, the prime minister somehow formed the impression that McConachie was the president of the whole Canadian Pacific Railway. The illusion was helped along by Grant fortuitously producing from his wallet his old union card, unused since the summer he had slaved for the CNR as a stoker. That a stoker could rise to become president of the largest company in Canada was immensely appealing to Chiffley, and he got up from the table and embraced "his brother of the tracks." Accordingly, the next day CPA was granted a "temporary" permit to fly into Sydney—one that it would use for two decades.[18]

McConachie returned to Montreal in Christmas 1948, full of his Australian success, his eagerness infecting the overworked, despairing Neal. Of course, the CPR president wanted to see the North Star in action and immediately had Grant arrange a demonstration flight, to Kingston, Jamaica. The airline's order would not be ready until next May, so borrowing an RCAF North Star from Canadair, Neal and McConachie flew to Kingston after New Year's Day. Neal didn't return with Grant a week later, excusing his absence by saying he was suffering from a cold.

Had he not been away, McConachie might have heard the rumours floating around Windsor Station about Neal drinking heavily and entertaining his mistress in, it was said, the sacred CPR presidential rail car. After four decades of bludgeoning his way to the top, the strain had caught up with the office boy and Neal was publicly unravelling. But such was the power of those who governed the CPR that even these indiscretions might have been overlooked had Neal, after a respectable holiday, returned to Windsor Station. However, it took an absence of three months and an impending annual general meeting before the board of directors forced him to resign. Unwin later said that McConachie took the news very hard, that he wandered around for days saying "he would do anything for that man."

The CPR press release that announced Neal's "retirement" on March 8, 1948, also introduced the team of George Walker as chairman and William Mather as president. Never again would the positions of president and chairman be held by the same person. Both men could have consoled themselves with the knowledge that the railway's air branch finally had some future; that year and the next, the airline was awarded the following new routes:

Vancouver–Whitehorse (direct)
Montreal–Val d'Or–Rouyn (at first with Cansos, then DC-3s)
The Pas–Churchill
Vancouver–Whitehorse–Dawson City–Fairbanks

The remainder of the domestic network consisted of:

Edmonton–Peace River–Yellowknife–Coppermine
Edmonton–Fort McMurray–Norman Wells–Aklavik
Montreal–Quebec City–Chicoutimi
Montreal–Quebec City–Baie Comeau–Lower St. Lawrence–Seven Islands
Vancouver–Port Hardy–Prince Rupert
Vancouver–Penticton–Castlegar–Calgary
Regina–Saskatoon–North Battleford
Winnipeg–Dauphin–The Pas–Flin Flon
Winnipeg–Kenora–Red Lake
Red Lake–Pickle Lake–Sioux Lookout

A sample of the fares on the domestic routes:

Edmonton–Whitehorse $75
Edmonton–Dawson City $120
Winnipeg–Pickle Lake $33.75
Kenora–Red Lake $8.00
Montreal–Saguenay $15.50

By April 30, 1949, the airline had grown to 821 employees and had a monthly payroll of $193,727.

That year the beaver on the CPR's logo was given a tree branch to chew on, but James Richardson's Canada goose continued to fly unchanged on CPA aircraft, to the chagrin of its vice-president, Dick Ryan. Before he and McConachie attended a directors' meeting at Windsor Station, Ryan suggested that consideration might be given to changing the flying goose emblem.

Now that we were flying at 300 miles per hour, a goose hardly seems an appropriate symbol. Should it not be stylized or changed to a speed bird? At the meeting I had an easel put in the boardroom and placed on it a sketch of a Canada goose, a stylized goose, and a speed bird. I pointed out that a wing-flapping goose flying at a top speed of 60 miles per hour was hardly an appropriate symbol for an airline company operating planes that flew at 300 miles an hour. I suggested that the goose be replaced by a speed bird.

175

During my presentation I noticed that Mr. Mather was becoming somewhat agitated. Suddenly he exploded. "Gentlemen," he said, "I have always been a great admirer of the Canada goose. I have often hunted it and I respect that bird. Gentlemen, as long as I am chairman of this board a goose will remain a goose." And he thumped the table with his fist to emphasize his remarks. There was dead silence for a few seconds and soon the board proceeded with other matters.

When Grant and I were able to get together after the meeting I said, "Well Grant, I sure made a mess of that presentation, but I had no idea that our chairman was such a strong supporter of the Canada goose. What are we going to do now?" I knew Grant was never at a loss for ideas.

"Think nothing of it, Dick," he said. "We will simply make the transition gradually." And that is exactly what we did."[19]

As sober as Walker and Mather might have been, it wasn't long before McConachie went to work on both of them, spinning his expensive dreams. Having first warned the new rulers that the services to the Orient could initially lose money, he said that CPA had to expand into Tokyo, Shanghai and Hong Kong while the options were open to it. The directors were still reeling from that one when Grant informed them that to be closer to the action, he wanted to move the airline out of Montreal to far-off Vancouver. It was obvious to all, he pointed out, that TCA had the Atlantic Ocean all sewn up and was itself transferring from Winnipeg to Montreal to concentrate on its European and Caribbean routes.

In January 1949, McConachie caught a Northwest Airlines flight from Seattle for the Orient—a trip he had dreamed of twenty years before. The rights to land at Tokyo, Shanghai and Hong Kong could only be granted by the three men who ruled the Far East at that moment in history. In order of importance, they were Supreme Allied Commander General Douglas MacArthur, General Chiang Kai-shek and Sir Alexander Grantham, the British governor of Hong Kong. To the mystification of many, McConachie took along with him Wop May, former bush pilot and now CPA's director for northern development.

The first stop was Tokyo. Four years into the American military occupation the ruined city was rebuilding from the war, and everything was rigidly controlled by the US armed forces. The two Canadians laboriously worked their way through the "desks," collecting permits for

In 1949, the CPR's beaver was given a log to chew on and the Lodestar was taken off the shield. CAA M4337

fuel, food, banking and radio transmission. But it gradually became clear that only The Great One, MacArthur, had the authority to guarantee landing rights. The problem was that no one was allowed an audience with the general in his eyrie in Tokyo's Dai-Ichi Building.

It was McConachie's good fortune that the Canadian ambassador to Japan was Herbert Norman. Posted to Japan before the war, Norman had endured house arrest in the legation building from 1939 to 1942. Because he was fluent in Japanese, on his return to Ottawa he was seconded to Special Intelligence and, when the Philippines was liberated in 1945, assisted in the evacuation of Canadian prisoners of war. Norman was now the Canadian liaison on the Supreme Allied Command staff—there was no one whom MacArthur respected more—and McConachie's audience with the general could only be arranged through Norman.

Even the diplomat's stature could not secure a meeting time longer than a tightly controlled fifteen minutes. Before he was to meet with MacArthur, McConachie went over his speech, covering all the main points—the fact that CPA was Canada's official airline in the Pacific, the reasons why the permits were necessary, and so on. Finally, he was led into MacArthur's presence, whereupon both lit up their pipes and puffed away "like two old pensioners." The general, on hearing the words "Canadian Pacific," reminisced about his rail trip to Banff and Lake Louise and rambled on about the wonders of the Rockies, while McConachie became ever more painfully aware that the precious fifteen minutes were ticking away. Suddenly MacArthur, looking at his watch, said he had an embassy dinner to attend. The audience was over. As he was shown out the door, McConachie managed to blurt out that CPA needed the permits to land in Japan. Ushering him out, MacArthur said that they would be ready at nine o'clock the next morning. And they were.

Landing in Shanghai, McConachie must have wondered if this time he had overreached himself. He had no idea of the chaos that China was experiencing. The huge port, preparing to face the onslaught of the Communist armies, was protected by unpaid, poorly armed Nationalist troops sheltering behind a thin wooden palisade.[20] Although Chiang Kai-shek was in the city at the time, he was packing to leave for Taiwan by gunboat. He was not even president of Nationalist China any longer, so there was no paramount authority for McConachie to approach. With bribery rampant—all the banks had been looted—only the most powerful of connections could have got Canadian Pacific the landing rights. This is where Wop May proved to be the smart choice as travelling companion.

Chiang Kai-shek's personal bodyguard was Morris Cohen, an Englishman exiled to the Canadian frontier by his despairing family. On the Alberta frontier he was at various times a pedlar, a cowboy and a gambler, acquiring the nickname of "Two Gun." At the turn of the century, Cohen was a hero to the Chinese community of Edmonton as a result of having knocked out a criminal who preyed on their restaurants. Thus, in 1911, when Dr. Sun Yat-sen came to Canada to raise funds for the Nationalist cause, Cohen was chosen to be his bodyguard. Dr. Sun made him a full general in the Nationalist Chinese army, and although Canadian and Jewish, Cohen became a sort of ambassador for China.

At their reunion, he and Wop May greeted each other as old friends and exchanged stories of their Edmonton days; the landing rights were assured. Later, at an embassy party, McConachie met Madame Chiang, who was about to fly to New York with her personal fortune. Thus it came as no surprise when Tom Davis, the Canadian ambassador, informed McConachie that the Nationalist government had approved CPA's application to serve Shanghai.

British-owned Hong Kong proved to be no problem—quite the opposite. Thousands of refugees were fleeing the Communist armies, so the resources of the tiny colony were severely strained. Food and water were especially scarce, and the British authorities were ecstatic that an airline might take as many refugees as it could away from Hong Kong. The fare from Hong Kong was $726. The Chinese community in British Columbia were responsible for sponsoring their relatives and friends to Canada, and one of the conditions of immigration from Hong Kong at that time was that the sponsor paid the fare of $726 in full, in advance

Before the CPA C-4-1s were ready, McConachie leased two RCAF North Stars for familiarization flights to Australia and the Orient. He is shown here inspecting the Merlins with W.G. Townley, general manager of operations, and C.H. Pentland, manager of Pacific operations. CAA 989-189-31

and in Canada. This sure source of income allowed CPA to make a fortune on the Hong Kong route for decades to come.

On April 19, 1949, the first CPA familiarization flight took off from Vancouver for Tokyo and Shanghai via Anchorage. For the airline, this was its first international flight as well as its first four-engined one. It used North Star CF-TEP, leased from the RCAF, which became the first Canadian commercial aircraft to cross the Pacific. Because the aircraft's Merlin engine was unavailable outside the country, a spare Merlin was carried on board.

Between Anchorage and Tokyo, Shemya was used as a refuelling and "overnight layover" by the crews of commercial aircraft bound for Japan. Because of Shemya's location, fog and poor visibility were the norm, and the CPA crew remember having to make a few approaches before landing. Coming into Shanghai's Lunghwa airport in heavy rain on April 25, the Canadians could see the civil war raging all around the city. When they landed, the advance units of the Communist military were a few miles from the airport perimeter. Canadian embassy staff met the aircraft and threw some classified files on board, and the North Star was hurriedly refuelled before taking off for Hong Kong.[21]

The familiarization flight for Australia and New Zealand was less eventful. CPA's first North Star, *Empress of Sydney*, left Vancouver on May 28 and landed in Sydney on June 3 and Auckland on June 8, returning to Vancouver on June 11. The official launch to Australia took place on July 10, 1949. This time the aircraft was packed with thirty-two guests, including influential VIPs like media baron Roy Thomson, former bush pilot Romeo Vachon, now with the Air Transport Board, and Clarence Campbell of the National Hockey League. One of the journalists selected was Ronald A. Keith, who was then reporting for

The first familiarization flight about to depart on April 19, 1949, using RCAF North Star CF-TEP for Tokyo and Shanghai via Anchorage. From top left: Capt. J.K. Potter, Capt. C. North Sawle (CPA's chief pilot), Capt. Louis C. Stevenson, Capt. M.D. Lee, Capt. George Knox, Capt. Len Fraser, Navigator P.D. Roy, Capt. Robert Goldie, Capt. Archie Vanhee, Navigator Fred Wicker. CAA 226

Canadian Aviation magazine: the "most pronounced impression of the flight was one of unreality. You boarded after breakfast in Vancouver, spent an indolent day lounging in the sky, then surprisingly found yourself swimming in the moonlit surf of Waikiki the same night." Because the Australian flight was linked with the airline's Fairbanks–Whitehorse–Vancouver service, the prospect of being in Alaska one morning and swimming in Hawaii's warm surf the next night was entirely possible.

> Two nights and a day at the Royal Hawaiian Hotel in Honolulu, then another day at 18,000 feet, lazing around at 240 mph and a night landing at Nadi Airport, Fiji Islands. [For the inaugural flight] some of the residents had laid on a lavish party at the airport's Mocambo cocktail lounge, an assignment which occupied the undivided attention of most of the party for some hours after arrival. The program included native music, ceremonies and dances which, we can guarantee, will rate high in tourist attraction.
>
> The swish of Fijian grass skirts and the peculiar rhythms of native music, breakfast served by barefooted waiters, these memories of the brief sojourn in Fiji are still vivid as you touch down at Sydney the next afternoon.
>
> Roaming through the streets of Sydney, strongly reminiscent of London, and listening to the distinctive accents of the Australians, you find yourself not quite believing in this modern transportation miracle.[22]

In Sydney, Mrs. Drakeford, wife of the Australian minister of air, christened the aircraft *Empress of Sydney*, and all returned to Vancouver on July 21. Ground servicing was done in San Francisco by United Air Lines; in Honolulu, Nadi and Canton by Pan American and in Sydney by Australian National Airlines (ANA). McConachie recruited Australian flight attendants in Australia as well; the first of many were Margaret Lineham, a "check" girl from ANA; Pamela Hookham, a model from Sydney ("the first Australian girl to take to water skis"); and Shirley Burgess, the ANA receptionist at the Australia Hotel in Sydney. But the airline emphasized that each plane would also have a registered nurse.

The first scheduled flight to Australia departed Vancouver for Sydney on July 13, with a single paying passenger: the bush pilot Pat Reid, who worked for Imperial Oil. The same crew worked the entire Vancouver–Sydney flight, a strenuous ten-day job. Navigation was fairly rudimentary by the later standards of the inertial navigational system (INS). The navigator used conformal conic projection charts and the navigation aids were LF range, ADF HF/DF, MF/DF, LORAN (the basic navigational aid for oceanic flights) and commercial radio stations. Between Honolulu and Canton, LORAN was used to check position every thirty minutes until it ran out. Drift sight readings double-checked the track, and position was also checked with sun position lines and in relation to the west leg of Palmyra Island. The final stage was completed using the ADF range tuned to Canton. The crew had to keep their headsets on at all times to listen for their call through all the radio traffic. The flight times were:

> Vancouver–San Francisco: 3:55 hrs.
> San Francisco–Honolulu: 10:25 hrs.
> Honolulu–Canton Island: 8:07 hrs.
> Canton Island–Fiji: 5:57 hrs.
> Fiji–Sydney: 8:33 hrs.

Covering 8,403 miles in 37 hours and 18 minutes, the Canadair Four flew at an average speed of 195 knots. Included in the price of a one-way fare of $685 was two nights' accommodation in the Royal Hawaiian Hotel, Honolulu, and one night in Fiji.

CPA estimated that the overall cost of flying one North Star worked out to $500 an hour; thus the total cost was $18,500 for the whole 37-hour trip. The break-even load factor was filling 25 of the 36 seats, but taking into account the mail and cargo revenue, an actual break-even

The *Empress of Sydney* at Mascot airport, Sydney, July 1949. Left to right: Capt. C. North Sawle, Capt. Bud Potter, stewardesses Helen McCracken and Pamela Hookham. The same crew worked the entire Vancouver–Sydney flight, a strenuous ten-day job. CAA (UNN)

level of 20 seats was more accurate. However, the Australian run, departing every two weeks, was fortunate if it had any passengers at all. In the 1950 winter season it carried a total of 125 passengers—48 of whom had been given or won complimentary passes—and although in December the following year the Nadi–Sydney trip was re-routed via Auckland, McConachie must have reflected that Howe and McGregor were right in their assumptions: despite Commonwealth solidarity and the swish of Fijian grass skirts, no Canadian wanted to fly to Australia and New Zealand. The South Pacific service was a money-loser.

It was the gruelling flights to Australia that led CPA stewardesses to consider joining CALFAA, the TCA flight attendants' association. Barbara (Ayliffe) Stewart, a former TCA stewardess who worked for CPA in 1949 recalled stewardesses getting off flights in tears from sheer exhaustion. Although the pilots had an eighty-five-hour monthly flight time limit, there was no such limitation for the stewardesses. "You have no idea what those girls looked like after some of those flights," said Stewart, who acted as adviser to the management. "All I could think of was that they were going to be old women long before their time."

One of the Australian stewardesses hired in April 1949 was Lizabeth Brewis, who felt that her colleagues in CPA were "underpaid and overworked." Although Australian attendants had the reputation of being militant "rabble rousers," at first Brewis was reluctant to take the lead in organizing the Canadian stewardesses. She was also concerned that "the family atmosphere" of the airline might suffer and that the Australians might be considered "ungrateful" foreigners in Canada. Nevertheless, led by Brewis, all thirty-nine CPA stewardesses based in Edmonton, Winnipeg, Dorval, Regina and Vancouver voted to join CAL-

The inaugural run to Hawaii rated a full traditional welcome with dignitaries and grass skirts at Honolulu airport. The airfare to Sydney included an overnight at the Royal Hawaiian Hotel. CAA 989-165-144

FAA. Negotiations started in 1950 between the airline and a CALFAA committee consisting of Jean Saddy, Liz Brewis, Barbara Ayliffe—and TCA's Muriel Peacock. Since the airline had no transcontinental routes then, CPA stewardesses rarely came in contact with their TCA colleagues and were grateful for Peacock's guidance.

The airline didn't particularly welcome this. The cherished "family atmosphere" was, in reality, little more than benevolent paternalism. Grant McConachie, for example, prided himself on knowing every stewardess by name, but if they were now going to be unionized his expansion plans, especially on the long Pacific flights, might be jeopardized. His old friend from the YSAT days, Domestic Manager Barney Phillips, put it even better. He warned the CALFAA committee that stewardesses could be "picked up on any corner" to replace the current employees if they did not like their salaries and working conditions.[23]

Fortunately more liberal heads prevailed. After patient negotiations, a one-year agreement, effective March 1, 1951, was signed with the cabin staff. It covered rates of pay that varied from a minimum monthly salary of $200 for registered nurses to $280 per month after six years. The in-charge stewardess received an additional $10 per month, and an overseas rate of $30 was paid on top of that. Flight-time

Touring the airline's new facilities are (left to right): CPA President G.W.G. McConachie, future president Ian Gray, talking to Buck Crump (a future CPR president) and an unknown. CAA 989-195-25

limitations were held at 1,020 hours per calendar year, with a quarterly limitation of 260 hours on domestic operations and 270 hours overseas. CALFAA or no, stewardesses still had to buy their own uniforms, had to keep their hair "up" past their ear lobes at all times, couldn't wear glasses but were expected to wear girdles and, of course, had to resign if they got married.

At Windsor Station that August, William Mather and George Walker were too engrossed in the Canadian Pacific Railway's slow, painful conversion to diesel power to bother much about the airline staff packing up and moving to Vancouver. McConachie had talked C.D. Howe out of the old war-surplus plant at Vancouver Airport, where Boeing had built Cansos during the war, and had paid Crown Assets Disposal $20,000 for it. CPA's fleet now consisted of 2 Norsemen, 4 Ansons, 3 Cansos, 4 Lodestars, 16 DC-3s and 4 Canadair Fours, demonstrating that—although it still serviced much of the bush country—the airline was shedding its frontier image.

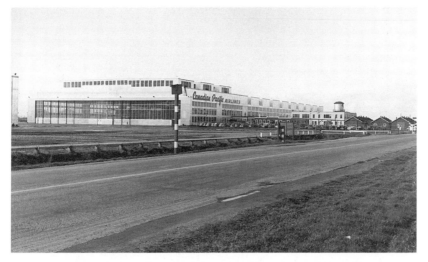

McConachie leased the old wartime Boeing plant at Vancouver airport from Crown Assets Disposal for twenty years for $20,000. It remained the main airline facility until the operations centre opened in 1969–70.
CAA (UNN)

Under Grant McConachie, CPA's focus was now international, perched on the edge of Canada and facing out across the sparsely travelled Pacific. Although Hong Kong was profitable, there was little hope of ever making a profit from Tokyo or Sydney. None of this seemed to bother the airline's president. The previous year, on September 6, 1949, he had been in England, at the Farnborough Air Show, and watched fascinated as de Havilland's Comet streaked overhead.

Withdrawing from its bush routes in 1959, the two de Havilland Otters CF-CZO and CF-CZP were the last of CPA's bush aircraft. Their direct lineage from the de Havilland Beaver, which Punch Dickins had helped design, is evident. CAA 988-31-26

CHAPTER SIX

Clash of the Tartans

I t was commonly thought that C.D. Howe would retire from politics in 1946 and succeed Herbert Symington as president of Trans-Canada Air Lines. The minister had expressed this wish many times himself, but when Mackenzie King asked him to stay on in Louis St. Laurent's cabinet, the question of Symington's successor was unexpectedly raised. Howe's reliance on Americans like P.G. Johnson to get the national airline running had been criticized so Symington had been careful to vet candidates who were not only native-born—preferably in a heroic mould—but who also had some administrative experience in aviation. In 1945 he had hired three group captains directly from the RCAF, one of whom was the air ace Gordon McGregor.

By coincidence, both Grant McConachie and Gordon McGregor were in England four years later at the Farnborough Air Show, and both fell in love with British-made aircraft. During the war the British realized that they were no longer able to compete with the United States in the field of conventional aircraft, and as a result de Havilland and Vickers-Armstrong had each created jet-powered and turbo-jet transports, the de Havilland 106 Comet and the Vickers Viscount.

The Comet surprised the world at the 1949 Farnborough Air Show. Although it had no serious competitors—Boeing did not roll out its 707 for another five years—the Comet 1 was rushed through its testing so that BOAC could begin scheduled services with it on May 2, 1952. Since the Comet's Ghost 50 turbo-jets were typically thirsty early jet engines and did not allow for flying the Atlantic, BOAC used the aircraft on the

Grant McConachie's motto was "If it's new, I'll take two." He hoped that CPA's Comet would gain maximum publicity by setting a speed record between London and Sydney.
CAA (UNN)

London–Johannesburg route, which allowed for frequent refuelling stops.

De Havilland's sales staff at Hatfield thought that the president of CPA was more enthusiastic about the Comet 1 than they were. Since the war the Americans had concentrated on building jet bombers like the B-47, which was hampered by its heavy fuel consumption. None of the other major aeronautical nations—not even the United States, the Soviet Union or France—had anything like the Comet, but McConachie had to have—two!

At the same time, in the neighbouring town of Wisley, home of Vickers-Armstrong, when Gordon McGregor took the controls of its new airliner, the Viscount, he thought it "handled as beautifully as [a] Spitfire." The type of aircraft that seduced each man at that Farnborough Air Show—McConachie's revolutionary Comet and McGregor's neat, nimble Viscount—demonstrated a great deal about their characters.

The two airline presidents shared more than their Scottish ancestry. Like Eddie Rickenbacker and Howard Hughes, both were experienced pilots—McConachie had even been checked out to fly the North Star—and both were the last of their profession to run major airlines in Canada. Eight years older than McConachie, McGregor was his apotheosis. McConachie had kept his various companies aloft through bravado, unpaid bank loans and sheer luck; McGregor thrived in the security of large organizations, beginning his working life in Bell Telephone before joining the RCAF.

Twice winner of the Webster Trophy for the best amateur pilot in the 1930s, at the age of thirty-nine McGregor had been the oldest fighter pilot to participate in the Battle of Britain. He survived a young man's war because of a natural caution that combined with a singlemindedness to

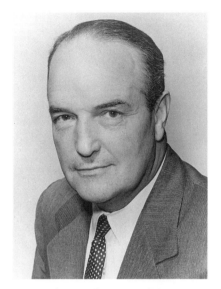

The prudent, assertive Scot Gordon McGregor, president of Trans-Canada Air Lines, 1948–68. He was appalled and irritated by McConachie's tactics and never failed to wonder why Ottawa allowed CPA to exist.
ACA (UNN)

make him the commanding officer of 126 Wing by D-Day. Herbert Symington had made him TCA's general traffic manager, and at the 1947 International Air Transport Association conference in Brazil, all the delegates were impressed by McGregor, who spoke sensibly for the standardization of air regulations. The prudent, assertive Scot was just the sort of president the government airline needed to get out of the red. Quite the opposite of the prolific McConachie, he rejected the Comet and the Canadian-made Avro Jetliner for what they were—untried and too advanced for a North American airline.

McGregor's infatuation for TCA was rivalled only by Howe's. Both men were ferociously prepared to protect it from competition. As a result of McGregor's stewardship, TCA climbed out of deficit and away from government meddling to record a modest surplus by 1951. Except for the routes to Britain and to the Carribean, however, TCA remained essentially a domestic airline, unlike CPA.

On the stormy night of February 2, 1950, at Haneda airport, the CPA Canadair Four CF-CPR was coming in to land from Hong Kong when the pilot misread the signals to go around, and the plane overshot the runway. The wet concrete made the Canadair Four's brakes ineffective, causing the aircraft to career toward Tokyo Bay. It jumped the sea wall and then buried its nose and engines in the water. Nine passengers suffered minor injuries and were quickly rescued from the shallow water. The Canadair Four, however, was a complete write-off, its Merlins having sucked in the salt water. Although the incident left McConachie short one airliner, at that time it didn't look like he would need another. Except for revenue from the Hong Kong emigrants, the Vancouver–Tokyo service, like the South Pacific one, was economically disastrous, the crew on both routes invariably outnumbering the passengers.[1] The Canadian public had no interest in visiting its recent enemy Japan, and under US military law no Japanese were allowed to leave the country.

That summer there was even less potential for tourism in the Far East, when North Korean tanks and troops struck across the 38th parallel on June 25, 1950. Mackenzie King had retired by then, and for Prime Minister Louis St. Laurent, Korea was a test. How much could Canada, now in a recession, afford to do? To help the United Nations forces hold back the Chinese millions pouring across the Yalu River, all that minister of defence Brooke Claxton could scrape together were two destroyers, the North Stars from the RCAF's 426 transport squadron and a single army brigade. But Washington saw Moscow's hand in this and, mobilizing every infantry division it could, aimed them across the Pacific Ocean.

McConachie was never short of the imagination or courage to let an opportunity such as this pass. He lobbied Ottawa to award CPA the contract to airlift American soldiers from Tacoma, Washington, to Tokyo. Since DND would pay the entire cost, it would enable Canada to help in a tangible though noncombative way. Thus between August 14, 1950, and March 30, 1955, when the service ended, forty thousand US military personnel were flown by CPA on 703 charter flights.

Anyone else would have used the troop-carrying flights to cut back on such luxuries as the first-class gourmet meals for the GIs, but McConachie understood the potential of advertising and insisted that they continue. He hoped that every soldier who had enjoyed the in-flight service might one day return to do so with his family. When word spread about CPA's standards, the story was that officers pulled rank to fly the airline.

It was a good deal all around: the troops were flown out to Tokyo, and Chinese immigrants were picked up in Hong Kong and brought to Vancouver. CPA made $16,839,790 on the Korean airlift, more than covering the losses from the unprofitable Vancouver–Sydney trips. Gordon McGregor, who would never admit to regretting his decision to turn down the Pacific routes, summed it up. "Jesus Christ!" he marvelled. "Only McConachie could be that lucky."[2]

Canadian Pacific Airlines showed a profit in 1951 and received a new domestic service from the Air Transport Board: Vancouver–Kamloops–Williams Lake–Quesnel–Prince George–Edmonton–Lloydminster–

For the Hong Kong flights the airline recruited their first Chinese stewardesses, having them pose in traditional costume on September 28, 1950. They are Gloria Wu, Lorraine Hgu and Joyce Lam. CAA 635

North Battleford. In 1952 it was awarded three more routes: Prince Rupert–Terrace, Edmonton– Uranium City and Rouyn–Toronto. Despite increased labour costs and a strike in the oil industry, domestic revenues continued to increase substantially.

The recession had ended by now and Canadians were taking to the air in unprecedented numbers, more than the designers of the North Star had anticipated. TCA's flights across Canada, to Europe and the Caribbean were filling up fast, and to squeeze as much profit out of the North Stars as possible, McGregor increased their capacity from 36 passengers to 55 and eventually to 62. CPA's Canadair Fours also seated a maximum of 36 (which featured in airline ads as having "great leg room"), but resulted in unprofitable flying.

The North Star, the aircraft that had been part of Howe's bargain with CPA, was unloved by both crew and passengers.[3] The controversy over the Merlin engines refused to go away. Designed by Rolls-Royce to provide military aircraft with sudden bursts of speed—for Spitfires to scramble or for Lancaster bombers to evade flak—the Merlins were ideal in combat situations. Their unreliability during prolonged use and high fuel and parts consumption were never considered until they entered commercial service in the North Stars. The Merlins also suffered from an overly sensitive fire-control system, which unnervingly flashed warning lights in the cockpit for no apparent reason.

The passengers who flew in a North Star never forgot the Merlin's noise. Caused by the engine's exhausts venting against the fuselage, this had not been noticed in the Spitfires because pilots always kept their earphones on. Eager for postwar exports, however, Rolls-Royce offered all customers an unconditional guarantee that they would modify the Merlins whenever needed, clinching their sale to Canadair.[4]

Sherry Chinn was born on June 22, 1952, in an immigrant-filled CPA DC-4 flying from Hong Kong over the North Pacific between Tokyo and Shemya. It was flown by Captain George Jarvis, who twenty-two years later, when Chinn returned to Hong Kong in a Boeing 747 as a guest of the airline, was also captain. As a memento of the occasion he presented her with an autographed photograph of the actual DC-4 she was born in. CAA (UNN)

McConachie could not wait to unload his three remaining North Stars onto TCA. He did so for a profit, replacing them in 1950— with five Douglas DC-4s from Pan American Airways. Unlike the North Stars, the DC-4s were unpressurized, but they were lighter and had a longer range, making them more economical to operate on the long Pacific runs. Better still, they were powered by four Pratt & Whitney Twin Wasp radial engines that had been well tested in the aircraft's service with the military.

The DC-4 had first flown in 1938 in response to a requirement by United Airlines to compete against Boeing. The prototype, which featured a triple tail, proved too complex for airline usage, and by the time Douglas had redesigned the whole aircraft the war had begun and it was pressed into troop-carrying as the C-54 Skymaster. In 1945, it was snapped up in large numbers as an interim airliner until the DC-6s were available. The CPA DC-4s remained in use until January 1953, when they were replaced by the DC-6B.

Returning from Farnborough in December 1949, McConachie had somehow talked the CPR's board of directors into buying two de Havilland Comet 1As for use on the Pacific route from Vancouver to Sydney. It was the very limited range of the Comet 1 that led de Havilland to develop the 1A, which had an additional fuel capacity of 1,000 gallons and room for forty-four passengers. The airline's chief engineer, Ian Gray, warned McConachie that the Comet's range was still limited to 2,000 miles (3200 km)—shorter than that of the North

The model of the second Comet, *Empress of Vancouver*, unpacked and about to be displayed in the concourse at Windsor Station. CAA (UNN)

Star—and that it didn't have the capability to fly between Honolulu and Vancouver. But McConachie could not be dissuaded. With refuelling stops at Canton Island and Fiji, the Comets could fly the Honolulu–Sydney portion; they could be based in Sydney to pick up passengers in Honolulu, who had been brought there by DC-4. CPA, Air France and the RCAF (which disdained the Toronto-made Avro Jetliner) placed orders for Comets. Their cruising speed was almost double that of contemporary piston-engined aircraft—500 versus 280 miles per hour (800 vs. 448 kph)—and, since Comets were the fastest passenger-carrying aircraft in the world, McConachie was sure that customers would love flying in them as much as he did.

BOAC began experiencing the first doubts about the Comet on October 26, 1952, when one was damaged on takeoff from Rome airport. The accident report indicated that if the aircraft's nose was lifted too soon, the engines didn't have enough power for it to attain flying speed. De Havilland subsequently fitted its later Comets with drooped leading edges to overcome this.

Hatfield on a misty March 1, 1953, the CPA crew about to embark on CF-CUM for Heathrow (which was also shrouded in fog) to take on more fuel. Left to right: Charles Pentland, C. North Sawle, John Cook and James Smith. CAA 988-10-3

By late that year the Canadian Pacific Comet 1A CF-CUN was nearing completion at Hatfield and had been christened the *Empress of Hawaii*. Typically, McConachie wanted its delivery flight to Australia to garner maximum publicity by setting a new speed record. The crew he chose were the best CPA had. Captain C.A. "Charlie" Pentland from Winnipeg had trained with Imperial Airways and BOAC and was the airline's director of operations overseas. Captain North Sawle was an experienced bush pilot and had flown on the survey flights over the Pacific. John R. Cook, the radio operator, had trained in RCAF transport command, and James A. Smith, the engineer, had joined Canadian Airways Ltd. in 1940. All were put through the course on flying the Comet at de Havilland's Hatfield plant, "unlearning" everything they knew about conventional flying.

On March 1, 1953, CF-CUN was taken to Heathrow Airport to fuel up for the flight to Sydney. Six de Havilland technicians were also on board to monitor the aircraft's performance. The first segment, to Beirut, went well, and the Comet made record time to Karachi. In the early morning darkness of March 3, the *Empress of Hawaii* rolled down the runway again to take off from Karachi Airport. Pentland raised the nose wheel of the heavily loaded jet, unknowingly increasing the drag on the wing and diminishing the thrust of the engines. The Comet remained on the ground but continued to hurtle down the runway. The pilot instinctively raised the nose sharply to get the aircraft into the air. Its tail scraped the ground, but the full fuel tanks would not let it lift off. Pentland, realizing his error, brought the nose wheel down, hoping the Comet could work up the flying speed to take off. By then it was too late—there was no more runway.

The undercarriage wheel hit the wall of the drainage ditch at the perimeter of the runway. The aircraft then swerved into the ditch, slamming the concrete wall on the far side. The explosion lit up Karachi. There were no survivors.

In Australia the minister of civil aviation, Hubert Anthony, asked for a minute's silence in the House of Representatives. In London a court of inquiry later concluded that the accident was caused by pilot error. On May 2, when a BOAC Comet crashed on takeoff from Calcutta airport, once more the pilot was blamed. Even before these tragedies, de Havilland was redesigning the Comet with more powerful Avon engines, prompting McConachie to go before his board of directors and obtain their permission to order three Comet IIs.

It took the mid-air disintegration of two of BOAC's Comets, on January 10 and April 8, 1954, for the British to withdraw the aircraft's certificate of airworthiness and McConachie to lose interest in the Comet. Exhaustive testing determined that metal fatigue, brought about by the aircraft's skin flexing, had caused the last two crashes. That this would have been the fate of the CPA Comets with a full load of passengers over the trackless Pacific occurred to many at Canadian Pacific. The other CPA Comet 1A, CF-CUM, was never delivered. The loss of the Comets and CPA's re-equipment with DC-6As and Bs and Convairs all combined to create a net operating loss for the company in 1953.

More mundane than the saga of the Comet was the addition of the Convair 240s to the Canadian Pacific fleet. The tubby little airliner had come about in 1945 when American Airlines was looking for a modern replacement of its DC-3s. Convair designed a forty-passenger pressurized aircraft with a cruising speed of 280 statute miles per hour (most aircraft are calibrated in nautical miles). Canadian Pacific bought five

Convair 240s from Continental Airlines in 1953 for use on its British Columbia routes. Harley Hatfield, a CPA flight attendant, recalled, "I was at Penticton airport when the first Convair came in, perhaps a familiarization flight, as one of the old DC-3s was also there. Somehow in making room for both of them in the loading zone, the DC-3 got the tip of one wing behind a fence post. As a DC-3 could not back up, the airport manager had to search out a handsaw . . . and cut off the top of the post, to the cheers of the onlookers."

Allan Clark's memories of the Convair were that "it was a fun airplane to fly, but operating it in BC's southern mountains required due diligence because of the icing problems." Wing de-icing was done by hot air heated by the engine exhausts, and the propellers were equipped with electric de-icers. It was said that the Convair "couldn't carry enough ice to cool a martini."

I was First Officer on one flight from Calgary to Vancouver with scheduled stops at Cranbrook, Castlegar and Penticton. The Calgary–Cranbrook leg was uneventful, but upon commencing the approach into Castlegar, when we noticed an accumulation of ice on the wing's leading edge, it started to get interesting. Weather at Castlegar was marginal and Gerry Lautch, the captain, thought it best to proceed to Penticton and requested a clearance at 16,000 feet. I was flying this leg and as the plane reached 14,500 feet it would not climb higher because of the ice accumulation. To prevent the Convair from buffetting due to "stall onset," a speed of 200 miles per hour [320 kph] was necessary, which corresponded to a 300-foot [90 m] rate of descent. This was a non-radar environment and we were navigating using the Crescent Valley, Carmi and Penticton radio ranges. To add to the din, ice coming off the props continually banged against the fuselage.

We crossed Carmi and Penticton well below the minimum altitude but turning north over Penticton towards the Naramata beacon the ice came off the Captain's windshield in one big chunk and Gerry was able to see the lights of the Okanagan Valley. With my own windshield still covered in ice, Gerry made a combined visual/localizer approach to a successful landing. After taxiing to the ramp, an inspection showed ice accumulations of up to 1½ inches on the nose, engine cowls and my windshield and a ridge on top of the wing where it had refrozen after being melted by the

CPA bought five Convair 240s to replace its DC-3s flying in the BC mountains. With its unique de-icing system, it was said that the Convair "couldn't carry enough ice to cool a martini." CAA 988-43-42i

de-icing system. The flight to Vancouver was cancelled until the next day.

A short (very short) runway had been built at Sandspit, allowing the first Convair 240 flight to fly the Vancouver–Port Hardy–Sandspit route on January 24, 1953, and replace the Cansos. That March, Convairs would be used on the Vancouver–Whitehorse run, and the following year between Penticton and Calgary.

Although the airline's letterhead proclaimed it to be "Wings of the World's Greatest Travel System," CPA was as fragmented as Canadian Airways Ltd. had been. It might fly to Australia and Hong Kong, but within Canada its network ended at Sioux Lookout, Ontario, and began again at Montreal with the Quebec City–North Shore runs. To travel between Vancouver and Windsor Station, Montreal, senior CPA staff still had to patronize TCA or take the train. It was obvious that as long as the government airline had the monopoly on transcontinental flights, CPA would never prosper. All it was doing was feeding customers from overseas to TCA at Vancouver.

If his airline was forbidden to carry people across the country, McConachie figured, there was no reason why CPA should not begin a transcontinental freight service. TCA had been careful not to compete for cargo with its CNR parent, especially in the latter's deficit years, and McGregor had assigned a single DC-3 as a freighter, which was just about adequate for the demands of the market. In late 1951, McConachie approached C.D. Howe and Minister of Transport Lionel Chevrier about CPA operating all-cargo flights on a Vancouver–Edmonton–The Pas–Toronto and Montreal route, pointing out the political advantages of helping small businesses in those cities. Competition gave the shipper a choice and would lead to the lowering of freight rates. From his own fish-hauling days, McConachie claimed that 21,000 pounds (9450 kg) of fresh fish could be flown every week from The Pas to the cities in eastern Canada. Careful not to compete with TCA, he even promised not to carry any freight between Toronto and Montreal. In truth, the idea of serving Edmonton and The Pas but not Winnipeg was because he hoped that the Air Transport Board would see that this northern route did not put CPA in competition with TCA. Not for the first time did McConachie overestimate his powers of persuasion. In anticipation of the route, he bought two brand new DC-6A freighters from Douglas.

Introduced in 1947, the DC-6 was a stretched version of the DC-4, identified by its square rather than round windows, and at first was available only in the "6A" cargo version. Like the Lockheed Super Constellation 1649A, the DC-6 would be the ultimate in four-engine piston airliners—stylish and economical. The classic beauty of both aircraft was never surpassed by the jets that were soon to displace them. Douglas extended the DC-6A fuselage another thirteen inches (32.5 cm), taking off the cargo doors, and put in seating for up to 102 passengers (although Canadian Pacific carried only 61–64 passengers in luxury). Passengers liked the DC-6s because they were pressurized and flew "above the weather"; these aircraft became the first true intercontinental airliners of globe-girdling airlines Pan American, KLM, SAS—and CPA.

Airborne courier services were twenty years into the future, and TCA staff knew that there was little market for a national air freight service, let alone two. Canny Scot that he was, McGregor saw into the ploy. The next step, he predicted, would be McConachie losing so much money that he would ask to fill the half-empty CPA DC-6As with passengers. Donald Gordon, president of the CNR (and also of Scottish ancestry), was also wary of Canadian Pacific's railway and airline

CPA DC-6B *Empress of Tokyo*, the star of the inaugural flight to
Auckland, New Zealand. Its bunks in the rear made it popular with
honeymooners going to Honolulu. CAA 988-235-17

encroachment. He wrote to McGregor, warning him that he had
watched with some admiration as CPA had taken TCA's cast-off routes
and, in his words, "woven them into an airline system which now
threatens to blanket Canada far more effectively than . . . TCA." Gordon
offered to lend his considerable weight to squash the possible incursion.

The cross-Canada cargo competition was dissected at the Air
Transport Board hearings in early 1953. As they would do time and
again, TCA's lawyers proved more skillful than CPA's and demolished
McConachie's arguments about providing shippers with cheap freight
rates. When TCA's legal team had finished, the board was satisfied that
CPA had entered into this scheme expecting to lose money—backed as
it was by the wealthy CPR—as a means of breaching the TCA transcontinental monopoly.

It was up to the Cabinet to make the final decision to reject
McConachie's air freight project, but the publicity the hearing garnered
was worth every penny. Newspapers from Vancouver to Saint John
called TCA "a dog in the manger." Support for CPA also came from

small operators like Russ Baker, who had just renamed his Central British Columbia Airways as Pacific Western Airlines (PWA). Although he had lost, McConachie did get some pleasure from knowing that all the controversy caused TCA to drop its cargo rates by 10 percent and increase its own transcontinental air freight service.

Temporarily stymied, McConachie now looked for other opportunities, both north and south of the country. South and Central America were eager for air services, and TCA was also interested in expanding to Mexico; it was negotiating a bilateral agreement by having the Department of External Affairs meet with the Mexicans through the normal channels. Rather than going through this laborious process, as he had in China, McConachie handled all the negotiations himself, once more going directly to the top. He flew to Mexico City and gave away Cadillac limousines "like playing cards" to secure the proper permits from Mexican officials. McGregor, aghast at McConachie doing business this way, was even more shocked in 1952, when Ottawa granted CPA permission to fly from Vancouver to Mexico City. Although few British Columbians might want to visit Mexico, it gave CPA a beachhead into Central America.

McConachie took delivery of his first DC-6B, CF-CUO *Empress of Lima*, on January 17, 1953. A week after it arrived at Vancouver it was on its way to Honolulu to pick up passengers, and a month later it was also serving Sydney. "Those flights on the DC-6 to the South Pacific with only two stewardesses were long flights," remembered former flight attendant Barbara (Barnes) Chambers. "One stewardess was always on duty during the night. She would tour up and down the aisles while everyone else was sleeping. The other stewardess would have a little rest, either in a vacant seat, or if you could squeeze yourself into a coat cupboard under all the garbage and coats and on top of the life raft, you could maybe take your half hour off there."[5] Between 1953 and 1969, CPA bought twenty DC-6Bs, using them first on the international routes and then switching them to flights within British Columbia.

Not content with having CPA fly to Mexico, McConachie secured landing rights in Lima, Peru and the first Vancouver–Mexico City–Lima flights began on October 24, 1953. Now the airline bordered the Pacific Ocean in an arc, from Sydney, Hong Kong and Tokyo to Vancouver and down to Mexico City and Lima. The hiring of two Mexican stewardesses provoked some resentment among the CPA stewardesses at the growing number of foreign nationals who would work for lower salaries. The airline maintained that, like the Chinese and later the Dutch crew, the Mexican employees were outside CALFAA's certification, and it only partially backed down when the seventy-five CPA stewardesses voted by 95 percent in favour of a strike over the issue.

In January 1954 TCA was given its Toronto–Mexico City service, a route that McGregor expected to be marginally economical but that he knew McConachie desperately wanted. The federal government solved the problem by brokering a trade. On November 1, 1955, it was agreed that in exchange for TCA's Toronto–Mexico City route, CPA would give up its Quebec City–North Shore network. McGregor had moved TCA's headquarters from Winnipeg to Montreal—a blow that the prairie city never forgot—and the high-density Quebec routes emanating from Dorval fitted neatly into TCA's domestic network. For McConachie, the trade was worth it. While CPA kept Montreal, it was also at last in Toronto, the fastest-growing market in the country.

Alcan's Kitimat construction project was slowing down in 1954, causing both Jim Spilsbury's Queen Charlotte Airways and Russ Baker's Pacific Western Airways to suffer heavy losses and approach bankruptcy. Fortunately for both companies, another air transport bonanza was about to begin. On February 22, 1955, Washington announced (without consulting Ottawa) that it was going to build a chain of radar stations across the Arctic to prevent the Soviet Union from launching a "Pearl Harbor" type of attack against North America. Extending from Point Barrow in Alaska to Baffin Island, the Distant Early Warning (DEW) Line was to be entirely paid for by the United States.

Although the American giant Western Electric would be the primary contractor, every small Canadian air carrier—and both the RCAF and the USAF—got into the act of ferrying supplies to the sites. Eastern Provincial Airlines, Associated Airways, Canadian Pacific Airlines, Pacific Western Airlines, Dorval Air, Maritime Central Airways, Wheeler, Boreal Airways—even Jack Moar's little Northern Flights—all had contracts showered on them to deliver goods to the DEW Line radar sites. Until it was completed in 1957, using aircraft ranging from Norsemen and DC-3s to Globemasters and Avro Yorks, an estimated 45,000 flights were launched into the Arctic.

For its part, in early 1955 CPA bought eight Curtiss C-46Fs from Flying Tigers, Miami Airlines and a Cuban airline, and by March Capt.

One of the eight CPA Curtiss C-46s at a DEW Line site in the winter of 1956–57. The double-bubble design of the C-46, in which only the top half was to be pressurized, was too controversial for the standards of the day, and the aircraft was restricted to freighting. CAA 988-137-1

C.N. Neal flew the first one to a radar site. The venerable (in aircraft terms) firm of Curtiss had built the fish-shaped C-46 in 1940 as a partly pressurized airliner, hoping to steal some of the orders away from Douglas and its DC-3. The DEW Line C-46s were bare shells containing only the tie-down equipment, but fortunately for the crews their cockpits had some lining in them and were heated by Janitrol heaters operating at full capacity.

There was no lack of pilots volunteering from CPA because for First Officers the steady hours of flying on the shuttle meant promotion to the rank of captain sooner than could normally be expected. Ground school was conducted at Vancouver, with Fort Nelson used for flight training. The loads usually were barrels of fuel oil and sometimes grossed about 48,000 pounds (21,600 kg) in total. Each DEW Line carrier conducted its operations out of a single base; Fort Nelson was the main operating base for CPA, but its C-46s also flew from Yellowknife and Hay River—wherever goods could be trucked in—and the radar sites were located between 300 and 600 miles (480–960 km) from the bases.

The pilot scheduling was simple: first in, first out, with two return trips daily, about ten hours in total. It was not uncommon for a pilot to be back at base after having flown ninety-five hours in ten days. Life was flying, sleeping—and more flying. In winter one could fly for months without ever seeing daylight, and in summer the long hours of daylight meant viewing the Arctic in all its splendour. Since engineers were not carried on these flights, the pilots refuelled or added oil when necessary, sometimes in a -40° temperatures. In summer the ever-present swarms of mosquitoes made flying uncomfortable, especially when they got into the aircraft. When the contract began in March 1955, hard-surface runways were nonexistent, except at Norman Wells, Aklavik and Cambridge Bay.

Because the C-46s had been bought from three different airlines, they were known for their varied cockpit arrangements. "It was not uncommon for a crew to board an aircraft and spend the first five minutes getting oriented. The Artificial Horizon and Directional Gyro were where they always are, but some of the cockpits had the 'Turn and Bank' centred on the glare shield above the engine instruments."[6] By May, some of the smaller sites had 4,000-foot (1200 m) gravel runways and low-powered nondirectional beacons, but navigation en route was still done by following the terrain, which had not been mapped too accurately—crews were always drawing in their own sitings of lakes and rivers.

Don O'Grady, appointed to the chief mechanic's position at Norman Wells, wrote that accommodations were also primitive. At first, the Norman Wells base consisted of three wooden structures: a combined ticket office and waiting room, a nose hangar and an outhouse. The ground equipment was an old International panel truck used to transport passengers and mail, a Herman Nelson heater and a small transportable power unit. Later, additional living and sleeping quarters were put up in a double-decker barrack block, in which the flight crews slept in rooms on the upper floor and the maintenance staff on the lower. "You would be surprised how imaginative and decorative each individual proved to be," Don noted.

Centrefolds were unheard of, let alone depicted as they are today, but we did manage some facsimile. I can even recall seeing one

Connie Brierley, a stewardess from Calgary, was chosen to make the presentation with McConachie of the airline's stylized Flying Goose crest to the Department of Transport to mark the construction of the new terminal at Dorval airport on December 15, 1955. CAA (UNN)

guy's room adorned with airplane pictures. Most important was the Mess Hall run by the chef, Vern Olsen, and the Native girls that the company brought in for the cooking duties. Eventually, we had the best-fed, best-housed, best-clothed (winter geared supplied) operation on the DEW Line, and it was up to one and all to deliver the goods. And deliver it we did. We operated night and day, weather permitting. Aircraft mechanicals plagued us incessantly at the start. Many causes, radio, electrical, power plant, airframe . . . but we learnt and with sure perseverence, guts and will, many of the causes were overcome."[7]

The bulk of the DEW Line contract was completed by October 1956, by which time the airline had transported 36.5 million pounds (1.6 million kg) of freight. The C-46s were then upgraded to a "super" version, and two were converted to carry passengers.

CPA was not the only airline to make money during the DEW Line operations. Before the DEW Line construction began, Pacific Western

Airlines (PWA) had puttered about with an incongruous fleet of seventy-five aircraft, a mix of floatplanes and ex-BCATP Avro Ansons, but it was still primarily a charter–bush carrier. When the DEW Line contract earned PWA an astounding $23 million, Russ Baker began to model himself on his former boss, Grant McConachie, and dreamed of expansion to more populous urban areas. Although the Alcan project might have lifted PWA off the ground, Baker felt that the experience that its staff got on the DEW Line gave them the expertise to expand into a conventional airline.

In 1955 Baker bought out Associated Airways and Queen Charlotte Airways for their routes, especially QCA's scheduled service to Kitimat. Since Kitimat had no airport, PWA used Grumman seaplanes to service the port—when the weather allowed. Unfortunately for PWA, the nearby city of Terrace had an all-weather airport that accommodated CPA's comfortable Convairs. Business there was so good that CPA intended to upgrade its Convair flights with the larger DC-6Bs. As he had done with the forest patrol contracts, Baker lobbied hard to have an airport built at Kitimat, promising to buy turboprop F27s, which were faster than the DC-6Bs, if it was built.

The Air Transport Board arranged a swap, to the satisfaction of both parties. In return for leaving Kitimat–Terrace to CPA in 1957, PWA received all of CPA's remaining DEW Line operations plus its Edmonton–Lloydminster, North Battleford, Prince Albert, Saskatoon, Moose Jaw and Regina service. Baker was ecstatic—at last he had moved inland from the coast. He celebrated by buying a DC-3 for the prairie route and a pair of DC-4s for the DEW Line. The Regina run would be a money-loser, but like his mentor Grant McConachie, Baker was already looking east to the population centres of Ontario and Quebec.

Many would-be airline entrepreneurs watched the DC-6B in Scandinavian Airlines System (SAS) colours at Winnipeg airport with some envy. Isolated during World War II by Sweden's neutrality, SAS had pioneered commercial flying over the polar region between Norway and North America as soon as the war had ended. The president of SAS, Bernt Balchen, had been one of the pilots employed by Western Canada Airways for the Churchill airlift and knew the Canadian North well.[8] Through Balchen's efforts, on November 15, 1954, SAS's inaugural flight touched down at Winnipeg on the airline's Copenhagen–Sondre

Stromfjord–Winnipeg–Los Angeles run. It didn't cross the North Pole, but the idea of flying over the polar wastes in such snug comfort captured the public's imagination.

Then, in December, TCA's acquisition of a Vickers Viscount made it the first airline to operate turboprop equipment in North America. For Grant McConachie, who mourned the recent loss of his Comets, the stakes had moved up in the contest between the battling Scots.

The Great Circle formed by flying from western Canada to western Europe had been one of McConachie's boyhood dreams, so once again he did his super-salesman act before an audience of CPR directors. Stretching the truth, he satisfied their concerns about the weather in the Far North (unlike the Atlantic, the Arctic was cold and clear), and assured them that there would always be an airfield within one hundred miles (160 km) of the flight route (derelict leftovers from the wartime Crimson Route, but the directors were not to know that). McConachie concluded with, "not only would a polar service bypass TCA's Atlantic preserve, but it would open up western Canada to European trade opportunities." The directors agreed to finance the polar route, and since London, Paris and Dusseldorf were already served by TCA, McConachie asked the Air Transport Board to approve the Vancouver–Yellowknife–Amsterdam route for CPA. As part of the PWA–CPA swap, Canadian Pacific already had refuelling rights at Yellowknife.

This effort became known in Ottawa as "McConachie's Folly." Even C.D. Howe, who admired McConachie, grumped, "CPA's planes will be as empty as the regions over which they'll be flying." Operating on the principle of giving McConachie enough rope to hang himself, the minister of transport, George C. Marler, announced on January 21, 1955 that Ottawa had designated CPA to fly "a northern Great Circle route from Vancouver to Amsterdam." Marler also stated that CPA "in its application had indicated that Vancouver was the only point in Canada which it proposed to serve; and that governmental authorization would contain a condition to this effect."[9]

But McConachie saw beyond Amsterdam itself. Inheriting the seafaring drive of their ancestors, KLM's directors had made Amsterdam's Schiphol airport the terminus for flights across Europe, Africa and the Middle East. Once CPA's passengers were at Schiphol, they could connect with myriad flights to anywhere on KLM. It worked the other way too; passengers from those KLM destinations would funnel into

Not allowed to fly to London, CPA made Amsterdam its European gateway with this DC-6B inaugural flight on June 4,1955. The polar route became known as "McConachie's Folly." CAA 989-71-187

Schiphol and catch CPA's DC-6Bs to the West Coast of North America. The accountants told McConachie that Great Circle passengers would pay the same fares as those for the flights across the Atlantic, giving the airline a higher revenue per mile flown. Thus, on April 11, 1955, CPA operated its first DC-6B charter flight to Amsterdam and then, on June 4, its scheduled inaugural flight.

The Vancouver–Amsterdam route also made the news in an unexpected way:

> World aviation history was made in Vancouver, Wednesday, by CPA when a father and his twin sons piloted a DC-6B airliner from Vancouver to Amsterdam. The unique flight over CPA's polar route was made by Captain Bob Randall and his 23-year-old sons Bob Jr. and Ted, all of 1755 West Sixty-eighth St. Vancouver. Ted and his father have piloted CPA aircraft together before, but because of intricate scheduling of pilots, there was only one chance in a thousand that Bob Jr. would accompany them as Second Officer. The boys joined their father in CPA in 1952 and last year were transferred to overseas operations.[10]

The sisters Nina and Madelene Youngman in 1958. Their hats were designed by Reg Sargeant from BOAC. Both would retire in 1997, after thirty-eight and forty years' service. Morrison family photo

Gaining Amsterdam did not, however, stop McConachie from continuing to seek another European destination. Blocked by TCA from landing in London or Paris, he obtained permission to begin services to Lisbon. Just as passengers for London could connect through Amsterdam, Lisbon would be the transfer point for Paris, Rome and Madrid. The first Montreal–Lisbon revenue flight took place on May 30, 1957, and because the returning DC-6Bs were usually empty, McConachie placed advertisements in newspapers in the Azores urging fishermen to hire on as construction workers at Kitimat in their off-season; CPA could fly them all the way from Santa Maria to Terrace.

That summer, everything changed, for McConachie, for Baker and especially for Gordon McGregor. Someone once observed that political parties don't win elections; governments lose them. Even so, few expected Louis St. Laurent's Liberals to lose the federal election on June 10. Later, C.D. Howe acidly commented that the Liberals lost because the party had run out of ideas. The arrogance of two decades in office had made them many enemies, especially on the prairies. Perhaps the grandfatherly St. Laurent made it all look so easy that he had bored the electorate and they wanted to hear the old-fashioned oratory of John Diefenbaker. Whatever the reason, never expecting to get in, the Progressive Conservatives had wildly campaigned on the benefits of competition. On June 11, all the small air carriers were gearing up to remind them of it.

For TCA the Tory era began badly, from the very first. Two days after the disappointing election results were in, Liberal ministers James Sinclair and Ralph Campney, intent on squeezing one final official privilege, requisitioned the government's VIP aircraft, the C-5, for a last trip at taxpayers' expense.[11] This forced John and Olive Diefenbaker to catch the regular TCA "red eye" overnight flight from Saskatoon to Ottawa on June 13–14. If the lapse in protocol wasn't bad enough, TCA staff managed to lose one of the Diefenbakers' suitcases at Prince Albert, causing "Dief" to go berserk. An aide who witnessed the incident observed: "He had a temper tantrum and jumped up and down like a little kid . . . 'Don't you know who I am?' he shouted at the TCA counter. 'I'm the new prime minister.'"[12] When the aircraft landed in Toronto, Diefenbaker's mood worsened: news of his tantrum was spread across the front page of *The Globe and Mail* (a reporter at the Prince Albert airport had phoned in the story)—a memory that throughout Diefenbaker's term in office did not endear the government airline to him.

Although CPA had obtained permission to abandon its service between Edmonton and Regina, it delayed cancelling until the service aircraft could carry Prime Minister Diefenbaker from his Prince Albert constituency to Regina. There he could appear on national television and thank Canadians for selecting him as the country's thirteenth prime minister. Neither the prime minister nor the airline was unaware of the symbolism of that flight.

The election victory had been such a surprise that there had been no time to mull over the selection of a cabinet. For his minister of transport, Diefenbaker unexpectedly chose George Hees. A better David to slay the Liberal monopolies' Goliath might have been George Drew, who had made a hobby of criticizing Howe's North Star and McGregor's TCA; but Hees ran his own business and represented Toronto's busy Spadina district. Although Hees was committed to competition in commercial aviation, he was held in check by his deputy minister, John Baldwin (a future president of Air Canada), who advised caution.

Counting on the new mood in Ottawa, PWA's Russ Baker filed an application with the Air Transport Board in September for a Vancouver–Windsor service, proposing to use the latest jet airliners on it. Baker put it about that de Havilland was practically willing to give PWA four Comet IVBs (a new, improved Comet) on easy credit terms. Gordon Ballentine once asked Baker why he was "such a bullshitter," and certainly Baker's claim seems like braggadocio. But Baker counted on McConachie's unhappy memories of the earlier Comet to prevent him from buying the only jet airliner available.

Russ Baker before the PWA Mallard amphibian used for the Kitimat run. His career in aviation began in 1928 as the office boy at WCA. Like McConachie, Baker loved wheeling and dealing as much as he did flying.
CAA 989-78-12

The IVBs were faster and more glamorous than TCA's lumbering Constellations and would, Baker knew, appeal to businessmen. Both the federal government and the public liked the idea that a spunky, third-rate Canadian airline would be flying jets before TCA or CPA or, for that matter, Pan American Airlines. Finally, there was the thought that Gordon McGregor would rather see the lesser threat of PWA on the transcontinental run than that of CPA. For the first time, in the eyes of the public, Grant McConachie was no longer the underdog.

That October, CPA was busy handing over its Quebec operations to TCA and its prairie routes to PWA. Not until November did management cobble together a proposal for some sort of matching transcontinental service using DC-6Bs. These would have 80 percent tourist capacity and 20 percent first class, compared with TCA's 1049s, which had 70 percent first class and 30 percent tourist capacity. It would mean, the CPA application concluded, "a considerable advantage to the public in vastly expanding the opportunities for low cost tourist air travel and the introduction of modern aircraft with larger (therefore cheaper) tourist accommodation." It requested three main transcontinental routes—Vancouver–Calgary–Saskatoon–Ottawa–Montreal, Vancouver–Edmonton–Montreal and Vancouver–Winnipeg–Toronto–Montreal—and two shorter ones: Edmonton–Regina–Toronto and Winnipeg–Ottawa–Montreal.

During the winter of 1957–58, McConachie, Baker and their wives flew to a holiday in Honolulu—on CPA passes. At the Royal Hawaiian Hotel, the two men played their favourite sport: using a transportation map, they divided Canada up between themselves. Both ran airlines that were, in a word that became popular thirty years later, "dysfunctional." CPA had routes on both sides of Canada and nothing in between, while PWA was a coastal airline with pretensions. Thus, when he returned from Hawaii on February 5, 1958, Baker amended his application to fly not into Windsor but to either Winnipeg–Toronto or Winnipeg–Montreal.

Two days later, in a speech to the Porcupine Chamber of Commerce in Timmins, the minister of transport, George Hees, made his move, stating:

> I have proceeded on the assumption that competition is a valuable stimulus to more efficient operation and to the provision of better service to the public. However, if business available is not sufficient or barely sufficient to provide an airline operator with a chance of operating on a self-sufficient basis, there is no point in attempting to create competition which can only result in either deterioration of service or an increase in rates due to a lack of enough business to cover costs.[13]

Like all of Diefenbaker's cabinet ministers, Hees mistrusted Ottawa bureaucrats. It was thought that they had been in power too long to be loyal to anyone except the Liberals, so Hees, rather than going to the Department of Transport, had commissioned British airline economist Stephen Wheatcroft to study competition on Canada's transcontinental routes and make recommendations. The Wheatcroft Report had considerable significance; it was the first comprehensive study of "the desirability and economic consequence of airline competition in Canada."

Wheatcroft gave Gordon McGregor's regime full marks for TCA's smooth running and compared it with CPA's deficits. His report found no justification for unrestricted competition on the main routes until sometime between 1961 and 1966. The exceptions were Vancouver–Victoria, Toronto–Montreal and Toronto–Winnipeg, which, it was believed, had enough traffic to warrant competitive services. As to the limited-frequency competition on the trunk routes, the addition of new daily transcontinental services as proposed by PWA and CPA "would be much too large for the mainline routes and would involve a loss to both TCA and the competitors."

The report concluded, "the maximum competitive capacity which could be permitted in 1959 without leading to a TCA deficit is approximately four daily transcontinental services." This was later halved to two flights "in order to ease TCA's problems of adjustment, particularly in terms of equipment commitment."

Replying on July 31, McConachie thought Wheatcroft's comparisons between TCA and CPA invalid because the report "did not take into account relative frequencies, fringe routes distinct from the trunk lines, complete dissimilarity of aircraft, or the unique cost conditions prevailing on the north–south routes." But the new president of the CPR, Norris Roy "Buck" Crump, refused to spend more money propping up McConachie's dreams. He needed all his company's resources to keep the railway from sinking under its own inefficiency and rusting stock.

The militancy of the labour unions, the spread of private automobiles, the 1897 Crow Rate (which froze grain-carrying rates) could all be blamed for the CPR's deficits, but Crump knew that all these factors paled in comparison with the cunning of the real adversary—the federal government. To tackle the visiting plagues of royal commissions and boards of enquiry, the favoured status of the CNR and unfriendly politicians, Crump made Ian Sinclair the CPR's general counsel. Both men were determined that CPA should start paying its way, and that meant getting it on the transcontinental route.

Armed with the Wheatcroft Report, Hees, before the Air Transport Board (ATB) hearings were to begin in October, invited the lawyers of both PWA and CPA to Ottawa on August 29 in order to examine their transcontinental applications. Grant McConachie must have been furious and a little envious of Russ Baker, who had stolen his best lines. He had been waging a war against TCA's monopoly for nearly ten years, and now there was a good chance that the Conservatives would grant the transcontinental routes to the upstarts at PWA because of their phantom Comet IVBs.

Russ Baker may never have had any intention of either buying the Comets or flying to Windsor. It may be that because he would not be perceived as a threat to TCA in either Toronto or Montreal, the ATB would grant his wish over McConachie's. In reality, Baker's application may have been a bargaining chip: the real prize was the CPR's northern network.[14]

After touring the country—"a waste of time and money," McGregor commented—the ATB hearings were to culminate in Ottawa at the Chateau Laurier hotel. Baker and McConachie got together there for yet another swap. On September 18, after a haggling session that would

CPR President Norris Roy Crump—
"Buck" to the media—the last of the
"railway presidents." Crump was
determined that the railway's airline
should make a profit and remain
independent of TCA. CAA 989-129-18

have done the prewar bush pilot profession much credit, it was agreed that PWA would withdraw its transcontinental application in favour of CPA, which would, if it got the routes, turn over all of its Mackenzie District network to PWA. If CPA were given permission to service London, England, it would turn over to PWA its Vancouver–Edmonton run and all of the Yukon services. Finally, if CPA received the jackpot of a Montreal/Toronto–London route and continental Europe, it would give PWA all its domestic services, including the Vancouver–Terrace run. Baker was also to take over CPA's remaining bush fleet and equipment in the Mackenzie District. On October 6, to the surprise of the government, PWA's lawyer withdrew his company's application to operate Comets across Canada.

The Baker–McConachie deal smelled of collusion, and when the ATB hearings began at the Chateau Laurier on October 20 they didn't go well for Canadian Pacific. Worse, PWA's legal counsel was shocked to discover CPA's lack of preparation. Though McConachie had filled a Convair with company personnel and flown them to Ottawa, he had brought no economic studies or market research to back up the testimony of the sympathetic witnesses he put on the stand. The old inflatable globe and song-and-dance act did nothing for the steely-eyed TCA lawyers, chief of whom was J.D. Edison, Q.C.

When Gordon McGregor took the stand, capitalizing on his years of speaking before parliamentary committees, he portrayed TCA as the public's protector from the sort of big-business connivance that all knew PWA and CPA had engaged in. Lack of competition for essential services like the cross-Canada routes, McGregor explained, was actually better for the customer—a lesson McGregor had learned from his Bell Telephone days. "Duplication of efforts and facilities from competitive

In the mockup cabin steward Bill Hoffman carves the decorated food for first class passengers. The airline was known for its high culinary standards, especially on the international routes. CAA 989-139-24

service tend to increase the end cost of the product and therefore increase the cost to the consumer." In his no-nonsense manner the TCA president asked, "Who would think of having a municipality with two sources of lighting supply, two sources of sewage disposal, two sources of garbage collection? It doesn't make good economic sense."[15] TCA, McGregor said, would expect to show a loss of $8 million in the first year of such competition, $12 million in the second year and up to $13 million in the third year of service. After the years of investment in the airline, no one, Liberal or Conservative, wanted to see a national institution humiliated into bankruptcy.

Then Edison questioned McConachie. His "ruthless, incisive cross-examination demolished the rhetorical super-structure of the CPA case to lay bare its fragile foundations of assumption, conjecture and hope."[16] By the end of the third day, McConachie ruefully admitted that they had no place to go but up.

The hearings were broadcast on the radio, and it wasn't long before Buck Crump in Montreal lost patience with the way things were going for his airline. Deciding to get involved, he ordered Ian Sinclair to take charge of the CPA case, parking the CPR presidential railcar at Union Station across the road from the Chateau Laurier for Sinclair while he did it. Without having had the time to prepare a strong defence, Sinclair turned the tide through sheer brilliance. He argued that a competitive transcontinental service would not cost either TCA or Canadian taxpayers anything. CPA forecast a 17 percent increase each year in mainline air traffic within Canada. For TCA this would mean a normal revenue increase of approximately $15 million in 1959. Assuming that one third of the business derived from the proposed CPA services were diverted from TCA, TCA's normal revenue increase would be reduced by only $4 million, leaving the government airline with an $11 million increase in 1959, despite the competition. Even if 100 percent of the CPA traffic on the proposed routes were diverted from TCA during the first year, the government airline would still enjoy a $3 million revenue increase for the year, based on normal traffic. Thus, the competition would not cost the taxpayer a penny.

Sinclair then referred to the precedent case of Colonial Airlines versus TCA. From 1927 until 1950, when TCA introduced Montreal–New York services, there had been a Colonial Airlines monopoly. Faced with TCA's competition, Colonial protested vigorously to Ottawa that this would undermine Colonial's position and have disastrous results. However, it was discovered in 1954 that despite the TCA flights, Colonial's passenger traffic actually increased in volume by 76 percent, representing newly created traffic generated by the competition. Sinclair's final argument was that after more than two decades of protection on the most lucrative routes in the country, TCA should by now be in a position to meet some degree of competition. Crump and all the CPR directors were so impressed by Sinclair's handling of the airline case that he was made vice-president in 1960.

The ATB hearings were interrupted on November 15 by news from Vancouver of Russ Baker's sudden death. Baker's influence in British Columbia was such that although not a Roman Catholic, a requiem mass was held for him at Holy Rosary Cathedral and two PWA aircraft took the body and the mourners to Fort St. James for burial. Pierre

Berton, who had shared Baker's early career on the Nahanni—romanticizing him in "The Headless Valley Expedition" story for the *Vancouver Sun*—called him the last of the great bush pilots.

Unfortunately, there seemed to be no heir apparent at PWA just when it needed one. As a result of the airline's rapid expansion, morale was very poor, especially among its eleven flight attendants. In 1958 PWA cabin crew, with the help of one of their members, Diane Jamieson, attempted to form a union to negotiate for better working conditions. When management threatened to fire them for doing so, the eleven attendants designated CALFAA as their bargaining agent. The airline continued to handle its labour contracts very poorly, and when mediation from the Department of Labour failed to change PWA's uncompromising attitude, CALFAA called the first strike in its history on February 27, 1963. The PWA stewardesses endured six long months of harrassment on the picket line. Only when organized labour in British Columbia, especially the loggers and construction workers, sympathized with the women and refused to patronize the airline, did PWA cave in.[17]

The ATB handed down its decisions on the last day of an eventful year. In summary, it approved a single daily flight for CPA on the transcontinental run, relating its approval not to the need for competition but to the airline's need for financial assistance.

1. The Board finds that additional transcontinental air services could not be introduced at this time without economic hardship and possible deficits for scheduled airlines.

2. The application of CPA in its original form is denied because the airline has computed its expenses on an unreasonable added cost basis with no projection on a full allocated cost basis. Its revenues were also estimated without a traffic study and schedules appear to be based on optimistic block speeds.

3. The Board finds that the position of CPA as a Canadian carrier operating international services needs strengthening and for that reason, proposes to issue to CPA a Class 1 scheduled commercial air service serving Vancouver, Winnipeg, Toronto and Montreal, restricting it to a frequency of one return flight a day.

It was a small victory for McConachie—Baldwin made sure of that— hardly worth the effort of reserving the three DC-6Bs needed for the

Rule Britannia. The last CPA aircraft in the postwar colour scheme, touted as the "Whispering Giants," were affectionately known to the crew as "The Brutes" or "The Anglo-Saxon Bunsen Burners." CAA 988-269-1

single daily flight. But it was a victory, nevertheless. Although CPA had not been awarded all five of the routes it requested, the TCA monopoly had been broken, and McConachie kept his bargain with the late Russ Baker to turn the Mackenzie District network over to PWA.

Gordon McGregor was predictably disappointed with the federal government's credulity. He knew McConachie had no intention of using the sixty-one-passenger DC-6Bs on the run—not when he was awaiting the delivery of six Bristol Britannias, which not only carried ninety-eight passengers each but were faster than TCA's Constellations. Between April and September 1958, having constructed its "Britannia Hangar," CPA took delivery of the first of six turboprop Bristol Britannias. The Britannia was an outcome of British wartime government recommendations for a medium-range, piston-engined airliner. What would have been a revolutionary transport aircraft if it had appeared in 1950, when it was faster than its closest competitor, the DC-6B, was instead delayed for technical reasons until the Boeing 707s were about to fly. By then it was hopelessly outclassed and outdated.

Fresh from Short's in Belfast, on April 1, 1958, the airline's first 314, CF-CZB, *Empress of Vancouver* in its special hangar. CAA 989-71-25

With a wingspan of over 142 feet (42 m), a length of over 124 feet (37 m), a height of 37½ feet (11.3 m) and a fuselage diameter of 12 feet (3.6 m), the Britannia was still a very impressive aircraft for the day. BOAC and subsequent customers were able to pack 139 passengers in it, although CPA carried only 89 on the long routes and 110 otherwise. Bristol, knowing that pure-jets were imminent, uncharacteristically (for a British concern) initiated an American-style promotion campaign including glossy brochures that identified its aircraft as "Whispering Giants" because of their size and quiet engines. In all, only eighty-five examples were produced of what must be the Rolls-Royce of airliners: handsome anachronisms, almost Edwardian in style, produced too late to make an impact.

To the Canadian pilots who flew them, the Britannias were affectionately known as "The Brutes." To the passengers and ground crew they were "The Anglo-Saxon Bunsen Burners" because of the peculiar smell emanating from their exhausts when the engines started up. Since Bristol was too busy with its military orders, it subcontracted the building of CPA's Britannias to Short's in Belfast.

Prior to operating the Britannias all CPA pilots were sent to ground school, which was regarded as more ardous than usual due to the complexity of the aircraft's electrical, hydraulic and fuel systems. For example, the "Smith Instrument Flight System" was so complicated that even one of the BOAC captains remarked, "You sure get a weird picture if you forget one of the steps!" On a "front" course the little aircraft symbol pointed straight ahead, and the pilot turned the Britannia so as to steer the aircraft symbol toward the localizer centreline. On a "back" course, however, the miniature aircraft on the screen appeared backward, so that if the localizer was displaced to the left the pilot had to turn the Britannia to the right to get on course; conversely, if the localizer was to the right the pilot had to turn to the left! It was thought that the manufacturer had to do this to avoid infringing on patents held by other instrument companies.

This eccentricity was nothing compared with the Britannia's ice protection system. While it used heat bled from the engines for most functions, for others it used electrically heated "Napier Spray Mats," which occasionally short-circuited in icy conditions. The result was spectacular, especially at night. The wires that were embedded in the mats would start arcing, and from the cockpit, the crew used to say, it looked like the "Northern Lights." Meanwhile, from the passenger's perspective, it looked like the tail was on fire.[18]

Other British features included handlebar controls on top of the column rather than a wheel; Capt. Cam Ross recalled pilots leaning into a turn as if they were on motorcycles. Ross liked the performance of the Proteus engine. Once he "took on" a TCA Super Constellation at Winnipeg, westbound to Vancouver. The Britannia was on three engines, having damaged a prop tip at Winnipeg, while the "Connie" was churning on all four. The Britannia had 16-foot (5 m) paddle blades, so ground clearance was critical; when taxiing at Winnipeg, they had clipped one of the taxiway lights. CPA unloaded the passengers; Ross feathered the damaged prop, put a brake on it and took off for Vancouver, still managing to beat out the TCA Constellation.

CPA used its Whispering Giants on the North Pacific run to Tokyo (beginning August 23, 1958), Honolulu (December 13, 1958), Hong Kong (September 11, 1959), Montreal–Mexico (December 18, 1959) and to Buenos Aires (May 10, 1960). Undoubtedly, the airline's proudest day with the aircraft was on May 4, 1959, when Captain Bud Potter flew one CF-CZX on the first CPA transcontinental flight from Vancouver to Montreal. C.D. Howe's monopoly had been breached.

The Britannias were the last CPA aircraft to have the postwar colour scheme, a broad red cheat line running the length of their fuselages, separating the upper white from the lower silver. Their all-white tails were distinguished by a series of red and white horizontal stripes halfway up, below which black lettering indicated the type of aircraft and fleet number.

The only Britannia crash occurred on July 21, 1962, when flight 301, CF-CZB, took off from Honolulu for Nadi. Capt. W.T. Jennings, encountering problems with one of the engines, feathered it and returned to Hawaii. On landing, the aircraft burst into flames, killing Jennings and all in the cockpit. Of the twenty-nine passengers only nine survived.

Having lost its cross-Canada monopoly—and soon contending with BOAC flights into Toronto—TCA and Gordon McGregor found the Tory years to be bad ones. McGregor always wondered why the prime minister haboured a bias against TCA and played an unaccustomed role of the injured party for all it was worth. He did secure from George Hees the promise of permission for TCA to fly to Tokyo and Hong Kong, a route that would put TCA in direct competition with CPA. But what distressed McConachie most was TCA initiating its own quasi-polar route: the "Hudson Bay Service" operated Constellations from Vancouver via Winnipeg and Gander to London. In 1959, when TCA opened services to Brussels and Zurich and between Montreal and Paris, McGregor even pressed for "Fifth Freedom" rights to carry passengers from Paris onward to Rome.[19]

When Buck Crump passed TCA's request for rights to London to the prime minister, nothing seemed to come of it, except that George Hees must have felt some repercussions—he called Gordon McGregor in to take back the promise of allowing TCA to fly to the Orient. CPA was in financial trouble, Hees explained, and could not stand the competition of another Canadian airline on its most lucrative route. McGregor's observation that giving in to CPA was like "paying hush money to a blackmailer" has become one of the classic quotes in aviation history.

Toronto had been a hard-won destination for CPA, and McConachie looked to where many of its Italian community wanted to go or bring

On May 4, 1959, with ribbon-cutting ceremonies in Montreal and Vancouver, Britannias CF-CZX and CF-CZW flew the first east- and west-bound flights. TCA's transcontinental monopoly had been breached. McGregor knew that McConachie had no intention of using the smaller DC-6Bs. CAA (UNN)

relatives from: Rome. He went off to Italy once more and negotiated for the rights himself. This time the Department of External Affairs had lost patience with his cowboy tactics and, through John Baldwin and the ATB, exerted pressure on Crump to admonish McConachie. His ill-timed blundering had tipped Canada's hand in bilateral negotiations with the Italians and according to McGregor had put it in a weak bargaining position. But when the bilateral agreement was signed with Italy that fall, as a consolation for being shut out of London, CPA was designated the official Canadian carrier to Rome.

Once more McGregor had been prescient. The Italian airline Alitalia was not only given reciprocal landing rights in Toronto and Montreal, but as a result of McConachie's cavalier approach Ottawa turned over to Alitalia "Fifth Freedom" rights onward to Chicago, Los Angeles and even Mexico City. No Canadian airline flew to these destinations then.

"It's our answer to a ridiculous ruling allowing us only one flight a day . . ."

A cartoon by Norris, published in the *Vancouver Sun*. CAA 989-71-174

In October 1959, CPA and TCA ordered from Douglas five DC-8-43 jets each, at a unit price of $6 million, all to be delivered in 1961. Because of the heavy cash outlay for the jets, McGregor warned Hees that TCA would go into a deficit. On January 22, 1960, he also apprised the prime minister of the situation. He suggested as a remedy that since both airlines were losing money, they should be merged. TCA had the domestic network that could feed into CPA's international one. Not only did the rationalization in aircraft and equipment make economic sense, it would present a unified stand against foreign governments, ending what McGregor called "giveaways" like the Fifth Freedom rights to Alitalia.

As occurred in 1937, there was no chance that the Canadian Pacific Railway would accept less than a controlling interest in whatever form McGregor's idea took, and the proposal went no further than the minister's office. When a reporter asked about the merger Crump is supposed to have grunted, "How much do they want for TCA?" Rather than have CPA disappear into the government airline, Crump had

McConachie buy two more DC-8s to compete with TCA and renewed his efforts to get the London route.

Hees moved on to become minister of trade and commerce, and Leon Balcer took his place. The advent of long-range aircraft like the Britannia had made airports like Gander, Goose Bay and Prestwick— once vital refuelling points for crossing the Atlantic—unnecessary. As a result, on September 6, 1960, the British and Canadian governments amended the 1949 bilateral agreement between both countries. This time, rather than a single airline being designated as the "chosen instrument," the wording was changed to read "airline or airlines of Canada."

Crump and McConachie were quick to seize on this change, and on February 20, 1961, three days before the first of the DC-8-43s were to arrive at Vancouver, CPA applied to operate a Vancouver–Edmonton–Gander–London route. This time, all seemed to go well for CPA and the ATB issued the licence on August 25. Airline press releases announced that the service was to begin in early October. But then the British government began asking for more information on the CPA flights—a traditional means of stalling. BOAC, like TCA, had gotten used to being protected by government monopoly, and on November 14 the British minister of air arrived in Ottawa for talks on re-examining the CPA licence. The British stated that a second Canadian carrier serving the United Kingdom "would be politically unacceptable at this time." Under pressure the ATB meekly withdrew the licence, and with it died McConachie's chances of tapping the British market. For once, his rival Gordon McGregor's hands were clean.

CPA took delivery of its first DC-8-43 on February 23, 1961. The jets inaugurated the airline's new colour scheme: although the upper fuselage was still white and the lower fuselage was silver, a red/white/red cheat line ran below the windows. The tail was still white but now had a vertical red line, bisected with a circle containing a flying goose and surmounted by the letters CPA. Lettering on the upper fuselage was still in script style, but later DC-8s would have the title in block letters.

On April 19 CPA used its second DC-8, CF-CPH, on a Vancouver–Ottawa–Montreal familiarization flight with seventy passengers, including the board of directors. For the transcontinental run the DC-8 carried 124 passengers, a fact not lost on Gordon McGregor, who pointed out that the traffic allowed CPA by the Wheatcroft Report and calculated on the capacity of the DC-6 had now jumped 58 percent.

The DC-8-43 CF-CPG, with its Rolls-Royce Conway turbofans at
Winnipeg, in November 1961. When tested by Douglas prior to delivery,
the *Empress of Montreal* became the first commercial aircraft to exceed
the speed of sound. CPA 26242

The DC-8s changed everything. They were so expensive that in
February 1962 Grant McConachie informed the CPR board that
because of them the airline was $7.6 million in deficit. TCA too lost $6
million that year, hardly a huge sum today but staggering then. With
his board's approval, McGregor applied to raise fares on the
Montreal–Vancouver flights. CPA did not follow suit, preferring to
absorb the loss in order to lure customers over to its single DC-8 flight
now being used on the run.

"Jet indigestion" was commonly blamed for the cost overruns that
all airlines suffered in the early 1960s. The immense amounts of capital required to finance and maintain them took airlines away from their
pilot founders and gave them over forever to bankers and insurance
companies. One by one, the men who had started the airlines with Ford
Trimotors and Fokkers retired or were pushed out—Howard Hughes left
TWA, Eddie Rickenbacker left Eastern Airlines, C.R. Smith bid farewell
to American Airlines. Finally came the turn of the American who had
influenced Canadian aviation so much—Juan Trippe retired from Pan
American Airways in 1969. The industry had grown up.

Grant McConachie himself began suffering from sharp pains in his
chest in 1962, and a year later he was admitted to hospital after having suffered a partial heart failure. On his return to the office, Buck
Crump ordered that he be relieved of most administrative duties,

The "blue uniform" of the early 1960s had a dress under the jacket because no slacks were allowed yet. The pillbox hat was inspired by hats worn by Jackie Kennedy.
CAA 989-55-48a

although typically McConachie continued his "burning the candle at both ends" lifestyle.

On February 4, 1963, Parliament was dissolved and Diefenbaker's days in power ended. All things considered, the Tories had been good for CPA, acknowledging the need for competition on the mainline and breaking down TCA's stranglehold on the transcontinental route. When the Liberals returned to power under Lester B. Pearson, CPA had climbed back to a profit of $347,000, after paying the CPR $1.3 million in interest on loans. Pearson's experiences with commercial aviation could be traced to his early career in the Department of External Affairs, when he had sat on committees that had sunk James Richardson's brave attempts to make Canadian Airways Ltd. the nation's "official instrument." The new prime minister was well acquainted with McConachie's aspirations, as was his minister of transport, a Manitoba professor of history named J.W. Pickersgill.

Called "Sailor Jack" because he was parachuted into the "safe" Liberal riding of Bonavista–Twillingate in Newfoundland to become part of the Liberal cabinet, Pickersgill had been the power (and, many said, the intelligence) behind the throne of Louis St. Laurent. Both Gordon McGregor and Buck Crump lost no time in lobbying him for their particular projects—the TCA–CPA merger or the opening up of London and Paris to CPA. On April 24, 1964, Pickersgill announced three principles as the basis of the Liberal government's civil aviation policy for the 1960s:

1. In the international field, air services provided by Canadian airlines should not be competitive or conflicting but represent a single integrated plan, which could be achieved by amalgamation, by partnership or by a clear division of fields of operations.

Reservations clerk uniform, January 20, 1960. CAA 989-55-46

After considering many possible divisions of the world between the two airlines it was finally agreed that the simplest approach would be to agree that each airline should continue to serve all points now served by it and that the areas . . . should be defined and extended so that practically the whole world would be open to service by one line or another.

As a result it has now been decided that CPA will serve the whole Pacific area, South and Central America and Southern Europe. Air Canada [TCA's successor] will serve Northern and Eastern Europe and the Caribbean. The only exception to this clear-cut division is that CPA will continue to serve the Netherlands.

In the case of the United States we continue to hope that negotiations for a new bilateral will soon be completed. Once they have been . . . a decision will be made as to which of the new areas is to be served by each of our national airlines.

The government has undertaken to regard Air Canada and CPA as its chosen instruments in the area of international operations allocated to it and to pursue vigorously with the governments of other countries extensions of the Canadian service to their respective areas of operations.

2. The second principle relates to the domestic field. Competition was not rejected but the government made it clear that any development of competition should not compromise or seriously injure the economic viability of Air Canada's mainline domestic operations which represent the essential framework of its network of domestic services.

3. The third principle stated that a definition of the role for regional air carriers providing scheduled service was necessary, including a relationship with the mainline carriers that would give them a reasonable chance to operate without subsidies. In

The lounge section of the DC-8 on the other side of the wall from the cockpit. The epaulettes on the white sharkskin blouse date the flight attendant's uniform as from the 1950s. CAA 989-71-65c

the discussions with the presidents of the two major airlines, it has been made clear that the government expects them to take some responsibility to assist in the working out of that policy.

The minister of transport then summoned the four presidents—Donald Gordon of the CNR, the CPR's Crump, TCA's McGregor and CPA's McConachie—to his office on April 27 to hammer out some sort of acceptance for his three principles. But the growling and hackles rising between two sets of sworn enemies produced nothing tangible, and all adjourned to the Chateau Laurier for a sombre lunch.

Besides acknowledging the status quo and giving CPA recognition,

When transport minister Jack Pickersgill formulated a role for the small carriers, Max Ward was able to supplement his pair of de Havilland Beavers (seen here with long-range tanks) with larger Bristol freighters.
Bombardier Regional Aircraft 7827

Pickersgill had also formulated a role for the small airlines known as "regionals": PWA, Eastern Provincial Airlines, Nordair, Transair and Quebecair. The availability of used piston-engined airliners and the growth of the leisure industry in the 1960s allowed small operators like Max Ward to supplement their bush operations with charter flights, both domestic and international. The two big airlines were told to encourage rather than tolerate the growth of the regionals. The small airlines could feed into the trunk routes from rural areas or run innovative services like PWA's "Airbus," a DC-4 that shuttled between Edmonton and Calgary. "There are many forms this assistance might take," Pickersgill would later tell the House, "ranging from route transfers and route operations to contractual relations and various forms of technical and administrative support."

Although the Liberals had returned to power, it seemed there were better days ahead for CPA. On February 28, 1965, Grant McConachie

The Flying Randalls (left to right): Parents Mrs. Hilda and Bob Randall holding the photo of a DC-8-43 cockpit, with Capt. Randall and sons F/O. H.E. Randall in front and F/O. R.B. Randall and S/O. J.Randall at the back. CAA (UNN)

reported that the airline had made a net profit of $4.8 million after payment of $2.3 million to the CPR. Operating revenue for 1964, at $61.5 million, was up 9.5 percent over 1963 and on international routes revenue was up 11 percent, due to the DC-8s. CPA flew 541,014 passengers an average of 1,817 miles each (indicative of its long routes) for a total that year of 983,067 passenger miles. This was an increase of 8.1 percent over the 909,407,000 passenger miles flown in the previous year. It was, McConachie would boast, the first time the airline had flown over half a million passengers in one year.

In March 1965, it seemed as if a spirit of inter-airline co-operation was about to dawn as Air Canada and CPA moved to implement an extensive program of co-operation. "The time has come to resolve certain of our differences," McConachie informed his staff, "and turn our combined energies to fighting the fiercer and less friendly foreign competitors." Both airlines signed a "statement of agreed areas of co-operation" in which it was agreed that if one carrier was unable to

Douglas DC-8-43 CF-CPJ sporting the new Canadian flag and airline colour scheme. Only thirty-two of this series were built; they were used by Canadian Pacific, Air Canada and Alitalia. CAA 989-51-58f

accommodate a customer, a place would be found for that person on the other Canadian airline before allocating him or her to any foreign operator. "In short," the CPA president hoped, "a customer will 'fly Canadian'."

But Grant McConachie would not live to see the outcome of Jack Pickersgill's pronouncements. On June 29, 1965, while in Long Beach, California to negotiate the lease of a Western Airlines DC-6B, McConachie was walking down the corridor of his motel when he had a fatal heart attack. His body was returned to Vancouver in a CPA DC-6B with three passengers on board: McConachie's son Don, his old friend Barney Phillips from the United Air Transport days and John Gilmer, the executive vice-president of CPA.

On July 3 about eight hundred people paid their last respects at Vancouver's Ryerson United Church to a bush pilot who had transformed Canadian aviation through courage and vision. The honorary pall bearers were Ian Sinclair, John Gilmer, John Baldwin and Gordon McGregor. The newspapers reported that the cavalcade of cars to Ocean

View Burial Park, South Burnaby, stretched half a mile. McConachie was laid to rest on a southern slope, just north of the approach path of the main runway at Vancouver airport.

One of the many newspaper tributes was from the *Ottawa Journal*: "In the hamlets of the North, in the weather-beaten fur trade buildings, the mourning will be as profound as in the glittering air terminal buildings where his jets come from across the oceans."

Old friend Maurice McGregor remembered sitting with Grant on a verandah in Acapulco one night in 1952, looking at the stars. He said, "We have to fly to the moon. We have to do this!" McGregor replied, "What are you going to do? You'll have to sit in a projectile!" His death at fifty-six severed the remaining ties that the CPA had with its pre-CPR origins. The consummate bush pilot, McConachie may have lacked the education, diplomacy and common-sense approach of future CPA presidents, but they could never hope to equal his charismatic personality. His public persona was that of a showman who enjoyed the brickbats of his detractors as much as the plaudits of his patrons. Like all gamblers, he was more passion than prudence. Without him, commercial aviation in Canada became a prosaic place.

The Class of May 1979. At this time there were 568 flight attendants based in Vancouver, 222 in Toronto, 82 in Montreal, 6 in Lima, 19 in Hong Kong and 6 in Amsterdam. CAA 989-55-38

CHAPTER SEVEN

Empress of Australia

Flying by Numbers

Pat Carney, the *Vancouver Sun*'s business columnist, expected J.C. Gilmer to be a typical "financial type"—dull and boring—when two weeks after Grant McConachie's death she interviewed him for the paper. As she sat in the sparsely furnished office that he had occupied as CPA's vice president since 1963, the only personal item she noted was a paper-knife that had a handle of coins set in plastic. Appropriate, she thought, for a chartered accountant who was also a coin collector.

The airline had been much in the news recently. The shock waves from McConachie's death had barely subsided when, a week before the interview, a CPA DC-6B exploded in mid-air near 100 Mile House, BC, killing forty-six passengers and a crew of six. After an investigation, the RCMP and the Department of Transport concluded that the tail section had been blown off by a bomb hidden in the toilet. The catastrophe only seemed to confirm that the days of the faithful DC-6s had come to an end.

"It's been said that your appointment signals a new era in the history of Canadian Pacific Airlines," Carney began. "There's an element of truth in that, or it is obvious that they would have looked for a president more in the mould of Grant," Gilmer replied. "The timing of his exit was fantastic. Things might very well have been an anticlimax for him in the future." "Then is the exciting era over for the airlines industry?" Carney asked. "It's never over," corrected Gilmer.

The excitement is in running the airline. In this business, if you're right, you're bright. If you're wrong, you're ruined. It's in taking hold of something that hasn't been done before and turning it into a profit. You see, we have a business in which the leverage factor is very high. Leverage is the way in which losses can be turned to profits. Unlike other industries, 90 percent of the costs of running an airline are fixed and do not vary with the volume of passengers carried. Of the fixed costs, 50 percent is associated with owning the planes, depreciation, insurance, hangar fees, maintenance—you bought the things; you have to look after them.

The remaining 40 percent of the fixed cost is associated with flying the airplane—landing fees, fuel and oil, salaries for the flight crew. Basically, it doesn't matter whether the aircraft flies empty or not. It still costs the same to run. That leaves only 10 percent—travel agents' commissions, in-flight food and bar consumption—to vary with the number of passengers carried. Consequently, once you are off the nut and sell enough to hit a break-even load factor, you stand to make a profit. Up to that point for every dollar of revenue, there is nothing coming back.

Unlike McConachie, Gilmer had applied the "leverage factor" all his life. Born in Northern Ireland in 1910, he had come to Canada at the age of four when his father, a veteran of the Boer War, emigrated to join a brother in Winnipeg. Jack graduated in business administration from the University of Manitoba at the height of the Depression, never contemplating a career in aviation. His first job was with the city of Winnipeg, punching out relief cards on an IBM system. "It proved to me that machine accounting was here to stay." In 1937, as local businessman James Richardson was trying to get Canadian Airways Ltd. recognized as the national airline, Gilmer went to Montreal to join a more substantial organization, the mighty CPR. He worked as a travelling auditor and was promoted to chief comptroller when McConachie moved CPA to Vancouver in 1949. Now, in 1965, he was president.

"This industry is not like producing sausages." Gilmer doodled on the pad and passionately swung into a brief explanation of airline economics. "We have the most perishable business in the world, the seat-mile. If it's not sold, we can't put it on the shelf." Gilmer was optimistic about the immediate future: bilateral talks with the United States were coming up, in which both Air Canada and CPA were expecting a share

John C. Gilmer succeeded McConachie as president of CPA. "In this business," he would say, "if you're right, you're bright. If you're wrong, you're ruined." He was fortunate that the airline, like the country, was in its seven years of plenty.
CAA (UNN)

in the spoils, and transport minister J.W. Pickersgill was thought to be sympathetic toward finding CPA more transcontinental flights. Both opportunities were gambles. Gilmer jabbed the air with his pencil. How was CPA going to work the US flights? Did it need another machine for them and, if so, what kind? Like the country's economy in 1965, the airline was financially sound, and its rank and file were adamantly opposed to any merger with Air Canada. Carney left the interview with the impression that, in his own way, Jack Gilmer was as intense as his predecessor.

Pickersgill was now House leader and at the height of his power. The consummate mandarin, he wasn't fooled by the smokescreen behind which both airlines hid their reluctance to merge. Since his meeting with the four transport-company presidents a year ago, little had occurred, aside from some literal window dressing, in which both airlines displayed each other's schedules at their sales counters.[1] To demonstrate its power the Department of Transport threatened to take Amsterdam away from CPA and to deprive TCA (renamed Air Canada on January 1, 1965) of any future prospects of routes to the Orient. Jack Gilmer and Air Canada president Gordon McGregor were also informed that the government would rehire Stephen Wheatcroft to reassess his 1958 report on competition in the airline industry. McGregor must have heard this with a sinking heart. The last time Wheatcroft had interfered in Canadian aviation, TCA had lost its transcontinental monopoly. What, he wondered, would Air Canada lose now?

When the second Wheatcroft report appeared later in the year, it was all that McGregor feared. The British economist didn't think that

The old terminal building, with its "Welcome to Vancouver" sign barely visible, was decidedly shabby by the 1960s. The Department of Transport was already planning a new one to be located north of the intersection of runways 12-30 and 08-26. CAA 989-51-51

CPA's single transcontinental flight had any effect on Air Canada's profits, but it had opened the door to giving the public a choice of carriers. Wheatcroft could not help but notice that CPA had taken advantage of the only flight permitted by increasing the passenger capacity of the aircraft used on the service: the DC-6B with 61 seats grew to the Britannia with 89 seats, which became the DC-8 with 141 seats. He thus proposed that the Air Transport Board limit the airline's seats per flight but increase the number of flights to two a day—whatever the market would allow.

As Gilmer had hoped, the Canada–US bilateral talks yielded Air Canada its first US continental route, Vancouver–San Francisco, and the inaugural DC-8 flight took place on January 30, 1967. However, lest it be forgotten that Air Canada was the government's airline, the Montreal-based company was awarded the much more profitable Montreal–Toronto–Los Angeles route. In March 1967, as a consolation prize, Pickersgill allowed CPA to apply for a second transcontinental licence; "subject to an annual review," the flights would be allowed to grow in number until CPA had 25 percent of the market. This second transcontinental service began on February 1, 1968, using DC-8s originating in Edmonton and from Calgary the following day. On the international scene, London remained unattainable, and Buck Crump wanted to exchange Amsterdam for it—all that CPA seemed to be doing was feeding customers into KLM's aircraft bound for Britain. Crump did not get his wish; instead, CPA was allowed to extend its southern European routes from Rome to Athens in 1968 and onward to Tel Aviv in 1971.

The euphoria of the country's centennial year was the ideal venue for transport minister Jack Pickersgill to launch his National Transportation Act. Using the Royal Commission on Transportation recommendations as a framework, Pickersgill was finally able to implement a variation on Mackenzie King's dictum: "competition if necessary, but not necessarily competition." Canada's road, rail, water and air systems, while still regulated, were to be allowed a modicum of freedom. The 1967 National Transportation Act was the next step to regulate commercial aviation, following the process begun in 1944 with the Aeronautics Act and the Air Transport Board. Responsibility for the airlines now passed to the Canadian Transport Commission (CTC), a regulatory agency that had the authority to grant routes and licences. Having integrated the country's transportation system as he thought fit, Pickersgill then wisely bowed out. The next generation of voters would not accept the backroom politics that "Sailor Jack" had orchestrated for three prime ministers, and the Canada that he and Lester Pearson had known was slipping away. Besides, the rising star of Pierre Elliott Trudeau foretold a new, non-WASP elite that was about to take power in Ottawa. So Pickersgill wrote his own job description, made himself first president of the CTC and, in September, resigned from the House.

Even before the historic year of 1967, the women's movement had gained some momentum in Canada, due mainly to coverage of similar events in the United States. Both Air Canada and CPA, and indeed all of the aviation industry, with its discriminatory practices regarding age, marital status and motherhood, were prime targets. Neither airline had any firm policies on the rights of its female employees because both counted on female flight attendants leaving their employment within a year or two. However, as the government's airline, Air Canada had to appear to set a good example, and in 1965 it had dropped the requirement for compulsory retirement at age thirty-two, allowed its stewardesses to continue working after marriage and provided maternity leave as well. In 1966 CALFAA had successfully negotiated the right to marriage for its CPA members, but it was not able to obtain maternity leave for them until 1969.[2] Both airlines had been guilty of putting recently hired male flight attendants, or "pursers," in charge of the more experienced females. However, besides being more experienced than the men, many of the women were university-educated, trained nurses and Canadian-born. Only when federal legislation was enacted in 1968 concerning sex discrimination was this most blatant form of inequality on CPA ended.[3]

The airline's industrial relations with its other employees were slightly less acidulous. Dick Ryan admitted that although "McConachie had many outstanding qualities, he wasn't an administrator." Industrial disputes, for example, were beyond him; although he was a pilot, he couldn't understand why CPA pilots were dissatisfied with their salaries or working conditions. Once, when they insisted that the company pay for their uniforms—as TCA did—McConachie threatened to supply them with coveralls that said "Aeroplane Driver" on the back. As he was quick to point out to them, the director in charge of flight operations received fractionally more money than they did, and while they flew eighty hours a month (which meant that in theory they only worked twenty days a month), he was permanently on call. Besides, pilots were management, weren't they? McConachie's standard reply to his officers who asked for raises in salary was that when the company was doing better, "they would be taken care of." He was so admired by all ranks that he could pull it off. There weren't more strikes was because "Grant was so popular with the employees that they hated to take this step against him."[4]

The dawning of the Age of Aquarius was also keenly felt in the boardrooms of both Air Canada and CPA. In 1968, to the dismay of Air Canada's old-timers known as the "Class of '37"—the original TCA employees—Prime Minister Trudeau appointed Yves Pratte, a Quebec City lawyer, to take Gordon McGregor's place as president. Across the way from Air Canada's Place Ville Marie offices, at Windsor Station, Buck Crump also prepared to step down as CPR president. He was fortunate enough to be able to choose as his successor Ian Sinclair, the legal counsel who almost a decade before had rescued McConachie from certain death at the ATB hearings. The shrewd Crump realized that the railway's survival lay in adapting to the modern era and in being governed by a lawyer who could deal effectively with politicians

Thigh in the Sky: This CPA ad, with both models posed in a DC-8 interior, depicted the attitude that permeated the airline industry in the late 1960s. Although CPA never took it as far as the "Hi, I'm Debbie, Fly Me" advertising that some airlines in the United States did, it was this image of stewardesses that the Canadian Air Line Flight Attendants' Association (CALFAA) was working hard to eliminate.
CAA 989-80-11b

and parliamentary hearings. However, it was ironic that in his final year the last of the CPR's "steam engine" presidents would initiate a massive corporate facelift for the Canadian Pacific empire.

Sinclair had already made a name for himself by diversifying the CPR's ramshackle portfolio of minerals, land, timber and rolling stock into assets more suited to the twentieth century, that is, less influenced by cyclical downturns in the economy. In 1967 his chosen vehicle, Canadian Pacific Investments, announced a public share issue of $100 million. Four million shares were sold in CPA alone so that it would from now on effectively be owned by 75,000 or so registered shareholders. To traditionalists, it must have seemed as if the venerable CPR of Van Horne and Beatty was no longer ruled from the stuffy, panelled boardroom above Windsor Station but had moved to a cheesy, Formica-furnished penthouse off Carnaby Street.

Most scandalous of all, Canadian Pacific hired the New York advertising firm Lippincott & Margulies to design a "corporate image makeover" that, it was hoped, would drag the company from its Edwardian roots into the Swinging Sixties. The poor old beaver, which since 1946 had been quietly chewing on its branch on top of a red shield, was displaced yet again, by "The Multimark"—a triangle set in a segment of a circle, and both set in a square box. The triangle signified motion, the circle global capability and the square stability. In addition, each branch of the CP family was to be identified by a single word—CP Ships, CP Transport, CP Rail, CP Hotels—and its own distinctive colour—CP Ships was green and CP Rail was red.

Ian David Sinclair, the CPR's legal counsel, succeeded Buck Crump in 1969. He loved the glamour of the airline and used to say that he didn't need a corporate jet like other CEOs—he had twenty-five of them.
CAA (UNN)

For Canadian Pacific Airlines, the image makeover buried the last remnant of the James Richardson days. The "Flying Goose," which had been stylized in 1953, now disappeared altogether, as did the "spaghetti writing" of the company name. It took a year for the Multimark scheme to be implemented, and airline staff were given until April 1, 1969, to begin using the new name "CP Air" (which could not be translated into French) on all stationery, advertisements and announcements.

Passengers first noticed the "new look" when the pursers gave up their light blue jackets and bow ties for bright red blazers and Multimark ties. The aircraft underwent even more startling changes. Each upper fuselage and tail were painted orange with a wide red stripe that started on the top, above the pilot's window, and ran back to the very end of the plane below the tail. The remainder of the body was silver, with a large CP logo in white and red oñ the tail and the black title "CP Air" on the forward fuselage. The registration and fleet numbers were displayed in black on the rear fuselage and the tip of the tail respectively. No longer "Empress" but "Big Orange," no one could miss a CP Air aircraft, however crowded the airport![5]

The advertising team came up with the "hookline" of "Orange is Beautiful" or "*Orange, ça va loin*," because, so the story goes, when the agency's chief executive was in San Francisco he saw an orange tree in full bloom and was deeply impressed by its vibrancy. Through the years, there were many who cast a jaundiced eye at the gaudy corporate image, and more than one staff member heard a passenger say, "Your orange is a lemon!"

The first aircraft to wear the orange livery was the company's initial Boeing 737, CF-CPB, which was accepted on October 22, 1968. The

The "Multimark" and orange livery on the CP Air Boeing 727-217. Too "fuel inefficient" for the domestic routes, the 727 did not have the range for international routes. All 727s were sold off by 1977; this one went to British charter operator Dan Air. CAA (UNN)

Douglas DC-9s and BAC One Elevens were already in airline service in 1965 when Boeing announced that it would compete in the short-haul jet airliner race. Its previous entry in the medium range was a tri-jet, the Boeing 727, which had been built since 1960. Boeing had developed its 727s for "shuttle" airlines in the United States to get in and out of inner city airports such as New York's La Guardia, Chicago's Midway and Washington's National. The aircraft climbed quickly and descended, as one passenger remembers, straight down "almost like a helicopter." This made them difficult to fly; pilots of that generation had been trained on piston-engined aircraft and on final approach tried to "flare" into a glide, hitting the runway so hard that the 727's landing gear and fuselage bottom would break off. CP Air bought four 727-117s for its cross-Canada service in 1970, only to discover that they were unsuited to the airline's network. They proved fuel inefficient for the domestic routes, but did not have the range for the international runs. None of the airports used were in the midst of a large Canadian city, and all four 727s would be sold off by 1977. In 1975, the airline would

The workhorse 737 first flew in 1967 as Boeing's answer to the Douglas DC-9. By using a 707 fuselage, it allowed for six-abreast seating. Few airlines in the world have not at one time or another flown these DC-3s of the jet age. CAA (UNN)

buy two 727-217s for the Vancouver–Toronto and Vancouver–San Francisco routes but leased them out after six years' service.

Boeing meanwhile had adapted its 707 fuselage to create the 737. This allowed six-abreast seating, as in the larger jets, and commonality for the airlines that already owned 707s with regard to equipment like seats and galleys. When Lufthansa, the launch customer for the 737, asked that the 737-200 series be extended to increase seating to up to one hundred seats, this gave Boeing a considerable advantage over Douglas. It was with this 200 model that CP Air standardized its fleet, flying it on both the second and all subsequent transcontinental routes.

Another user of the versatile 737s was Vancouver-based PWA, which in 1968 had taken over CP Air's Calgary–Kamloops–Vancouver route. The previous year, Russ Baker's old company had graduated from the bush carrier business by selling off its Beavers and DC-4s and operating its first turbine-powered aircraft. PWA's president, Donald Watson, had been hired by Baker as his assistant just before his sudden death in

1958. When the major airlines moved out of Edmonton's downtown airport to a new international airport out of the city, PWA introduced its downtown "shuttle" service to Calgary. The "Chieftain Airbus" offered low fares, hourly service and guaranteed seating. It had, as future president Rhys Eyton pointed out, identified a need in the market; and the airline, from management down to the frontline staff were prepared to take risks in "an atmosphere of fun and excitement that stimulated employees to give that extra effort." In the next two decades, PWA's entrepreneurial corporate culture led it into ventures that concentrated on a high aircraft utilization, usually of a single type of aircraft, blended with management's enjoyment of risk-taking. To compete with Wardair for the vacation charter industry, for instance, PWA had begun flying vacationers to the Grand Cayman Islands in 1964, investing in Boeing 707s[6], 727s and the trouble-plagued Lockheed Electras. Watson's six-year presidency coincided with a boom in oil and mining exploration, when an innovative airline could not but flourish. In 1966, when PWA was contracted to airlift drilling equipment into the North, its pilots discovered that the thickness they needed for heavy aircraft to land on ice was about forty-eight inches. In 1967, after seeing Alaskan Airlines use Hercules freighters, PWA bought the first of four Lockheed 100-10s for $4 million each. At last the Canadian Arctic had an aircraft that could deliver 46,000 pounds (20,909 kg) onto a rough airstrip at below-zero temperatures—the turbine-powered engines were easier to start than pistons in the subzero climate. In the summer, the PWA "Herc" freighters tramped around the world in true bush pilot fashion, touting for cargoes. Through eighty countries, they carried D-8 Cats, oil rigs, housing, vehicles, relief supplies, military airport hardware, a ship's propeller shaft, space components, cattle and sheep, currency, gold and silver bullion, crates of fruit, fireworks, sheep's casings in brine—anything that could be fitted into the holds.[7]

By 1974 this entrepreneurial spirit had positioned PWA as Canada's third-largest airline and the largest regional carrier in the country. Its potential to grow even bigger attracted the attention of Alberta's premier, Peter Lougheed, who saw PWA as a vital part of his strategy to diversify his province's wealth away from oil and gas and connect Alberta with export markets that the two main airlines did not serve. "For a landlocked province, transportation is the key—particularly air freight transportation," Lougheed emphasized. "Almost 80 percent of Pacific Western's revenue originates or terminates in Alberta. We want to assure that such a vital part of the transportation system in our province would continue to reflect the needs and interests of the people of Alberta." On behalf of the provincial government, on August 1, Lougheed bought 99.5 percent of PWA's shares for $37 million—preempting, it was thought, the NDP government in British Columbia, which was also interested in purchasing the airline. Prime Minister Trudeau, at odds with Lougheed over his energy policy, had the CTC take the province to court for a determination of whether the federal government or the provinces could own an airline. When the Supreme Court ruled in favour of Alberta, Lougheed had PWA's headquarters moved from Vancouver to Calgary. Don Watson then resigned, holding that if the aggressive, gutsy airline he had built up became a crown corporation, it would fall heir to the evils Air Canada was then known for: annual deficits, a bloated work force and poor customer service. Lougheed, on the other hand, was content to see Watson leave and Rhys Eyton take his place.[8]

Like Rhys Eyton, Jack Gilmer had the good fortune to take over an airline when it was in the black. In 1968, CP Air's revenues had exceeded $100 million for the first time and more than one million passengers had been carried; 17 percent more than in the previous year. Despite a national postal strike, mail revenues had increased to $6.5 million. Operating revenues totalled $106.7 million, 12 percent above the $95.2 million of the previous year. "Despite these milestones," said Gilmer "our 1968 net income is $2.3 million, compared with $3.2 million in 1967." One didn't need to be an accountant to know why. As a service industry, commercial aviation is susceptible to any downturn in the economy, and an international financial crisis, the unofficial "stay at home" policy in the United States, increases in wages and in the cost of materials—especially those used for the construction of the operations centre then being built at Vancouver airport—all affected the airline's revenue.

The last of CP Air's Britannias and DC-6s had been sold by April 1969, and except for the single DC-3 the airline achieved the milestone that McConachie had hoped for: it became "all jet."[9] The $24 million Vancouver airport operations centre was opened a year later, on April 17, 1970. Its floor space of 870,000 square feet included the main hangar, maintenance, stores, training classrooms, an administrative centre and a 32,000-square-foot flight kitchen. In 1971, CP Air sold its old hangars on the south side of Vancouver airport to PWA. It now had

In 1970 the airline was still small enough to be a family: Captain Ken Sorko, who had been flying since 1950, and his daughter Janyce, recently recruited, inevitably served on board the same flights. CAA (UNN)

1,300 employees in technical services who, besides servicing the airline's own aircraft and the 737s of Wein Air Alaska and PWA, as the new plant opened, were about to begin an engine overhaul contract for Canadian General Electric.

The fleet's diversity—Boeing 737s and Douglas DC-8s—presented "some interesting problems for maintenance," according to Ian Gray, the vice-president of technical services, who wanted to see standardization of equipment "sometime in the 1980s." Most of CP Air's markets consisted of traffic that peaked in the summer, so all major maintenance checks were done in the winter. The average daily use of the 737s was eight hours, and the DC-8s averaged fourteen hours in the summer and twelve in the winter.

The airline had grown to 5,148 employees, which necessitated the first use of the computers housed in the new operations centre in Vancouver. The airline's computer department had been formed in 1960 as a methods research office with three employees—and no computers. All such work was then done by the CPR-owned computers at Windsor Station. When in 1968 the airline put an IBM 1130 into service to begin automating reservations, CP Air became the first computerized Canadian carrier. Under W.R. Ellwood, computer services began with the rental of an IBM 360/40, used initially to process the bi-weekly paycheques. "It takes six hours to process the total payroll—that's about five seconds per employee," said an operator. Soon the IBM would also be used to provide figures for revenue accounting, billings for other airlines and passenger sales. In February 1975, an IBM 370-14 took over all of CP Air's manually operated reservations system and an intra-company competition was held to choose a name for it. The employees' favourite was "The Happy Booker," but management decided on the safer

"Pegasus." Every CP Air reservations desk had its own "television" display for an instant readout, while a few in the rest of the world were connected by teletype. The airline was particularly proud that not one of the reservations staff was laid off as a result of computerization.

Two of the departments that needed computers in 1970 were "Flight Crew Scheduling" and "Operations Planning, Control," both of which were still being done "by hand." The first involved scheduling the airline's 1,100 pilots, navigators and flight attendants to meet the demands of its timetable—without exceeding the maximum allowable daily-duty hours of any crew members. Scheduling Department manager Derek Staniforth and supervisor Art Jackman went to work as soon as the company's routings were approved. They paired flights, considering the safety regulations of rest periods and the most efficient use of staff with regard to wages. All federal transport regulations had to be met, problems of flight delays due to sickness and weather solved, and lodging, transportation and meals arranged for crews. This required pairing an outbound flight from Canada with an inbound flight from overseas. For example, Flight 204, due to leave Toronto at 6:30 p.m. on Wednesday and arrive in Rome at 7:45 a.m. on Thursday could be paired with Flight 203, which was due to leave Rome at 11 a.m. on Friday, giving the crew a twenty-four-hour rest before returning to Canada.

The demands of the North American service further complicated the procedure. For instance, the crew that returned to Canada on Flight 203 could go only as far as Montreal and had to "overnight" there to make a Vancouver flight the next day. The final pairings plotted by Staniforth and Jackman were then presented to the representatives of various employee unions, with whom they had formed committees, to draw up working periods or blocks—three-month blocks for overseas flights and monthly blocks for domestic. When this was completed, copies of the blocking schedule went to all flight personnel, who bid, in order of seniority, on the various work periods, until all the aircraft had crews assigned to them.[10]

Before World War II, decisions on planning and aircraft movements had been made by individual stations. As the airline grew, some central authority had to be established for the common good; hence the need for "Operations Planning and Control." Once a flight schedule was approved by "upstairs," director Wilf Cameron and superintendent Val Vaillancourt were in charge of it. They took into

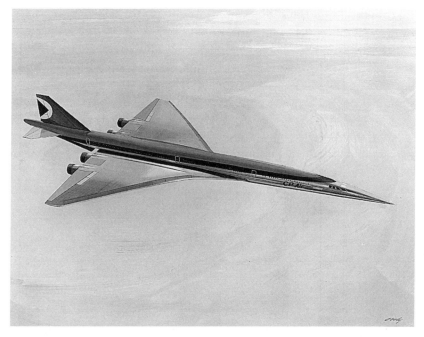

Like Pan American and TWA, CP Air considered the Boeing SST, which would be in service by 1974. Despite strong backing from the White House, Congress was fearful of subsidizing the aircraft's production indefinitely and killed off the US supersonic transport program in 1973. CAA 989-71-58

consideration the varied requirements of maintenance, crew routing, daylight time, customer service and payload control and assigned specific equipment to the routes. The timetable was then sent to "Ops Control," where supervisors Cliff Forsythe, Lawney Scown, Len Dixon and Roy Geddes—the "watchdogs of the air"—mulled it over. Like medieval alchemists, the four worked with indecipherable figures to come up with the magic formula. If in the pre-computer period CP Air had a "brain," it was their Ops Control blackboard which contained all aircraft movements for a twenty-eight-day period.

Finally there was Flight Dispatch, which dealt with flights in an immediate forty-eight-hour period. Under the guidance of J. Ferguson, CP Air flight dispatchers were responsible for preparing the technical information required by the flight crews immediately before their departure. The crews would come to Flight Dispatch to obtain their flight plans and briefings. The aircraft's capabilities and passenger load were matched with weather reports along the route, wind velocities, runway conditions and any special radio communications.

It might be that the old saying "Be careful what you wish for . . ." launched CP Air's "Executive Jet" program. By April 1, 1970, the airline's allotted 25 percent of the market had become seven daily transcontinental flights, for which it had bought the 737s. Faced with the problem of filling those extra seats and competing for a market that, east of Winnipeg, hardly knew that CP Air existed, the company embarked on a unique promotion exercise. If most Canadians still flew only once a year for a vacation—during school holidays and on weekends—another small segment of the population flew and flew . . . and flew again on weekdays. As early as 1964, *Time* magazine noticed that no fewer than 86 percent of domestic airline passengers in the United States were businessmen. Like the travelling "drummers" of a century before, they bounced between the major cities—New York, Boston, Chicago and Los Angeles—on expense accounts. Their effect on the architecture of North American airports was considerable. Rental-car companies catered to their needs by positioning cars in vast carparks nearby, and hotel chains built soundproof bedrooms adjoining the terminals. Some companies were moving out to the suburb nearest the airport. Every airline bought short-haul jet airliners to get these people in and out of airports and home for dinner. Eastern Airlines had pioneered this system by developing its famed "shuttle" between New York and Washington, keeping a spare aircraft ready for times when the regular one was filled.

Market research in Canada showed that in 1970, 80 percent of the cross-Canada air travellers were businessmen, and this was the group CP Air targeted. When a consultant study was undertaken to determine what executives wanted from an airline, the first two answers were predictable: a safe flight and on-time service. But the third reply was vague. Businessmen (it seemed from the survey that business travellers then were all male) "craved something different while in the air." The marketing department's studies disclosed that the average three-piece-suit-and-briefcase wielder was part of "a sophisticated and serious breed capable of being influenced by high-calibre customer service." As a result, "Executive Jet" was launched on what H. Bryan Renwick, vice-president, marketing and sales, described as a *vive la différence* basis.

The first impression the businessmen got of the new corporate strategy was that the stewardesses had put on the latest fashion in "non-uniforms"—predominately ankle-length midis. The CP Air flight attendants didn't wear miniskirts but had several different standard uniforms—which made, as a male passenger commented, the best "view seat" on CP Air aircraft "the one on the aisle."[11] But the press announcement of the new uniforms enthusiastically stated that "as the seasons change, so will the stewardess's clothes. For the first three months, pant suits in a variety of colours, maxi coats and mod glasses will be featured." Only a certain group of flight attendants volunteered to wear these as a marketing ploy, and one stewardess recalled that the feeling among them was that the experiment was "very positive and that it was fun."

The other features of the Executive Jet were more functional. Lest the businessmen become restless,

> there will be portable dictating equipment and speedy secretarial services, up-to-the-minute financial reports and current business magazines on board. Those who do use the hand-sized, battery-powered dictating machines available will be able to drop off the tapes at the next CP Air counter, from where they will be taken to the nearest Office Overload for typing. They will send it to the address requested by the customer.
>
> For the salesman who just wants to unwind there is a library on board, which includes hardbound bestsellers[12] and magazines such as *The Harvard Business Review, The Economist* and *Playboy,* as well as colouring books for children—and any adult that might be so inclined.[13]

The "non-uniforms" did attract public attention, but not the kind that the company's marketing department hoped for. Never in CP Air's history did any ad campaign elicit such a response from the public. More than two hundred letters, all from businessmen, flooded Gilmer's desk in the first ten days of the change of uniform on the "transcon." Cabin staff and airline offices were deluged with what the marketing department called "written and verbal feedback," including such comments as "You have succeeded in making Canada's most beautiful stewardesses look like Russian Army corporals," from one distressed male passenger. Another wrote, "I flew with my grandmother as a child. Why do I have to continue now that I am a lecherous old man?" And they continued: "Adoption of this fashion labels your executives as "party-poopers";

Stepping out on April 1, 1970, in the "non-uniforms" worn on Executive Jet transcontinental flights are (from left) Mrs. Trouda Holmes, Shirley Hazel and Jennifer Buckland.
CAA 989-55-55

"When your idiot vice-president put your stewardesses into midis, I issued instructions to my staff that henceforth nobody in the company was to fly CP Air" and one final query, "Can my mother become a stewardess at CP Air? She is 60 and can supply her own uniform."

The airline capitalized on what the media dubbed the "Thigh in the Sky" controversy by taking out two-page advertisements in all major newspapers on October 26, asking readers to clip a ballot about the "non-uniforms," tick off the "I Hate It / I Like It" boxes and mail it in. By the second week of November, 10,972 ballots had been received, of which 9,796 (91 percent) voted to "hate the midi." Eventually thirty thousand "anti-midi" ballots were received. Their replacement was a compromise mini-dress in a floral pattern, which made its appearance in March 1971. It was many years before anyone, passenger or staff, forgot about the "non-uniform" campaign.[14]

That month, CP Air became the first Canadian carrier to be authorized for inertial navigation system (INS) operations. The company's increasing dependence on INS meant that its navigational team would eventually be disbanded. It was only fitting that the first to make use of the INS on a flight from Vancouver to Honolulu and back was the airline's chief overseas pilot, Willbert George Melvin Knox. Like many of Canada's first generation of airline pilots, "Mel" had been a barnstormer on the prairies, getting his licence at a flying club, before joining CPA by way of Prairie Airways and the RCAF. He had been one of the nine pilots chosen by McConachie to fly on the company's first overseas flight to Shanghai in 1949, and when Mel retired in 1971 he estimated that he had flown 23,500 hours and more than 11 million kilometres in 28 types of aircraft. Inducted into Canada's Aviation Hall

"High Altitude Approval." A happy male passenger shows his delight at stewardess Shirley Hazel in the new CP Air mini dress. Customer dissatisfaction with the previous "non-uniforms" led to the airline receiving thirty thousand "I hate the midi" ballots. CAA 989-80-19

of Fame in 1973, Knox was noted for having made 204 Pacific flights, 84 Atlantic crossings and 56 polar flights to Amsterdam.

In the early 1970s, the airline was again negotiating with the Chinese government, for the first time since 1949, for routes from Vancouver to Beijing and Shanghai. On August 1, 1972 after a twenty-three-year absence, a Canadian commercial aircraft returned to the People's Republic of China when CP Air carried a federal government trade mission to Beijing via Tokyo, Hong Kong and Canton. This time the Chinese had learned a few capitalist tricks. They wanted *quid pro quo* rights for their national airline, CAAC, to fly from Beijing to Vancouver, and to be allowed to pick up passengers on the Tokyo–Vancouver portion. Since Japanese airlines were not allowed to fly into China, Japan refused to allow CAAC landing rights in Tokyo. Then Beijing upped the ante. Although Vancouver had a large Chinese population and CP Air connected with Chinese communities in South America, this wasn't enough. China demanded that CAAC serve Toronto as well. After two years of frustrating negotiations, J.C. Gilmer was asked by the *Financial Post* whether it was all worth it. Would there be enough traffic to China to justify this?[15] Not initially, he replied, but one of the tenets of his predecessor, Grant McConachie, was that there existed a great pent-up demand in North America to see China, and when the authorities allowed tourists in, CP Air would be well placed to provide the flights. On March 9, 1973, the last of the diplomatic hurdles were cleared and the Shanghai and Beijing routes were finally awarded to CP Air. By now, however, the recession had set in and it was thirteen years before the inaugural scheduled flight to China took place.

Captain Mel Knox, CP Air's chief pilot, signs the aircraft log book for President J.C. Gilmer at the close of his final flight. Knox had begun flying in 1929, and he estimated that he had covered seven million miles (11.2 million km) in a total of 23,500 hours during a forty-two-year career.
CAA 989-134-13

On the other side of the world, another human disaster was unfolding. Having been assigned the continent of Africa as its "sphere of influence," CP Air was asked to provide aircraft for refugee flights from Uganda. From October 17 to November 4, 1972, its DC-8s made ten charter flights via Madrid and Nairobi to Entebbe to pick up the thousands of East Indians made homeless by Idi Amin. The average passenger load on each flight was 150.

Until the 1970s, airlines in Canada had traditionally recruited their pilots from two sources: the RCAF and the bush carriers. Thus, asked if they had considered hiring female pilots, they could truthfully say there just were no women with the necessary qualifications from either of those male-dominated fraternities. When pressed to do so by the liberalization of the times, Air Canada and CP Air admitted that they were reluctant to hire women pilots because of the risks involved: after expensive training women might leave for family reasons, or they made the male pilots (or their wives) uncomfortable or worse, passengers would refuse to fly with the airline if they knew who was at the controls. The factors that led to a change in the airlines' attitude was the regional and charter boom of the 1970s, which precipitated a need for pilots, and the power of the Royal Commission on the Status of Women.

Transair is remembered as the airline that hired Rosella Bjornson, the first female commercial pilot on a scheduled airline in North America. As a child, Rosella's playhouse had been an abandoned Anson Mark II, and at sixteen, she had logged many hours in the family aircraft. In 1969, the Winnipeg Flying Club hired her as an instructor. While upgrading her qualifications there, she kept applying to Air Canada, CP Air and PWA. The local regional carrier, Transair, was suf-

On August 1, 1972, after an absence of twenty-three years, the first Canadian passenger aircraft, CP Air DC-8 CF-CPF, landed in Beijing carrying a federal government trade mission. Cabin crew on that first flight were (left to right): steward Gary Grant, Colleen Campbell, Linda Lee, Suzanne Dore, Colleen Woodman and purser Wilhelm Gerig. CAA (UNN)

ficiently impressed to employ her in 1973 as the first female First Officer in North America. Under the glare of such publicity, Bjornson's career in Transair, PWA and later Canadian Airlines was constantly scrutinized by the press and male pilots. The strong sexual bias against women in the cockpit was part of the attitudes of the era, although Bjornson admitted that most of the male pilots she encountered were unfailingly helpful. Her biographer noted, "Rosella simply bided her time. She neither tried to convert the anti-female faction nor did she complain. 'I always tried to see their side of things and to keep things in perspective,' she said. Having a sense of humour was a definite help. 'On one of my first flights into Toronto I was working the radio and ATC asked if Transair's first officer's jockey shorts were too tight since his voice was so high!"[16]

Air Canada hired its first female pilot, Judy Evans Cameron, in 1978. Nordair hired Stefanie Crampton at the same time and, in 1979,

Sandra Lloyd was taken on by CP Air. Lloyd had obtained her private pilot's licence at the age of seventeen and her commercial and instructor's licences in 1975. Like Bjornson, she instructed until her big break came—Parsons Airways Northern in Flin Flon took her on. Then she was interviewed by CP Air: "They had dropped their height requirements. When I first applied you had to be at least five foot seven; I was five foot five. After my first interview they gave me a strength test. I found out later, that was not usual. I had about 2300 hours when they hired me as a second officer on the DC-8 in November 1979."

Kathy Davenport Zokol, the second woman pilot hired by CP Air, was an Australian who had arrived in Vancouver in 1969. She had been so entranced by flying across the Pacific that she took flying lessons, paying for them with two part-time jobs. She eventually found work with Futura Aviation: "I had to prove that I could handle a fairly sophisticated airplane—an Aero Commander, single pilot—and could hoist fifty- to a hundred-pound mail sacks. Yet I never tried to hide my femininity." Zokol's philosophy was that she wanted to be treated like a pilot when she was flying but as a woman when out of the plane. She joined CP Air in 1981.

Despite its success with the 707 in the 1960s, Boeing was still viewed primarily as a manufacturer of military aircraft like the B-47 and B-52. Its Stratocruiser airliner had been a failure, and even the 707 had begun life as the KC-135 aerial tanker. It was Douglas that airlines went to for transports, and they had done so since 1934, when the DC-2s revolutionized the industry. To recapture its predominance over Boeing and its success with the 707, Douglas decided to stretch its basic DC-8 design as far as it could go, bringing out the "Super Sixty" series in 1965. "To stretch it into the 63, a 20-foot 'plug' was inserted in front of the wing and a 20-foot 'plug' behind the wing," explained Capt. Bob Randall Jr., who for 90 percent of a forty-one-year career with CPA had flown Douglas products. "It was a fine airplane, although when equipped with the standard Pratt & Whitney JT-3 engine was somewhat underpowered. CP Air pilots flying nonstop to Tokyo were treated to many thrilling takeoffs out of Vancouver, often breaking ground with only several hundred feet of runway remaining.

"The Douglas Aircraft Company then came out with a long duct version of the JT-3 called the JT3-7. This was rated at 21,000 pounds thrust and enabled the Series 63 DC-8 to be certified at a take off gross weight of 356,000 pounds. Handled with care, this aircraft presented no problem. The DC-8-63 was a refinement over the earlier Rolls-Royce-powered DC-8-43, which at a weight of 300,000 pounds required 38 pounds of fuel for each air mile, whereas the Series 63, powered by the long-duct JT3-7, required only 30 pounds of fuel for each air mile traveled. CP Air flight crews were taught to use a five-second rotation technique to enable the aircraft to lift off with full oleo leg extension."

Airlines were continually demanding aircraft that could carry greater passenger loads. Although the Series 40 had been considered exceptional in 1960 because it took 179 passengers, five years later the arrival of the stretched DC-8-63 that could seat 259 passengers in comfort, named "Spacemasters" by CP Air, hardly provoked a murmur, perhaps because to the public it was still the same DC-8 design but with an aisle that seemed to go on forever. The engineers at Douglas knew that their colleagues at Boeing couldn't do the same with the 707. Because of its landing gear and wing sweep, if the 707 were stretched, the tail would scrape the runway each time it took off. The stretched DC-8-63 did have some "tail strikes" too, but as Captain Randall recalls, they only proved that it was an exceptional aircraft:

In early February 1975 I was flying a DC-8-63 from Vancouver to Amsterdam, stopping at Calgary. Normally, because of reduced engine power and less dense air at Calgary's airport (altitude of 3557 feet) a full gross takeoff weight of 356,000 lbs. was seldom possible. This day, because of a very strong headwind on Runway 34 and a cold -25 degrees, we were able to plan a take off at a full gross weight of 356,000 lbs. It went smoothly, utilizing the five-second rotation technique, and the aircraft lifted off just prior to the intersection of Runway 10-28 and Runway 16-34. After approximately two seconds, a loud "bang" was heard in the cockpit. We had suffered a "tail strike" on our takeoff. The aircraft's tail skid had come in contact with the raised crown of Runway 10-28 as we were airborne. (The raised crown at runway intersections is to assist in rain water runoff.) En route from Calgary to Amsterdam, I repeatedly checked the performance and takeoff calculations but could find no error. I did discover, however, that this particular aircraft had suffered two tail strikes in the previous week—while flying the same route.

On arrival in Amsterdam I informed the CP Air base engineer of the tail strike, and he said that this aircraft would be remaining in Amsterdam for our departure back to Canada. Two days later, when we arrived at the airport for the return flight, the base engineer told me he had replaced the tail skid and that he had also found the reason for the strike. When he led me out to the aircraft, I was surprised to see it sitting in the centre of a small lake approximately 250 feet wide. It seems that the water had been melting out of the aircraft's belly, and the engineer had discovered that it was from a cracked fresh-water line. This had been leaking for the last ten days and freezing every time it was exposed to winter conditions in Canada and while in flight. Our engineers later confirmed that we had been overloaded by an accumulation of frozen water in the belly weighing at least 10,000 lbs. Simply put, we had taken off from Calgary at least 10,000 lbs. over the maximum sea-level gross weight—366,000 lbs! This to me was ample proof that given the chance, the DC-8-63 was capable of tremendous performance.

When the US military asked for proposals from aircraft manufacturers for an oversized freighter with a front-loading hatch for airlifting tanks and armoured personnel carriers, the Lockheed Aircraft Corporation won the competition with its C-5A Galaxy. But planning on such a large scale, Boeing's engineers dropped the idea of trying to stretch the 707 and warmed to the concept of a civil aircraft with a cabin twenty feet wide and seating for 452 passengers. Juan Trippe was so impressed by the new aircraft's specifications that in 1966 he ordered twenty-five of them for Pan American Airways, and with such a firm commitment Boeing invested in a new plant at Everett, Washington—the largest building in the world—to construct the "Jumbo" jet. The first 747 was rolled out in the fall of 1968. Its maiden flight was planned to coincide with the sixty-fifth anniversary of the Wright Brothers first "hop" on December 17, 1903, but a problem with the JT9D engines delayed this to February 9, 1969—the day that in the history of the world, the masses began taking to the air.

The only Boeing 707 that CPA had ever used was "wet leased" from Standard Airways of Seattle on October 15, 1967, for the Vancouver–Hawaii runs. Because it had no flight crews cleared to fly it, only cabin crew were provided by the airline. On February 7, 1968, with visibility at zero, during landing at Vancouver airport, the aircraft ran off the runway and collided with parked aircraft, vehicles and two buildings. Purser Martin Verhoef and an airport employee were killed, and seventeen passengers were injured.

CP Air's first Boeing 747 arrived at the Vancouver base on November 15, 1973, its gleaming hull reflecting the curiosity of hundreds of the airline's employees.

"Wonderment and fascination matched their thinly disguised feelings of satisfaction that this orange giant was their very own ticket to the jumbo era," wrote the *CP Air News* editor. "Work stopped. People started. What little was said spoke volumes as veteran Ops. Centre tour guide Archie Rixon declared that it was a bigger 747 than anyone else's, while Bill Genereux, director of properties and procedures, announced that it was the cleanest machine he had ever seen. A blustery day kept the rain away until the spotless aircraft made a textbook landing and taxied straight to the sanctuary of its hangar to undergo customs clearance.[17]

The airline bought four 747-200s from Boeing, at $26 million each, for its long-haul flights to Tokyo, Hong Kong, Sydney and Amsterdam, to be delivered between November 1973 and December 1974.[18] As the flagships of the fleet and potent symbols of the airline's prosperity, they were christened and rechristened: C-FCRA, *Empress of Asia/Japan/ Italy*; C-FCRB, *Empress of Canada/Japan*; C-FCRD, *Empress of Australia/Japan/Canada*; C-FCRE, *Empress of Italy/Australia*.

Captain Rick Wiley never forgot the first time he climbed the circular staircase to the upper deck of a 747-200 and sat in the pilot's seat. He remembered thinking, "This is a big airplane!"[19] When the aircraft was on the ramp, the pilot's eye level was twenty-six feet above the ground, which was much like sitting on the top of a two-storey apartment building. It was not uncommon in those days, Wiley recalled, to be able to see the gravel on the roof of some of the older terminal buildings when parked at the boarding gate.

The cockpit, or "flight deck," as Boeing referred to it, was surprisingly small for an aircraft of this size. Whereas the main cabin was more than twenty-two feet wide, the flight deck located on top of the aircraft over the first class cabin was only seven feet wide. On entering it, one's attention was immediately drawn to the myriad instrument switches, control knobs and warning lights that covered the central pedestal, forward instrument and overhead panels and glare shield panel. Behind the captain's and first officer's seats, facing the right side wall was the second officer's station. Mounted on the sidewall was another large

Winged majesty: the mighty Boeing 747-200 was "bigger than big, brighter than bright" to the hundreds of CP Air staff who greeted it on November 15, 1973. CAA 988-222-61

panel with the indicating instruments and controls for most of the aircraft systems, including electrics, hydraulics, air conditioning and pressurization, pneumatics and fuel systems. In total, the flight deck had more than one thousand lights, gauges and switches on all its panels.

Taxiing the 747-200 could pose problems for pilots new to the aircraft. The pilots sat approximately seven feet ahead of the nose wheel and ninety-five feet ahead of the main wing gear wheels. However, the height of the flight deck gave the pilot the illusion that the taxiways were much narrower than they were. The 747 also seemed to be moving much slower than it actually was. New pilots not only had to make frequent references to the ground speed readouts to avoid taxiing too fast, but they also had to be constantly aware of the aircraft's 200-foot wingspan (the wingtips are barely visible from the flight deck) and the 231-foot fuselage behind when manoeuvring in a congested area.

The 747-200 was powered by the latest in engines, the Pratt & Whitney JT9D. To achieve the maximum gross takeoff weight of 770,000 pounds, the engines were equipped with a unique water

The airline bought four 747-200s, at $26 million each, between November 1973 and December 1974. Only three were originally delivered; the fourth, C-FCRD, was leased to Braniff. All plied the long-range routes to Tokyo, Amsterdam and Sydney until they were sold to Pakistan International Airlines in 1986 in return for its DC-10-30s. CAA (UNN)

injection system. Tanks within the engine pylons contained a total of 600 gallons of distilled water, and immediately prior to the takeoff roll, the pilots turned the four water-injection pumps on. Once the engine had reached a preset thrust, the water was injected in to reduce operating temperatures and increase thrust. Once the aircraft was airborne, any water remaining in the tanks was jettisoned overboard to prevent it from freezing at high altitudes. In flight, new pilots were pleasantly surprised at how light on the controls and responsive the 747-200 was. It cruised at a speed of Mach .86 (86 percent of the speed of sound) at altitudes of up to 41,000 feet. The aircraft had a still air range of 5,000 nautical miles or approximately ten hours in flight.

The launch of the CP Air 747 revenue flight to the Orient on December 16, 1973 was comprehensively covered in *The Snag Sheet* by the editor, Moe Morel.

Marking the beginning of a new growth period for CP Air, the sun saw fit to break through the dismal cloud over Vancouver's International Airport on the day of our inaugural Super Orange

The forward section of the 747-200 in 1979, with the flight attendants serving the main course. Their uniform was apple-green and navy.
CAA 989-55-50

Service to the Orient. This was my first flight in the Boeing 747-200 and my first visit to Tokyo. At 15:26 power is applied to the four Pratt & Whitney JT9D-7As with the response being very good. Acceleration is far more noticeable than I'd expected from this huge flying machine . . . and soon we had left behind the light-scattered cloud and entered the bright sunlight, blue sky and cotton-wool clouds of the pilot's world. As we continue to climb out over the Queen Charlottes, time is 15:32, it is difficult to imagine a trip such as this taking up to 15 days aboard the original *Empress of Asia*, a giant of a vessel in her time too, right up to 1940.

Lunch was excellent, well presented, nicely displayed, as was the wine which was European. What a pity we don't carry Canadian wines, but then I'm no connoisseur. Take off weight on this flight is 758,000 lbs. of which I believe 310,000 lbs. was fuel. There were 240 Economy and 12 First Class passengers plus a crew of 14. This represents the greatest number of passengers ever carried by our company on a single flight, the previous being 242 in a DC-8. General appearance of the cabin is excellent, lighting, colour scheme, centre wall lighting, side paneling all blend well. Seats while appearing to be the same as on the DC-8 are by far more comfortable and wider. Fabrics are well chosen and passenger service and in-flight entertainment controls are very nicely placed in the right inner seat arms. A disappointing feature is the lack of literature in the seat pockets; no postcards, pamphlets, brochures or maps and nothing in Japanese. Hopefully this will be remedied as time permits.

The flight was delayed for 30 minutes due, I was informed, by connecting cargo from Eastern Canada. Another delay was an unserviceable hydraulic quantity transmitter which didn't show up until after leaving the gate. I was invited by the Purser to view the first-class lounge, an invitation gladly accepted, having heard so much about it. One gets the impression of grandeur even before setting foot on the spiral staircase, which is quite something when you consider that this is an aircraft. It is not possible to feel like anything other than a VIP as one walks into this penthouse in the sky. Plush red carpets, coloured lighting, the orange, red and brown shades that go to make up the executive seating layout, complemented by wall fixtures of brass steam-engine gauges, lamps and other adornments. An attractive rear area divider with black motif on glass panels accentuates the elegance that is the upper lounge.

Dinner was served in royal manner, as is the custom on our international routes; once again the cardinal rule of fine food— good service, presented in gourmet style was achieved. Listening to Channel 6 and savoring a most pleasant glass of Cointreau, it was already 22:30. How time flies (no pun intended). The view from the starboard cabin window is one of a peaceful sunset. This must be Japan's setting sun, the tone being that of near burgundy. At 01:08 overhead signs appear, indicating the near end of a wonderful adventure for those aboard and the beginning of a new chapter in the history of CP Air. It is all a far cry from those days of the late Grant McConachie's fish hauling. But then from humble beginnings a great country was made.

An 01:17 touchdown at Tokyo is the climax of the inaugural flight of the *Empress of Asia*. (Times are mine, so are not official.)

Landing the 747-200 was quite an experience for new pilots; one described it as not unlike "landing an apartment building." Rick Wiley explained why:

> The pilot flies the aircraft down an approach path of approximately 3 degrees to cross the threshold of the runway with the main gear wheels 50 feet above the ground. The radio altimeter was designed to read out the altitude below the main gear wheels and because the flight deck was 50 feet above the main gear on landing, it was the second officer's job to call out the aircraft altitudes in 10 foot increments, from 50 feet down to 10 feet. At 30 feet the pilot would raise the angle of the nose slightly to reduce the rate of descent and bring the engine thrust back to idle. As the speed reduced, the main gear settled onto the runway. At this point the pilot's eye-level is still the height of a four-storey building above the ground [hence the analogy of landing an apartment building]. The pilot would then lower the nose wheels onto the runway and select the engine thrust reversers to slow the aircraft. The aircraft's automatic braking system would apply the wheelbrakes after touchdown to slow the aircraft at a preset rate of deceleration.

Moe Morel did not review the in-flight movie because the service only would be offered the following month. CP Air began showing in-flight movies when its first 747-200s were equipped with cumbersome projectors installed in the centre of their ceiling. The machines had to be brought down and the passengers that sat under them relocated. For the stewardesses who climbed onto the seats to load the film reels, it meant embarrassment and frustration. The whole exercise, it is said, "was one of the biggest reasons that female flight attendants wanted to wear slacks."

The first North American airline to ply its passengers with alcohol was Pan American Airways, in the 1930s, by virtue of the fact that it operated exclusively overseas. Because the restrictions on serving alcohol varied with each state, other US airlines were reluctant to follow suit; in any case, to numb themselves against the noisy, lurching, freezing aircraft of the day, many passengers brought their own supply. Horror stories of drunken, frightened passengers mistaking the cabin door for the toilet were part of every early flight attendant's repertoire. To compete with European rivals, Pan Am's stewards offered a selection of wines on board the Martin M-130 flying boats, and after the war they continued to do so on the DC-4 landplanes. CPA started serving wine and liqueurs on its first international flights to Australia, but not until August 1966 did Canada's draconian liquor laws permit alcohol aboard all aircraft on the transcontinental route. At first it was served in miniature bottles; later, because of the economizing, the flight attendant poured each drink from a standard-size bottle. When the volume of passengers made this impractical, the miniatures were returned once more.

If in-flight entertainment contributed to maintaining the high load factors for the airline, it was what kept the parent company, Canadian Pacific Railway (Canadian Pacific Ltd. or CP Ltd. since 1971), happy. In 1973, CP Air had a system load factor of 62.5 percent, and significantly, the transcontinental route, in which CP Air competed with Air Canada, had a 65.4 percent load factor. The "thinnest" market was still the Pacific route to Australia, with a 44 percent load factor. Yet, despite the fact that the airline prided itself on its intercontinental routes across the Pacific and North Pole, a remarkable 71 percent of its total revenues was earned within Canada. The heavy growth up to 1973 in anticipation of the Boeing 747s was reflected in the total number of the staff, giving the airline 6,469 employees and a payroll of $65 million the following year. No longer just a "people mover," the airline was also a major national employer.

The following year, like all other industries that used gasoline, the airline found that its progress was checked. The Yom Kippur war of 1973 led to the decision by Arab oil-producing countries to increase the price of oil by 70 percent. The price of 1,000 US gallons of Jet A1 fuel in July that year was $150.08; by July 1974 it took a quantum leap to $335.70, and a year later to $429.41. For CP Air, problems came in battalions. There followed a sixty-eight-day machinists' strike (the first in the airline's history), the recession spiral and below-forecast traffic. Although the total number of passengers carried in 1974 had risen from 1.76 million to 2.28 million, the load factor had fallen to 61.4 percent from 1973's 62.5 percent.

What really hurt the airline wasn't planned in Tehran or Riyadh but in Ottawa. On August 2, 1974, transport minister Jean Marchand awarded Air Canada the bulk of the route allocations under the new US–Canada bilateral air agreement. Of the seventeen routes made available, the only one given to CP Air was Vancouver–Los Angeles. At the same time an American company, Western Air Lines, was allowed to compete with it out of Vancouver to Hawaii, and Wardair was given the green light to operate a weekly Boeing 747 charter service between Canada and Amsterdam, which would skim off the cream during the peak season.

On that black day, many at the airline's corporate offices in Vancouver bitterly remembered Prime Minister Trudeau's recent election speech: "This government will be a government of the whole country and not only of Ontario and Quebec." As soon as the allocations were announced, cabinet minister Ron Basford was sent out to explain the unfairness to reporters, saying: "The allocations have to be looked at in balance with the whole international point of view in which CP Air has recently been granted routes to Milan and Beijing. These are two important routes." A prominent senior British Columbia Liberal (who asked not to be named) put it succinctly: "An appalling decision of the federal government aimed directly and squarely at free enterprise in favour of the socialist airline and squarely at Vancouver in favour of Montreal." Summing up the feeling of the airline's employees, CP Air's marketing and sales vice-president, H.B. Renwick, said, "The word 'disappointed' is not really enough. It should have been 'aghast' or 'astonished.' It's just unbelievable to me, quite frankly."

Competition with Air Canada dominated the 1970s. A shift in focus to eastern Canada began with the construction of a $12 million hangar facility at Toronto in March 1975. Gerry E. Manning, vice-president of customer services, said that 40 percent of the cabin staff were men, and that company policy was also to have at least one attendant who spoke the destination language on international flights. A French-speaking attendant was available on all domestic flights and "after Bill 22, all CP Air literature to be used in the province of Quebec would also be in both official languages." The airline was making a concerted effort to improve on its customer services; china and silver were used even in economy class, and no frozen food was served anywhere.[20] Manning calculated that 20 percent of the airline's passenger revenue dollars went toward customer service. That was very close to the bone because "if I spend 23 percent, CP Air will have no profit at all."

The costs of running the airline continued to climb in 1975 because of the air traffic control strike by French-speaking employees and a strike in Mexico by local staff. Although an all-time high in operating revenue of $332 million was reached, for the first time in thirteen years, the airline recorded a deficit: a $6.4 million net loss. A glance at the balance sheet explained why:

1975 Revenues

International passengers	$156 million
Domestic passengers	$124 million
Cargo	$28 million
Mail	$10 million
Charters	$7 million
Incidentals	$7 million
Total	**$332 million**

1975 Expenses

Salaries and personal expenses	$130 million
Fuel	$69 million
Materials, supplies and passenger meals	$32 million
Depreciation	$26 million
Sales commissions	$17 million
Purchased services	$14 million
Interest on borrowings	$15 million
Advertising	$9 million
All other costs	$32 million
Total	**$344 million**

Within two years, CP Air's financial reserves had been devastated by the increases in fuel prices. In 1973, its fuel bill was $24 million; in 1975, it totalled $69 million. Although more miles were flown, of the $45 million increase, $29 million was directly attributable to higher fuel prices. The deficit was especially embarrassing for Gilmer, who was about to retire, because all other CP transportation branches reported substantial profits that year. Ian Sinclair had succeeded in bullying the remainder of the Canadian Pacific empire into paying its way. Although he loved nothing better than to chat with CP Air flight crew, Sinclair often said the airline had been running too long on a mixture of gasoline fumes and glamour. With the CP Air Operations Centre flowing with red ink, the telephone calls between Windsor Station and Vancouver must have become decidedly chilly.

Before fuel prices forced airlines to squeeze more passengers into their aircraft, the upstairs bar of the early 747s came equipped with a cocktail lounge, some with pianos. Rediscovering its railway origins, CP Air had one of its 747s fitted out in railway memorabilia with locomotive lamps amid the orange carpeting and upholstery.
CAA 989-71-176a

To reduce fuel expenses, Gilmer had one DC-8 taken out of service in September 1975 and once more asked the CTC for permission to raise fares. It wasn't bad management but simply the oil crisis that had made it too expensive to run an airline that wasn't subsidized by the government. Costs had outstripped income. Both Air Canada and CP Air had enjoyed a steady annual growth of about 12 percent each year until 1974, when aircraft fuel prices shot up by 80 percent and both were plagued by labour strikes. It was little consolation to CP Air that in 1975–76, its rival at Dorval suffered a net loss of $23 million. Still, travelling by air in Canada, when compared with the rise in the consumer price index, was a bargain, despite the massive fuel price increases.

CP Air Economy Fares from Vancouver
to Prince George, Toronto and Amsterdam

Year	Consumer Price Index	Prince George	Toronto	Amsterdam
1961	100.0	$35	$99	$386
1965	107.4	$35	$109	$343
1970	129.7	$39	$128	$374
1975	184.8	$51	$157	$428
1976	198.7	$57	$178	$428

The consumer price index had almost doubled between 1961 and 1976, yet the CP Air fares had gone up only by 55.6 percent in that time. The load factors for the year told the story better than any other statistics:

Route	Percentage Needed to Break Even	Actual Percentage	Result
Polar	63.5	68.4	Profit
Orient	31.6	49.8	Profit
Hawaii	67.1	72.3	Profit
South Pacific	41.9	52.5	Profit
Western Canada	56.8	59.8	Profit
Atlantic	64.2	54.7	Loss
California	80.4	64.2	Loss
Transcontinental	64.8	58.9	Loss
Latin America	54.6	48.7	Loss
East Canada/Mexico	86.6	61.8	Loss
West Canada/Mexico	93.0	68.4	Loss

Besides the ever-increasing prices of landing fees and fuel, staff paycheques took a sizable chunk of the budget. The airline's 6,950 employees were paid a total of $119,648,000. There were 1,390 reservations clerks, 1,600 clerical and accounting staff, 1,790 maintenance staff, 45 in commissaries and 620 in the United States and overseas. The aircraft were flown by 570 pilots and 33 training captains. On any given day, 212 pilots and flight attendants were on crew layovers, at a cost of $4 million annually. In addition, there were 568 flight attendants based in

Vancouver, 222 in Toronto, 82 at Montreal, 6 at Lima, 19 at Hong Kong, and 6 in Amsterdam.

The fallout from the oil crisis only underlined to the airline's management that CP Air had to stand or fall on its own profitability. Those who expected the coffers in Windsor Station to support the airline indefinitely only had to ask where the fleet of great white Empress ships was now. Ian Sinclair might love the glamour of running an airline—he used to say that he didn't need a private corporate jet like other CEOs because he had twenty-five of them—but he also had a well-deserved reputation of taking no prisoners.

The standard remedy for a business in trouble is to implement a cost-cutting program. Gilmer put up for sale the three DC-8-40s that had Rolls-Royce Conway engines ("not quite as efficient as the Pratt & Whitneys we have on the Series 50s, and with many airports now demanding noise suppression they would be difficult to refit"). Other savings included delaying engine starts as long as possible, reducing cruising speeds while flight levels were flown as high as practicable and shutting down one or more engines immediately after landing. Instead of adding extra flights on high-density runs, first-class seating was removed on 737s to increase load factors. Services to Madrid and Tel Aviv were suspended (management finally acknowledged that no foreign airline could compete with El Al for traffic to Israel), two destinations in Mexico were dropped and frequencies to other cities were lowered. The Orient route could have been more profitable, but Japan refused to allow CP Air to operate more than three flights a week in the winter months.

On the point of retirement at the year's end, Jack Gilmer concluded, "in the final analysis, whatever we do, our bottom line depends on how the government treats us. Whatever happens . . . beyond doubt CP Air has shown the Canadian people the benefits of competition." On February 1, 1976, Ian Gray, the engineer who had once advised Grant McConachie against the Comets, became president of CP Air. In one of his first interviews Gray announced, "Our share of the market is now immensely worse. We believe we were shafted on the allocation of routes. We came out minus on it." He also reminded Ottawa, "If the government wants certain international routes flown for policy reasons, then it must subsidize them. There is no way that passengers flying on other routes should be asked to carry those losses." With the 25 percent capacity restriction in transcontinental flying, the major problem was

The caption on the archive print of this publicity photo reads: "'Dear Boss, sorry to report I missed the assignment of the new CP Air Boeing 747 *Empress of Italy* due to circumstances beyond the control of any young photographer.' The photos, taken at Vancouver International Airport recently, point out the problems of proper exposure, composure and focus facing photographers employing petite assistants." CAA 989-55-6

the turnaround requirement at points west of Edmonton or Calgary and east of Montreal. This meant "deadlegs" between cities at unacceptable hours. CP Air had to have the same licence as Air Canada in order to have the same transcontinental schedule flexibility. It was the original charter carrier in Canada, Gray maintained, and had formed Transpacific Tours in 1973 in partnership with a Japanese firm; "but it did not maintain its prime position and thus allowed Mr. Max Ward to develop a very fine charter operation, Wardair."

The charter market exerted a powerful pull on two groups who had time to spare: the students and the retired. "Snowbirds" jetted to Florida, university sophomores backpacked around Europe and "affinity" clubs sprang up everywhere to circumvent government regulations. Rather than begin its own charter company to take advantage of the boom, in 1973 Air Canada had embarked on an unsuccessful courtship of Wardair. In February 1977, as the recession hit, both major airlines introduced "Charter Class Canada" fares on their domestic routes by reserving a few seats on each flight. The public could fight over these seats, provided the trips were longer than seven hundred miles, the reservation was made sixty days in advance and the passenger remained at least ten days at the destination. Needless to say, this wasn't entirely successful; consumers demanded that all the seats on the aircraft be made available, prompting the president of the CTC, Edgar Benson, to conduct hearings on cheap fares. Advance-bookings charters

(ABC) encouraged Gray to develop CP Air's own charter division. In its first year, CP Air flew 960 inclusive tour charters (ITC) and 283 ABC. Later CP Air Holidays was established, concentrating on ABCs, especially to Britain, a profitable market still serviced only by Air Canada's scheduled flights.

Max Ward had never looked back after buying two 189-passenger Boeing 707s (christened *"Punch" Dickins* and *"Wop" May*) in 1968 for the Hawaii charter market. In 1973 he purchased a 455-passenger 747, and in 1976, with the purchase of a second 747, Wardair International Ltd. was incorporated. Now he had too many seats to fill. In May 1978, as the scheduled airlines brought in rock-bottom-fare programs like Air Canada's "Nighthawk" and CP Air's "Courier," the charter balloon burst in Canada and the regionals abandoned their large aircraft operations. Wardair recorded losses in 1978 and 1979 as it faced competition from Ontario Worldair, higher fuel prices and the grounding of two crucial DC-10s for twenty-five days after a Chicago crash. Max Ward campaigned to run multiple-stop charters within Canada but was so hampered by CTC restrictions that, as he once wrote, he wished he had been born American.

Whatever his woes, Ward never forgot his bush pilot origins or mentors. CP Air christened its aircraft "Empresses," but Ward took great pleasure in naming his white-and-blue fleet after Canadian Airways Ltd. pioneers like Herbert Hollick-Kenyon, Romeo Vachon, H.A. Oaks, Jack Moar—and Grant McConachie.

One of Ward's two 707s had been sold to him by Quebecair, which in the early 1970s was struggling to survive. In typical regional airline fashion, it bought time by latching onto a giant construction project—as Kitimat had saved PWA, so the James Bay Hydro project put Quebecair in the black as it shuttled workers back and forth. The 1974 boom in charter travel gave Quebecair some breathing space but as it tried to compete with Air Canada and CP Air, especially on the Quebec City–Montreal–Toronto triangle—even buying 737s for it—the airline remained financially shaky, and the government of Quebec took it over in 1981. With the rise of Quebec nationalism, the airline assumed greater emotional importance than its existence warranted. It did not help that Montreal's Nordair, considered "anglophone," had been given routes that shut Quebecair out of the province's north. The provincial governments in both Toronto and Quebec City expressed interest in merging Quebecair with Nordair or Air Ontario, thus creating one strong regional airline that could compete with the two main carriers or the encroachments of thriving PWA.

Because of its strategic location, at various times Nordair attracted a number of suitors, including Air Canada, CP Air and Quebecair. Like PWA, the airline made money where it could, taking Muslim pilgrims to the Hadj and El Al passengers from Montreal to Miami. Nordair had gone public in 1972, and in 1978 Air Canada bought 86 percent of its stock with the idea of making Nordair its charter operator and gradually merging it, but political opposition in the House of Commons prevented this.

From CP Air's point of view—conveniently forgetting its own origins—the freedom given to regional airlines during the Trudeau years was nothing short of ridiculous. Set up after the war to service bush communities with DC-3s and Otters, encouraged by Jack Pickersgill, the regionals began operating jet flights with DC-8s and Boeing 707s far beyond the original intent: Winnipeg–Toronto (Transair), charters from Toronto to Belfast (Quebecair), charters to Cuba, London and Paris (Nordair) or Vancouver–Seattle (PWA).[21] Transport minister Jean Marchand had allowed the bush airlines to continue indulging in charters as a means of paying for unprofitable routes at home, but now they threatened to become competing transcontinental carriers. CP Air wasn't the only worried airline; in 1977 when PWA bought out Transair especially for its valuable Winnipeg–Toronto route, Air Canada didn't need to compete with yet another transcontinental airline and struck a deal with Rhys Eyton. The Toronto service was dropped for route authorities between Calgary and Winnipeg through Regina and Saskatoon. It would be three years before Eyton was presented with another opportunity to expand to the golden gateway of Toronto. Then he chose a route between Calgary through Brandon to Toronto, one that had not been considered viable by CP Air. Russ Baker's dream of flying between British Columbia and Ontario had come true.

By 1980, the CP Air DC-8s were between fifteen and twenty years old. They had given the airline good service, but now rising fuel costs and noise restrictions at airports were rendering them unpopular. The previous year, Ian Gray had ordered four Boeing 767-200s with an option to purchase four more, all to be delivered between December 1984 and June 1985. He told the press that the airline required a minimum of eight new long-range aircraft. Now that he had the power to

The CP Air flight kitchens in 1975—the unsung heroes of commercial aviation. In-flight catering is a science posing as art. Food must be prepared in advance, remain unaffected by transportation, pressurization and reheating, contain nothing that would inadvertently affect the religious and dietary sensibilities of 380 people, or have bones or paprika to cause choking at 30,000 feet—and be enjoyed with limited elbow room and under the scrutiny of fellow passengers. Phew! CAA 989-134-32e

rationalize the airline's fleet, he decided on the Boeing 737 for domestic services and the McDonnell Douglas DC-10 on international routes.

Capt. Tony Archbold remembers the DC-10 as "a true pilot's aircraft, and everyone who has ever flown them loves their stability in virtually all conditions and the relative ease at which a smooth landing can be made. Many former DC-10 pilots who have gone on to the 747-400 will state that the DC-10 is a superior aircraft in turbulence with its smoother ride."

Like the DC-8s, the DC-10s could also be extended in size and range. Once more the requirements of the US military boosted commercial aircraft development when it asked aircraft manufacturers for a replacement for its now-ancient Boeing 707 tankers. Boeing showed scant interest in adapting its 747s, and since the Lockheed C-5s were no longer being made, Douglas came up with a version of its DC-10 called the KC-10A "Extender." It won the military contract, which allowed it to build heavier, longer-range DC-10 models. The Series 30 and 40 had more powerful engines and were almost twice as heavy, and to support the additional weight extra landing gear were fitted into the belly of the aircraft.

Before they could come into widespread use, the early DC-10s (the 10 Series) suffered a series of catastrophes connected with their new, forward-opening cargo doors, introduced to maximize the volume for the containerized cargo system. A faulty latch mechanism led to the DC-10 nickname, "The Douglas Death Cruiser." On March 3, 1974, a Turkish DC-10-10 crashed outside Paris as a result of depressurization in flight, caused by the blowing open of a cargo door, killing all 346 people on board. Five years later, on May 25, 1979, 279 passengers died aboard an American Airlines DC-10-10 when it lost an engine on takeoff from Chicago. It had literally come adrift from the airframe. The US Federal Aviation Authority grounded all DC-10s for thirty-six days until engine mountings had been replaced, but passengers were too frightened to fly anything with an engine in the tail. The aircraft's unpopularity was by then so bad that CP Air flight attendants were told to identify them on the public address system specifically as DC-10-30s, not 10-10s.

CP Air placed an order for four DC-10-30s and took delivery of the first two in March and April 1979, but they were immediately leased to Varig Airlines in Brazil until June 1980. As a result the first DC-10 to enter service with the airline was the third aircraft delivered, arriving in November. In 1981 and 1982 three more factory-ordered DC-10-30s entered service with CP Air, for a total of seven. An eighth was acquired from Singapore Airlines, which was divesting itself of DC-10s in 1982. In all the airline would acquire twelve DC-10s, leasing three from United Air Lines and obtaining four from Pakistan International Airlines in exchange for its 747s.[22]

Despite their earlier evil reputation, the DC-10s remained in service with Canadian Airlines twenty years later, and in the late 1990s many were still in use worldwide, with Northwest, Fedex and Continental Airlines adding them to their fleets. In 1998, total airframe hours on the Canadian DC-10s ranged from a high of 81,500 hours to a low of 66,000 hours, and they could have probably lasted another twenty years. The only situation that might have brought about their early retirement would have been a sudden sharp increase in the price of oil, for fuel efficiency was not one of the DC-10's strengths. With its payload capability of 90,000 pounds plus, a full passenger load could be carried in combination with considerable cargo; the longest route it was used on with Canadian Airlines was Toronto–Tokyo, although it was versatile enough to fly stage lengths such as Vancouver–Calgary–Toronto. On the longest flights an extra or augment pilot was assigned to the crew, thus providing four pilots, so that crew members could be relieved of duty in rotation. The off-duty time was usually spent in the crew bunk; a role that DC-10 crews call "dozing for dollars."

In 1983, in a cross-leasing arrangement between CP Air and United Airlines, each exchanged three DC-10s. United had been awarded a new route between Seattle and Hong Kong and needed an aircraft with considerably more range than its DC-10-10, whereas CP Air was wasting its long-range DC-10-30s on shorter routes such as the "transcon." The authorities in both countries agreed that the aircraft would retain their original registrations, which meant that all of CP Air's DC-10 crews had to be issued with the appropriate American commercial licences, leading to a bureaucratic nightmare for the airline's administrative staff. The leased aircraft were returned to their original operators in 1987, with some complications. United had modified the cabin interiors of the CP Air aircraft and added an additional fuel tank in the aft cargo compartment in order to enable a Seattle–Hong Kong flight. These modifications were extended to two other CP Air aircraft; the five modified aircraft became the DC-10-30ER model, with a heavier gross takeoff weight, slightly more engine thrust and longer range. All five were still with the airline in 1998.

Gray was able to go on a shopping spree for the DC-10s because of Claude I. Taylor, Yves Pratte's replacement at Air Canada. For years CP Air had attacked Ottawa concerning Air Canada's special status. Some of the transport ministers, such as Jack Pickersgill, had sympathized, gradually decontrolling the skies overhead. But since C.D. Howe's day the federal government had been committed to protecting the government's investment in its airline and ensuring that there would never be an oversupply of carriers, as had happened with the two railways early in the century. Every Canadian Pacific Railway boss—Beatty, Coleman, Crump, Sinclair—had argued with Ottawa for a level playing field for CP Air, but ironically only when Air Canada's Claude Taylor informed

Otto Lang, the Liberal minister of transport, in 1978 that the government's child had grown up and could fly on its own, did the federal government start to cut the umbilical cord. After forty-one years, Ottawa began to extricate itself from the airline business through the Air Canada Act, which removed the special advantages (and burdens) that the national carrier had as an instrument of government policy. Air Canada had now to operate with "due regard to sound business principles and in particular the contemplation of profit." It was also to be regulated by the CTC, like every other carrier in the country.

Taylor admitted that he didn't do this out of altruism. Consumer pressure and the wave of deregulation sweeping the United States had played their part. "The public mood was demanding more competition," he later explained. "If I had opposed the Open Sky policy publicly, I would be a dead duck, PR-wise. I think the timing was just right, although I had a little trouble convincing the more socialistic members of my board."[23] On March 23, 1979, Otto Lang made the "Open Sky" proclamation that would usher in unfettered competition between Air Canada and CP Air on the mainline routes.

Taylor and Gray were very alike in temperament and came from similar backgrounds. The fifty-four-year-old Taylor had been born into a New Brunswick farming community. His first job as a teenager had been carrying water to the workers building the airports for the Commonwealth Air Training Plan. In thirty years, he had worked his way up the ladder from reservations clerk to president of Air Canada. At fifty-nine, the CP Air president was an aeronautical engineer with thirty-six years' experience in aviation. Gray combined what Ronald Keith called "a slide-rule mentality with an intense and sincere concern for the human side of the business." Driving a small red station wagon even as president, he flew economy class on his own airline and absolutely hated to be met at airports. Unlike their predecessors or the prima donnas that ran airlines in the United States, both men looked and behaved like friendly family bankers. At the dawn of the Open Sky era, their charges compared thus:[24]

	Aircraft	Number of seats	Employees	Revenues, 1978	Profits after taxes, 1978
CP Air	27	4,975	7,496	$465.8 million	$20.9 million
Air Canada	110	15,412	20,964	$1,322.6 million	$47.5 million

When he heard that CP Air was no longer to be confined to 25 percent of the cross-Canada market, Ian Gray didn't break out the champagne, nor did anyone do cartwheels on Grant McConachie Drive. If anything, the CP Air staff just knuckled down to work harder—after all, the airline had just lost its identity as an underdog. Gray admitted that "we're as happy as larks, as for the first time we've got a whack at fighting for as much of the market as we can get, but let's not kid ourselves. It's going to take time for us to rev. up to speed. Profits, not proliferation is the key to the future." Too much competition, Gray said, might have consequences similar to those that followed the building of the transcontinental railways.

If the CP Air president was cautious at the turn of events, Claude I. Taylor was positively beaming on the fortieth floor of Place Ville Marie. "I'll tell you what delights me the most," he laughed. "At last the crutch had been yanked out from under those scoundrels. For years CP Air has squeezed every drop of juice they could at playing the underdog. But from now on, they and the Canadian Chamber of Commerce can look for another scapegoat."

Gray did, however, concede that the bad days were finally over. His analysts were forecasting an overall market growth of 13 percent a year through to 1984 and a 43 percent share of the transcontinental market by then. Given the new freedoms, Gray warned that although the skies overhead were not going to be blanketed with CP Air aircraft, there were several routes that the airline wanted, especially in the Maritimes. On March 29, CP Air applied to the CTC to add Halifax to the "transcon" and on December 14, Victoria as well. Like Air Canada, the airline also experimented with cut-rate flights. That spring, Air Canada had kicked off its "seat sale" by slashing two thirds off its economy-flight tickets. This proved so popular that CP Air followed on June 1 with its "SkyBus/Aerobus," a "no-frills" weekly flight between Vancouver–Toronto, Calgary–Toronto, and Edmonton–Toronto commencing in July. There were no refunds for missed flights, limited baggage, no allowances, no newspapers, no meals or movies and a $20 service charge for changing your reservation. But the price was right: $94.50 for a one-way Toronto–Vancouver flight, compared with the $212 normal economy fare. The airline's switchboard was jammed with so many calls that by October the SkyBus flights were increased to three weekly for Vancouver–Toronto and Winnipeg–Montreal. The Skybus was so successful—by 1982 Skybus fares were available on most of CP Air's

Ian Alexander Gray became CP Air president on February 1, 1976. The airline, he said, was being "shafted by Ottawa on the allocation of routes" and would never be profitable if confined to 25 percent capacity in transcontinental flying.
CAA 988-182-13

regular flights—that Air Canada offered the same to its customers but with free nonalcoholic drinks. The discount concept became something of a Frankenstein's monster when CP Air, Air Canada and Wardair were operating in the red and the CTC began a detailed investigation into Skybus. The service ended on December 14, 1982 and was succeeded by CP Air Saver Class fares.

Ironically, it was the businessmen who now complained as their regular fares rose for flying on the same route as the Skybus clientele. To keep them as customers, on September 15, 1980, CP Air introduced "The Company Jet/Premiair." Its 737s flew four times a day on the Montreal–Toronto corridor with low-density seating and pre- and in-flight services dedicated to executives. Unfortunately, Air Canada had cornered the market with its hourly, no-reservation service called "Rapidair," and "The Company Jet" lasted only one year.

At almost the same time, to placate the business sector, the airline introduced its "Empress Service." Starting October 20 on transcontinental and certain international routes, this featured a separate on-board cabin, wider seating, a choice of entrees, complimentary liquor and movies. First class accommodation on the 747 on the Orient route was moved upstairs to accommodate twelve fully reclining sleeperettes, while the forward cabin was reconfigured to carry twenty-eight passengers in "Royal Canadian Business Class" for "a modest premium over the regular economy fare." On October 29, 1984, its equivalent in long-haul domestic flights, "Attaché Class," made its debut. Three specially modified 737s in an ivory and grey livery reconfigured with twelve seats in first class but fifty-four in business class made for an office in the sky.

Although D.A. Colussy was president and CEO only from 1982 to 1984, he realigned the airline's domestic network from the linear to the "hub and spoke" system, purchased EPA and Air Maritime, bought CP Hotels and introduced the "frequent flyer" program. CAA 990-26-8

With unaccustomed speed, the CTC granted CP Air approval to serve Halifax from Toronto on April 9, 1980. This time, a lobby group from the Maritimes sabotaged the airline's expansion. It pressured cabinet to overturn the decision and award it to Eastern Provincial Airlines (EPA), which locked into Air Canada's reservation system and was already flying between Montreal and Halifax. So desperate was CP Air to break into the Maritimes that when EPA dropped its Montreal–Halifax route in March 1981, CP Air grabbed it. EPA's president was Harry Steele, a former career naval officer who could never resist dealmaking. Described by *The Globe & Mail* as "an ornery, hard-driving manager who took a hard-nosed approach to labour relations," he once called striking pilots at EPA "a group of oversexed, overpaid bus drivers." Steele recognized that one of EPA's problems was that as a subsidiary of his Newfoundland Capital Corporation (NCC), it was expected to be anchored at Gander, on the very edge of the country. To tap into the busy Ontario market it had to move closer to central Canada, or at least get on the mainland. But since EPA had made a profit of $5 million in 1981–82, mainly on the Toronto route, Steele was busy expanding his NCC empire in other directions.

Ian Gray's last years at CP Air were marred by deficits. A second worldwide oil crisis hadn't finished with the travel industry yet, and airline traffic continued to plummet worldwide. By 1982, the year that Gray retired, the company's losses totalled $39 million. Attempting to "better structure its financial position," in November 1982, four million common shares were sold to CP Ltd., raising $20 million. A major recapitalization program in January 1983 improved the company's debt–equity ratio. Twenty-one aircraft with a net book value of $501.6

million were sold to CP Ltd. and as part of the agreement CP Air assumed debt and loan obligations of $262.4 million associated with the aircraft. Daniel Colussy, the company's new president, put into effect a cost-cutting program so that between 1982 and 1984 the number of employees decreased from 8,860 to 7,207. Then, in September 1983, the airline altered its fleet plans by cancelling the 767s and ordering ten 737-300s for delivery by April 1985.

The choice of aircraft was a clue to Colussy's strategy. In 1983, he imported the "hub and spoke" system to Canada from the United States. Jimmy Carter's presidency and Congress's passing of the Airlines Deregulation Act in 1978 were some of the main reasons why US airlines like Delta (from Carter's home state of Georgia) and American Airlines fortified their bases into "hubs"—in this case Atlanta and Dallas–Fort Worth. The "hub and spoke" strategy replaced the old linear-route pattern by ending the crisscrossing of routes to funnel all flights into a single airport, thus allowing home airlines to retain their customers as long as possible. In the words of one PWA executive, "you have to ensure that once you get a passenger on your plane, he stays on your plane." By controlling most flights in and out of a big city, airlines can charge a premium for fares and fight any incursions from competitors by flooding contested routes with frequent flights.

To be successful, the science of "hub and spokes" depended on high-speed computerized reservation systems like Apollo, which was marketed in Canada by Gemini and equally owned by Air Canada, PWA and Covia, a travel distribution jointly owned by United Airlines and American Airlines.[25] These systems locked 3,500 travel agents around the country into making customer reservations not only on aircraft but also for hotels and car rentals. By storing in their memory banks details such as the numbers of business and leisure travellers, air-mile totals, smoking or nonsmoking and meal and seat preferences, the computers could analyze historical travel patterns on individual routes and determine how many seats to sell and at what fare level. They also provided the hub airlines' operations centres with precise timing to deal with the increased frequency in flights by scheduling incoming aircraft from the "spoke" airports up to the "hub's" gates at exact intervals, giving the passengers and their baggage adequate time to transfer to outgoing flights on the same airline. Known as "spaghetti junctions," "blocked drains" (especially when airports were closed down due to weather) or

Affectionately called the "Douglas Diesel" by younger pilots, the DC-10-30, shown here in later livery, delivered to CP Air in 1979, is known to be a true pilot's aircraft. It had a very strong airframe, was extremely stable and, unless there was a sharp increase in oil prices, could remain in airline service forever. CAA (UNN)

"holding tanks," Atlanta and Dallas–Fort Worth are the focal point of their airlines' vast regional webs. Delta ensured its supremacy in Atlanta by appointing Henry A. Biedenharn, the former CEO of Coca Cola, and Andrew Young, the city's ex-mayor, to its board. So successful was Delta's operation of its hub that it was said no one knew where they were going to go when they died, but sure as anything, to get there they would have to change in Atlanta.

Unfortunately, having traded away its domestic routes to TCA and PWA in the 1960s, CP Air had retained only its transcontinental service and a few British Columbia destinations. Colussy, seeing what the lack of a domestic network had done to Pan American Airways, initiated a program that was to dominate CP Air for the next four years. In a complete reversal of the international expansion of the McConachie–Gilmer years, under Colussy and his successor, Don Carty, the airline single-mindedly pursued the acquisition of a domestic network to feed into its hubs at Toronto and Vancouver. Both presidents knew that if CP Air did not do so, cash-rich, aggressive PWA, which had already made commercial arrangements with Quebecair, would.

In Ottawa there were those who looked with some doubt on the advantages of deregulation, fearing that it might lead to greater concentration of the airline industry and less competition and higher fares for consumers. Although in 1982 a House of Commons standing committee on transport had recommended a "go slow" approach—continued regulation of the transportation industry, allowing for some competition—none of the airlines were fooled. The aviation industry knows no political boundaries, and deregulation was coming to Canada whether the federal government liked it or not. Air Canada, CP Air and PWA began hunting for regionals to buy or to ally themselves with.

Colussy began with overtures to Nordair and EPA, initially getting them to integrate their timetables and reservation codes with his own. He signed a co-operative agreement with EPA on April 24, 1983, luring it away from Air Canada at the Toronto, Ottawa, Montreal and Halifax airports and obtaining the contract to overhaul its 737s as well. On October 3, on the West Coast, CP Air entered into a commercial arrangement with Jim Pattison, the owner of Air BC, to synchronize its Dash 7 flights from Victoria, Comox, Kamloops, Kelowna, Seattle, Penticton and Cranbrook to feed into its Vancouver "hub." Finally, on May 10, 1984, the minister of transport grasped the nettle and issued a policy statement allowing carriers to compete on routes anywhere in Canada, rather than restricting them to specific geographic regions.

When building the transcontinental railway a hundred years before, the CPR had discovered that west of Calgary, because of the steeply graded lines through the Rockies, it could not equip its trains with dining cars. To cater to the healthy appetites of Victorian travellers, the railway erected a series of palatial hotels at Field, Glacier and North Bend, where its passengers were sumptuously fed and rested. Later, more hotels were built at the resorts of Banff and Lake Louise and then opposite the railway termini in Calgary, Ottawa, Toronto and Quebec City. The imposing architecture of the "chateaux" became, with the Mounties, part of Canada's romantic image. But the decline in passenger traffic on the CPR in the 1970s meant that these Gothic piles had recorded a deficit of $1.4 million in 1983 and now could be filled with airline passengers. Thus, on December 1, 1983, for $125 million, the ownership or lease of twenty-two CP hotels was transferred from Canadian Pacific Enterprises to CP Air as an autonomous division. The company now controlled seventeen hotels in Canada, three in West Germany and two in Israel.

In July 1984 another US import, the "frequent flyer" program, was introduced into Canada. It rewarded passengers' loyalty by providing free or reduced-cost travel if sufficient air mileage had been accumulated with the airline. The CP Air "Travel Bonus" program was so successful that Air Canada immediately followed suit with its own. Both airlines had seen their customers turn to US carriers to take advantage of the lower fares that deregulation had brought about.

The first step in the strategy to create a regional network came sooner than expected. On August 31, EPA moved to Halifax and was plagued by labour disputes. It was now in deficit, and Harry Steele—who once said "No one has ever made any money on a sustained basis in Canada operating an airline" sold it to CP Air for $20 million. The sale included EPA's subsidiary, Air Maritime, the first airline in Canada to ban smoking on board its flights. Most passengers would not have noticed the changeover because the 737s kept their EPA colours and Harry Steele remained as the airline's CEO. The more observant might have noticed the CP Air stickers on the fuselage and that some of the aircraft were being called Empresses, but complete integration was not expected for another three years, when all the planes were to be repainted in CP Air colours.

Colussy resigned as president in December to begin his own business, although he retained his position as CP Air chairman for a year after. The board of directors interviewed several candidates for president, including some from Air Canada. They made the most suitable choice: Donald J. Carty, a Canadian who had returned a creaky giant, American Airlines, to profitability by implementing cost-saving measures such as slashing jobs, negotiating wage controls and buying the more efficient aircraft.

Carty was appointed president and CEO of CP Air in March 1985 and took the helm on the 11th. Born in Toronto on July 23, 1946, and educated in elementary and high schools in Montreal, Carty graduated from Queen's University and completed his formal education at Harvard Graduate School of Business Administration. He worked briefly for CP Ltd. and Air Canada, becoming general manager of part of Celanese Canada, and since 1979 had held a number of positions in American Airlines, including operations research, information, financial planning and restructuring the airline. Carty had also been the chief

President and CEO in March 1985, Canadian Donald J. Carty was head-hunted from American Airlines. He returned to them two years later and was offered the crown at Air Canada in 1992 but turned them down to succeed Robert Crandall at AMR in 1998. CAA 988-269-19

financial officer of Americana Hotels, which at that time was a subsidiary of American Airlines. As the senior vice-president and controller, Carty was available when CP Air headhunted him because the president and CEO positions were then occupied by the flamboyant Robert Crandall, only eleven years his senior.

Carty arrived in Vancouver in time for CP Air's Annual General Meeting, pleased that thanks to Colussy, the airline could report a modest profit in 1984 of $5.6 million after a loss of $77 million the three years before. But hardly had Carty settled in when on May 10, 1985, CP Air took a direct hit from its old nemesis—the federal government. Transport minister Don Mazankowski announced that Wardair, then about to start its first scheduled services to Puerto Rico, was to be given permission to serve London (Gatwick) and Paris as Canada's second scheduled airline to Britain and France. The British capital had been sought by Canadian Pacific as far back as 1949, and now not only was it blocked out of both London airports (Heathrow was where Air Canada landed), but also Manchester and Prestwick, because Wardair had been given rights to those as well.

Max Ward had almost lost his company in the recession of 1981–82 and had clung on through a bitter strike by his stewardesses until the era of deregulation. In 1984, Wardair was firmly in the black, with profits of $9.7 million on revenues of $410 million. He had done this by offering hundreds of novice Canadian flyers those extra touches that Air Canada and CP Air provided only in first class: steaks cooked to individual taste and served on Royal Doulton china. High public praise, the former bush pilot discovered, carried powerful political clout in Ottawa. To bureaucrats whose idea of living dangerously was choosing bran muffins over doughnuts in the government cafeteria, Ward was revered as a romantic throwback, a pioneering hero who, it was said, still answered his own phone. (And how many airlines still bore the name of the founder?)

"The government says it wants competition and then grants a licence to an airline limited in scope." CP Air's senior vice-president of projects and planning, Ted Shetzen, wondered whether Wardair had the fleet to fulfill the responsibilities of a scheduled service. An Air Canada spokesman put in better: "Wardair now flies the 'cream' routes. It's a lot harder to fill a plane and make money on runs to Frobisher Bay." Even Ward admitted that with deregulation, "You have to be careful what you go for, because you'll get it."

Air Canada had come off badly under deregulation. The former monopoly carrier had expected inroads into its mainline routes from CP Air, but now third- and fourth-ranked carriers like Wardair and PWA were also entering its market. The government had been firm in its instructions to the airline not to fight the encroachments, and indeed, after May 1984 Air Canada, with admirable restraint, did not file a single intervention in route proceedings before the CTC. The fear in Montreal was that CP Air would merge with PWA—a combination that would threaten the viability of the government's airline, especially when it tried to appear attractive to potential investors.

"Freedom to Move—A Framework for Transportation Reform" was Mazankowski's opening salvo to deregulate transportation in Canada. Released on July 15, 1985, the paper completed what Pickersgill's National Transportation Act had begun in 1967. With the passage of new legislation, it would no longer be necessary for a carrier to establish that its service was required by "public convenience and necessity." As long as an airline was "fit, willing and able" (had adequate liability insurance coverage, complied with the Department of Transport safety regulations and was Canadian), it was legal. The bill allowed for the establishment of a new, smaller National Transportation Agency (NTA) to replace the CTC and, with some restrictions concerning foreign ownership—a phrase that would grow in importance—was given royal assent on August 28, 1987.

As EPA had been Colussy's goal, so Nordair was Carty's object of desire. The Montreal-based airline was everything a regional airline should be: employee-oriented, competitive and with a profitable route structure. In the same province, Quebecair was suffering and the

The great-grandson of the Convair 240, the Allison-powered 580 was the last hurrah of pre-STOL turboprops. It was also Nordair Metro's last aircraft, bought in 1985 for services between Quebec City, Montreal, Ottawa, Bagotville, Gaspé, Iles de la Madeleine and Sept-Iles. CAA 988-222-81

provincial government in Quebec City continued trying to merge it with Nordair. By 1984, Air Canada had reluctantly divested itself of Nordair, and 65 percent of it was now owned by 680 of the airline's employees, who invested $2 million in it, and Innocan, a venture capital company, while the Quebec government took 35 percent. On September 20, 1985, the Quebec government offered to buy the other 65 percent toward a merger. The Innocan investors and Nordair staff did not want their livelihood put in the hands of what they considered a politically managed Quebecair. The result was on November 8, Carty shared a press conference with Jean Douville, the president of Nordair, announcing that they had sold their shares to CP Air for $17 million.

Public reaction, especially in recession-hit Montreal, was highly favourable to the merger. Aviation analyst Steve Garmaise, of the stock brokerage firm of Wood Gundy, said: "CP Air was strong in the East and West with nothing in the centre. Nordair had carved out a niche for itself in Ontario and Quebec, so the merger is a natural." Carty was pleased because the combined schedule of both carriers greatly enhanced CP Air's presence in the "Golden Triangle" of Montreal, Toronto and Ottawa (sometimes called the Bermuda Triangle because of the many small airlines that have disappeared in it), which was then Air Canada's home turf. This left 35 percent of the airline still with the Quebec government, which did not want to relinquish control to a company from British Columbia.

In December, Nordair Metro was formed by Nordair and some Quebec investors as a commuter airline. It operated Convair 580s between Quebec City, Montreal, Ottawa, Bagotville and Gaspé, Iles de la Madeleine and Sept-Iles. Although Nordair Metro's parent, Nordair, was gradually being integrated into CP Air, the struggle over Quebecair would not culminate until July 31, 1986. Following a change of government in Quebec City, Nordair Metro was allowed to buy Quebecair for $21 million. Quebecair's four 737s (two of which had been leased to Pan American Airways since 1983) were sold to CP Air for $50 million, and Nordair absorbed the airline's debt of $64 million. The Quebec government accepted defeat and gave up its Nordair shares to CP Air for $13.5 million.

Strangely, although Nordair was swallowed up by January 1987, Quebecair lived on as a holding company—as part of CP Air and later Canadian Airlines, its livery and new name reflected the overall change. In the new Canadian colours, the name was now "Inter-Canadian"; the last "a" in the name was replaced by ">" to make the word bilingual. Nordair Metro, the last vestige of the airline that could be traced to Boreal Airways, disappeared into the holding company Quebec Aviation, operating under the name of Inter-Canadien. The Convair 580s were traded in for Fokker F.28s and painted in Canadian Airlines colours.

In a major policy speech to the Vancouver Board of Trade on October 16, Carty announced that the airline's wide-body fleet was to be expanded from eight to twelve DC-10s with the purchase of four Model 30s from Page Avjet Corporation, Orlando, Florida. The last of the Boeing 747-200s were to be withdrawn from service on October 25. This fleet standardization would enable CP Air "to achieve numerous benefits by operating only one type of wide body, including a 10 percent increase in aircraft utilization, improved flight training, reduced inventories and lower expenses in maintenance." The company was also going to buy three additional Boeing 737-200s and negotiated with the GPA Group of Shannon, Ireland, to exchange its remaining 737-300s for 737-200s.

Financially, the year 1985 ended badly for Carty because only CP Air's hotel division made any profit. Although its operating income of $34.7 million was one of the largest in its history, the airline posted a net loss of $8.9 million on revenues of $1,395.3 million, compared with a net income in the previous year of $13.7 million on revenues of $1,215.7 million. The weak Canadian dollar, inflation and continued downward pressure on fares were blamed. The airline now employed 8,500 staff, of whom more than 70 percent were covered by collective agreements, which had been ratified that year.

CP Air was working feverishly to reinvent itself, and Carty wanted his employees to be the first to hear that it was reverting to its old name. On December 17 he used the Christmas issue of *CP News* to announce the phasing out of the orange livery, before it was made public in January. He was careful to pay tribute to Buck Crump's choice of corporate makeover in 1968:

> Orange was no doubt the right colour for the 1960s when it was introduced. It was appropriate as an agent of change at that time and it served its purpose well. But as experts say, a corporation's wardrobe retains its luster for only 15 to 20 years, and after 18 years in orange it was felt that Canadian Pacific Air Lines should reposition itself in this critical time of its development. With the acquisition of the Nordair and the EPA 737s and the repainting of the Pakistani DC-10s, it is a cost effective time to change . . . We are proud of being a Canadian airline and it makes sense to tell the people of the countries we serve who we are.[26]

To formally introduce the new name—Canadian Pacific Air Lines (CPAL)—and new colour scheme on January 12, 1986, one of the former Pakistan International Airlines DC-10-30s was registered as C-FCRE and christened *Empress of Canada*.[27] Its lower fuselage and tail was painted in "Pacific Blue" and the upper body in "Sky White," both separated by a stripe of "Corporation Red"—to underline the company's aggressiveness, the president said. The orange of the elongated "Multimark" gave way to corporate grey and although the triangle and semicircle were retained, four pinstripes signified speed and five rectangles symbolized the five continents that the airline served. On the upper body, forward of the wings, each aircraft carried the name "Canadian Pacific" on one side and "Canadien Pacifique" on the other.

This 1985 uniform was the personal favourite of many flight attendants. Designed by Hugh Garber, in keeping with the company's "business" look, the colours are grey, rose and burgundy. CAA 989-71-112b

Wardair too was undergoing a transformation. After years of drinking from the cup of sorrows, Max was getting everything he wished for. On May 4, 1986, Wardair was given permission to operate scheduled domestic flights between Toronto–Calgary–Edmonton and Vancouver, something he had put on his "wish list" since 1978. Passengers could not help notice that the Wardair DC-10-30s were now configured into first and economy classes—like every other airline.

Colussy and Carty had accomplished what they set out to do. The major regionals in eastern Canada now fed into CPAL's Montreal and Toronto hubs, and through its expansion and alliances, the Vancouver airline was the largest it had ever been. The service to New Zealand had been restarted after a lapse of sixteen years, and there were nonstops to Hong Kong and new routes planned to Brazil and Singapore. On April 29, 1986, CPAL introduced the first nonstop between Vancouver and Shanghai, and it was to extend this to Beijing soon. The fleet standardization that Gray had planned was almost completed.

Was it a Pyrrhic victory? Unlike airlines in the United States, Canada's operated on an east–west axis in a very narrow linear market, which was less suited to a hub system. There were only two urban areas—Toronto–Hamilton and Montreal–Ottawa–Quebec City—that resembled anything like the American idea of "hubs," and both were within driving distance of each other. Toronto was too far south to become a hub for Ontario, and without a high-speed train link between its Mirabel and Dorval airports Montreal was handicapped. In the United States, airlines maintained several secondary hubs that they could pull out of or retreat to as the market dictated; American Airlines, for example, would drop its hubs at Raleigh–Durham, San Jose and

Hostesses, winged waitresses or guardian angels—the lot of a flight attendant is hard enough without having to wear uniforms that are sex-oriented or at worst pose a serious safety problem. Here Debbie Harland models the airline's sensible, cheerful approach. CAA 989-71-112a

Nashville but strengthen Miami. Northwest Airlines retreated from Washington (National) and shifted its resources to Detroit and Minneapolis. Airlines in Canada had no such options. With most of the Canadian population living along the US border, within reach of Buffalo or Seattle, airlines in Canada did not have that luxury.

Besides, the country was too underpopulated to support more than two transcontinental carriers, one of which was a crown corporation—a notion inconceivable to the Americans. Air transport was most needed in the bush, but as James Richardson discovered, the North's scattered communities could never hope to generate enough traffic. Finally, except for the provincial governments and Canadian Pacific, no Canadian entrepreneurs had the capital to sustain airlines in economic downturns.

Colussy and Carty staked much on a strategy that made sense for CP Air at the time: they acquired or allied it with regionals that served a similar purpose. But as the CPR discovered in 1942, throwing together diverse aircraft fleets and discontented employees who thought their pensions and seniority were in jeopardy and then hoping for harmony was asking for trouble.

Whatever the future labour problems might be, more tangible was CPAL's debt, which in 1986 had ballooned to $600 million, something that the parent company, CP Ltd., viewed with trepidation. It had received no cash flow from the airline in the 1980s, and Sinclair's creation Canadian Pacific Enterprises was itself in poor financial health. He had retired at the end of 1984 and despite what he said about the

The Boeing 737-300 was a stretch version of the earlier 200, allowing for twenty additional passengers. CP Air took delivery of six in 1985 before deciding to standardize on the 200 series and selling them off within a year. The "Attaché" livery was ivory and grey. CAA 988-222-52a

airline, would never have considered selling off the jewel in CP's crown. Perhaps it was the memory of how he had first come to the attention of Canadian Pacific's directors after his quick-witted defence of CPA in 1958. The present chairman, Fred Burbidge, and president and CEO, William Stinson, had no such qualms. CP Ltd. (combined with CP Enterprises in 1985) had lost $275 million in the first six months of 1986—the first such loss in 107 years. Stinson tried to recoup some ready cash on September 30, 1986, by selling the railway's oldest sub-sidiary, money-losing Cominco, for $397 million in cash and a total price of $472 million. Bits of the Canadian Pacific empire then fell off as quickly as Sinclair had bought them: CP Trucks, the flight kitchens, Maple Leaf Mills, the insurance companies and most of the Bermuda-registered CP Ships. Stinson explained the dismantling by saying he was laying the foundations for a new Canadian Pacific. By 1988, the company that had once toppled a government was investing in school buses. Watching it all, the last mogul of Windsor Station, Bill Sinclair, concluded, "It's a lot easier to sell assets than it is to build or buy them. Burbidge and Stinson never had any identification with . . . mining or air."[28]

The CP Ltd. portfolio managers watched as Don Carty pursued his aggressive growth strategy through expansion into eastern Canada, and the question they must have asked each other was, Will it work? Whoever ran it, under whatever colours or name, the airline had a record of poor financial performance, and whatever Carty did in the jungle of deregulation was risky. As with Cominco, selling CPAL would generate cash for specific investments.

Another alternative was to take CPAL public and sell off the 44 million shares that CP Ltd. owned in it. The earliest this could be done, because of the Nordair complications, was thought to be March 1989, when their market value was estimated at between $6 per share and $9 per share, which would net CP Ltd. between $255 million and $390 million. Either way, CP Ltd. was preparing to leave the airline business, and Stinson contacted Rhys Eyton.

Even before Quebec and the Maritimes purchases were sown up, Carty had been negotiating with the Deluce family of Timmins, Ontario. They owned Austin Airways, a small provincial feeder airline that could be brought down to Toronto and integrated into its hub-and-spoke system, thus completing CPAL's domestic network from coast to coast. On October 22, 1986, he performed his magic and shook hands with the Deluce family over a merger before flying off to Vancouver the following morning. Then, as he remembers, "On Monday and Tuesday they didn't return our calls and we knew something was wrong." The Deluces had also been talking to Air Canada's president, Pierre Jeanniot, and on October 30 announced that they had sold to him for $40 million. Air Canada too wanted to move Austin Airways to its Toronto hub. CPAL threatened to sue, but events were moving too fast by then.

When he was president of CP Air, Colussy had broached the idea of a merger between his battered airline and overfed PWA, but Eyton had bigger fish to catch. The analogy was not far from the truth; Eyton fished as he did business—a fishing buddy and PWA board member remembers that fishing excursions with him were a competition for the largest catch—and Eyton usually won. PWA and Air Canada already shared an interest in Air Ontario, in which PWA held 24.5 percent, and in Time Air, in which PWA owned 44 percent. But the merger scenario was predicated on the immediate privatization of Air Canada, which would be followed by its absorption of PWA. This windfall would not only give PWA shareholders 20 percent of the new megacarrier's stock

The terminal at Hong Kong's Kai Tak airport was four years old when CPA started using DC–8s on the route. Historically, the Vancouver–Hong Kong flights had subsidized other, less lucrative destinations. CAA (UNN)

but make it the controlling shareholder of mighty Air Canada, a situation that Russ Baker, Gordon McGregor and C.D. Howe would never have dreamed possible.[29] Eyton also suggested that Air Canada concentrate on the international and national routes; PWA would be its regional carrier. For PWA, the summer of 1986 was high noon. Deregulation meant that if it didn't merge with or buy out other airlines, PWA's fate might be the same as that of the other regionals. There wasn't room in Canada for a third large airline. "We looked at all the airlines, including Wardair," said Murray Sigler, PWA's executive vice-president then, "but there was nothing to gain by buying Wardair."

Prime Minister Brian Mulroney, who had campaigned on the promise that Air Canada was not for sale, shunned all talk of the subject in the summer of 1986, and the deal to privatize the "People's Airline" looked likely to drag on into the following year. Not until April 1987 did the government announce its intention to issue Air Canada shares to the public, starting in 1988. That August, Rhys Eyton, now

CEO of PWA, crossed the Rubicon and engineered the sale and lease-back of sixteen of PWA's Boeing 737s, amassing a war chest of $250 million. If there was a single moment in Canadian aviation history when one man held its destiny in his hands, this was it.

At the annual general meeting of the Air Transport Association in November, it was rumoured that a major acquisition was about to occur. Was it to be Air Canada or Canadian Pacific Air Lines? When PWA divested itself of its Air Ontario shares, Air Canada knew the romance had ended and ran for shelter. On November 27, guessing that it would not be able to use Time Air as its "connector" anymore, Air Canada bought 100 percent of Air BC. It seemed that it had also been talking to Jim Pattison. The Empire had struck back.

Overleaf: In the grim month of November 1996, the employees had heard it all before: their president asking for wage cuts, concessions from government and shareholders, the closing down of routes . . . This time, Benson warned them, there was no Plan B. Rick Sloboda/Canadian Airlines

CHAPTER EIGHT

"Though He with Giants Fight"

On March 8, 1976, when Donald Watson, president of PWA, resigned suddenly, the shock waves reached from the Chateau Lacombe Hotel in Edmonton to a deserted Mexican beach. Roderick R. McDaniel, who had just been appointed chairman of PWA, cast around for a replacement. The criteria he set had certain provisos: "What makes some people really good at business, and what doesn't? You have to have a competitive spirit, intelligence, confidence, compassion, energy, discipline, foresight and a sense of humour." Rhys Eyton came to mind. Years later, McDaniel added, "Rhys had all these attributes, and more." At that moment, Eyton was sitting on a beach in Mexico, contemplating his future, when one of the officers called with the news.

There are two versions of how Rhys Eyton entered the airline business. John Condit has it that having obtained his chartered accountant (CA) designation in 1966, Eyton had planned to enter the University of British Columbia to complete a doctorate when his friend Robert Samis, son of PWA's board chairman, suggested instead that he take a job in the airline's accounting section. Another story is that as a chartered accountant with a private practice in the same building that housed PWA's head office, Eyton used to ride the elevator daily with Bruce Samis, his friend's father, who one day invited him to join the airline.[1] Whichever version is true, Eyton's change in career ensured that commercial aviation in Canada would never be the same.

Of Welsh stock, Rhys Tudor Eyton was born in Vancouver on September 23, 1935, and had a sheltered, middle-class upbringing.

Although his father was a General Motors executive, Rhys paid his way through university with summer jobs, working amid the filth and stench of canneries and whaling stations in northern British Columbia. "I learnt a lot about getting along with people," he recalls. "Some of those career whalers didn't have time for college boys." He attended UBC and the University of Western Ontario, graduating in 1958 with honours in business administration. Then followed a stint at London Life, and working toward his CA designation while at Heliwell MacLachlan, at that time the largest accounting firm in Vancouver. "I wanted the CA training, but I had pretty well determined that once through and with a little bit of experience, I would move on. I thought that maybe I would test my skills in other areas as well—general management had always interested me."

In 1967 Bruce Samis, the Vancouver investment dealer who had been responsible for keeping Russ Baker's dreams airborne, invited Eyton to join PWA as its treasury accountant. "Samis said he needed young men like me in the airline," recalls Eyton. "it was an interesting little company, and I was young enough to take a chance for a few years." The airline was twenty-one years old, still a quasi-bush operation. Its clientele sighed that PWA stood for "Please Wait Awhile" or, more savagely, "Pray While Aloft." But it had an annual revenue of $10 million and was poised to emerge from its chrysalis. "It was a company that could have gone either way. It was not financially strong and had no money to buy aircraft." The boom in oil exploration and drilling and CPA's Mackenzie routes, which Baker had exchanged with McConachie, meant that PWA would be able to expand away from the British Columbia coast, where CPA was, toward the North. But the airline was a risky investment, particularly after Baker's death. "You can't be an entrepreneur and avoid startup situations," Eyton admitted. Within nine months of joining, he was shipped off to Edmonton as assistant director of northern operations—the first time he had ever flown PWA. When in 1970 PWA experienced its first deficit in eight years ($250,000), Rhys implemented stringent cost controls, resuscitating it back to profitability.

He resigned a year later but, instead of accepting it, PWA gave him a special assignment: to turn its money-losing trucking division, Byer's Transport, around and sell it. Told that it was impossible, Eyton somehow managed to do so, selling to its employees. Modestly, he says that "We found a way to do it using tax advantages. It was a win–win situ-

Rhys Eyton's financial discipline had allowed PWA to buy Transair, CPA and Wardair in order to take Air Canada on in an equal footing. He might have succeeded had the Persian Gulf war not caused exorbitant fuel prices and resulted in empty aircraft worldwide. CAA (UNN)

ation. We got all our original investment and more." Later he would use the same techniques to sell the Transair helicopter subsidiary to its employees. He returned to the airline offices in Vancouver as executive vice-president in 1974, just as the Alberta government bought PWA. Rod McDaniel remembered the sudden resignations that the impending move from Vancouver to Calgary caused. "It was quite traumatic. This is a tough business, and you needed a first-class person to fill the role of president." Eyton barely had time to enjoy Don Watson's old office at Vancouver airport or his new home when the relocation to Calgary took place. His attitude was, "You go where the job is, and I can't think of anything tougher than trying to move a company when the politicians and the public are trying to stop it."

Once in Calgary, Eyton first rationalized the company's fleet, selling off the old Boeing 707s and Electras and standardizing on one type of aircraft, the Boeing 737, which was ideal for many of the short runways that PWA operated out of. His management style earned the attention of *The Financial Post* in October 1978: "Eyton runs the airline with business school preciseness . . . It is a case study in modern management methods." Eyton's first priority was to expand past the regional stage, to show Ottawa that PWA could be taken seriously and given transcontinental routes, as Air Canada and CP Air had. Buying Winnipeg-based Transair in 1978 and creating a prairie base to build on, Eyton then reached an agreement with Hal Cope to own a fraction of Time Air, another money-losing Manitoba carrier. Eyton then went on a carefully planned fleet-acquisition exercise, purchasing additional 737s, four Boeing 767s, a Boeing 767 simulator and support equipment. By 1981, revenues had

Mrs. A.J. O'Hanlan, formerly Jewel Butler, the airline's first flight attendant, returned to service to commemorate the fiftieth anniversary of flight attendants in Canada. Wearing an early uniform, she won a round of applause from passengers on the Edmonton–Whitehorse run when she distributed carnations with stewardess Lise Eneborn. CAA 989-71-180

risen to $300 million and stock was at a respectable $6.47 a share.

The oil recession of 1982 delayed his expansion plans, and options on two of the 767s were not taken up, but not for long. Attributing it to his CA training, Eyton displayed a real sense for what made profitability. In the first half of 1983, PWA was the only Canadian airline to make any money on scheduled flight operations. That year, as the Alberta government sold its interest in PWA, he masterminded one of the biggest privatization moves in Canadian history, limiting the individual shareholder to 4 percent of the stock, thereby discouraging future takeover attempts. His cost-cutting caused unrest among the PWA flight attendants, baggage handlers and maintenance crews, and the airline endured a four-month strike in late 1985. At one point, the union even picketed the Eyton family home in Calgary. But Rhys held firm, and the unions finally agreed to negotiate job-saving measures.[2]

Critics within the aviation industry sneered that Rhys Eyton lacked the bush pilot heritage of Russ Baker, Grant McConachie and Max Ward. Even Air Canada's Claude Taylor had worked his way up from the reservations desk. But by the end of 1985, Eyton's CA instincts had turned PWA's debt–equity ratio around, from 74–26 in 1982 to 22–78—the nirvana of CP Air, Wardair and Air Canada. In contrast to CP Air, PWA had no long-term debt, just under $300 million in shareholder's equity and an asset base of $400 million. Fred Wright, the president of Pemberton Securities, described the PWA president in glowing terms: "Rhys is at the leading edge of financing and the moves he makes are cloned all over the world."[3]

Perhaps it was the CA in Eyton that recognized the equity tied up in his Boeing 737s at a time when their value on world markets was high. In what the *Air Finance Journal* referred to as "a landmark deal presaging the future of airline equipment funding" in October 1986, PWA sold sixteen of its twenty Boeing 737s to the Irish group Guinness Peat Aviation (GPA) for a net cash gain of $255 million. PWA would also buy 25 percent of equity in GPA Jetprop, a GPA company formed to lease commuter aircraft and, two years later, 25 percent of GPA Airbus A-320 Ltd. The 737 aircraft were to be turned over between 1990 and 1995, but PWA leased them back from several banks until then. The $255 million, combined with cash on hand of $100 million, gave PWA $355 million and no debt. The sale/lease-back occurred at the best of all possible times, the US dollar was at its peak and all prices were in that currency. The deal set the rumour mills in Canada into full gear, grinding out speculation that the money had been earmarked by Eyton to buy up Air Canada after its privatization.

With a profit of $29.8 million, Eyton confidently continued on his blitzkrieg of the two major carriers' turf by buying 45 percent of Time Air and 25 percent of Air Ontario. In 1986, he was made president and CEO of the PWA Corporation and Murray Sigler became president and chief operating officer of the airline. Eyton's right-hand man, Harvard-educated Sigler was a Yellowknife lawyer who shared his boss's unobtrusive lifestyle and lived a block and half away from him in Calgary. Both, it was said, planned their headline-making deals on early-morning jogging runs. In its fortieth year, the airline that Russ Baker had begun for forest fire patrols was wealthy enough to consider purchasing Grant McConachie's Canadian Pacific Air Lines, something that both founders would never have thought plausible in their lifetimes. Eyton's thoughts about the purchase appear in the following document.

There are three main ways by which to value CPA:

Book Value

For the most part this is irrelevant. It is important to note that PWA had a market value of 80 percent of book value until the increased cash balance in 1986. This would suggest that CPA could be worth, say, 60 percent of its book value of $340 million, or about $200 million.

Net Asset Value (NAV)

The purchase price of most airline acquisitions in the United States have been stated as a percentage of their NAV. CPA currently has a NAV of about $450 million for a going concern. Most airlines are worth more dead than alive, but in CPA's case, liquidation to realize the extra value is impossible. NAV cannot be relied on for valuation here.

Packaging the Offer

The following points should be considered when drawing up the bid for acquisition:

1. CP Ltd. will need a visibly "good deal" to help minimize the perception that they lost a subsidiary to a much smaller competitor, or simply sold too cheap.

2. CP Ltd. may be interested in a large initial cash payment. In the recently completed Cominco deal, they received 84 percent of the purchase price in cash. On the other hand, the large cash inflow from the Cominco deal may take the pressure off CP Ltd. and they may now be more inclined to accept a note or preferred shares.

3. If they can't have a majority position, CP Ltd. wants out immediately or by a pre-defined schedule.

4. CP Ltd. may not want a "profit participation clause." If publicized, it may appear that they are banking on PWA to run CPA more effectively than they did—(This is speculation).

Utilizing these guidelines, I would suggest an offer (based on $250 million present value) of:

$150 million cash
$176 million note, due five years hence (non-interest bearing)
Total: $326 million

The published price of this deal would be $326 million or about $7.41/share. CPA (net of the hotels) would be selling for about 72 percent of net asset value. If a present value selling price of $250 million is agreeable to both parties, it probably makes even more sense to let CP Ltd. suggest the payment schedule which makes the most sense.

The major assumptions used in this analysis are as follows:

1. 1986 expected results are used as a "normal" year.

2. Three 737-200s are freed-up from domestic routes in each of 1987 and 1988.

3. Six wide-bodied aircraft replace 10 narrow-bodied on the transcontinental routes in 1989.

4. Present value purchase price of $250 million.

5. Equity injection from PWA in 1987 of $150 million.

6. Approximately 2,500 staff, 26 percent of CPA's workforce, can be laid off to receive a ratio of 130 staff/aircraft by 1989.

It can be easily argued that assumption (6) is not achievable. Subsequent sensitivities will deal with this possibility.

The assumptions used result in a return on investment of about 12 percent to PWA. This indicates a potential for quite good returns if more areas of creating value are found.

On November 30, Eyton and William Stinson were in Vancouver for the premier event in the football season, the Grey Cup game at BC Place Stadium. The odds-on favourite Edmonton Eskimos were to lose to the Hamilton Tiger Cats, but neither fan got to his seat. "We were too busy with the CPAL thing," Eyton remembers, and the deal was worked out before a television set. There were three meetings with Stinson. "It was difficult to find places . . . because the town was all booked up for the game. The real issue was price."

On December 2, 1986, PWA purchased CPAL for $300 million, assuming approximately $700 million in additional debt. By that single act, Eyton transformed a medium-sized regional airline into the twentieth-largest in the world, propelling it into the big leagues:

■ PWA now served 89 points on five continents.

■ Between Rome, its most easterly destination, and Hong Kong, the farthest west, the airline spanned 270 of the globe's 360 degrees latitude.

■ PWA crossed 19 of the world's 24 time zones.

■ Its employee ranks swelled from 2,700 to 13,000.

■ Its total revenue grew from $362 million to $1.87 billion.

■ PWA's 24 Boeing 737s were now added to CPAL's 55 Boeing 737s and 13 DC-10s—a total of 92 aircraft, compared with Air Canada's 101.

Up to the month before the purchase, Eyton played his cards close to his chest. PWA was looking at aligning itself, he told the *Calgary Herald*, with either Air Canada or CP Air, or perhaps Wardair, but didn't at the moment think of "going out and acquiring other carriers." On the day the news of the purchase broke, Eyton was typically unavailable for comment. Rather than basking in the media spotlight of one of the biggest corporate coups in Canadian aviation history, Eyton preferred company spokesman Jack Lawless to brief the media. "That's how quickly it came about—just in two weeks," Lawless explained to the press. "At present there were no plans to merge," he continued; Donald Carty would continue to run CPAL from Vancouver and Murray Sigler would head PWA from Calgary.

The deal still had to be ratified by the CTC, but no problem was seen there. In Ottawa transport minister John Crosbie and Don Mazankowski, who were pushing to have Air Canada privatized, told reporters that the takeover did not represent a threat to federal ambitions for a more competitive environment. Historically, Canadian Pacific Airlines (or CP Air or CPAL) had done better during the years the Progressive Conservatives were in power; notably, TCA's transcontinental monopoly was breached during the Diefenbaker era. Many Conservatives, particularly those in the western provinces, still harboured suspicions about C.D. Howe's government airline and voted for increased competition in air travel. But by the time deregulation had hit Canada, local members of Parliament were more concerned about the health of their local regional carriers and about getting cheaper airfares for their constituents than in championing the cause of what was considered a Vancouver-based airline.

In a move that Howe would have considered inexplicable, Liberal transport critic Andre Ouellet took Crosbie to task in the House, pressuring him to ensure that competition and reasonably priced air services would be maintained across Canada. The NDP transport critic also thundered about free enterprise, predicting bankruptcies for small carriers everywhere. Crosbie reminded Ouellet that it was the previous

Liberal government that had launched the country on the road to deregulation and that "anyone who feels the deal will not be in the public interest should take his concerns to the CTC."

Eyton confirmed that if PWA hadn't jumped, it would have been pushed. "I see nothing but 'upside' despite the fact that we've taken on a very large load of debt," he told the *Calgary Herald*. "With deregulation, we had to have the ability to compete toe-to-toe with Air Canada. In 1986, Air Canada's profit was a comfortable $40.4 million. The smaller carriers were being consumed by them or CP Air. We'd been chatting with the latter for ten years and the CP operation was the most natural fit." Recalling what had happened when the Alberta government had bought PWA, Eyton made it clear that "there will be no great shifts of people between Calgary and Vancouver." Besides, he had just built a home in Calgary, and that was where he and PWA were going to stay.

At Windsor Station, CP Ltd. was tight-lipped about the sale of what had been its crown jewel, but spokesman W.G. Murray issued a press release: "cutting a money-losing limb from the corporate tree would mean an extra $175 million for its treasury." "It was a good deal for both sides," wrote Tom Bradley, analyst for Richardson Greenshields. "The $300 million was a higher price than the company was worth to CP, but lower than it was worth to PWA. The company is worth more to PWA because CP doesn't have the ability to trim costs or rearrange routes." Most of the market also felt that it was a better deal for Canadian Pacific than for PWA. One observer put it this way: "When you take a good balance sheet (PWA's) and add debt to it, in an industry that is sensitive to interest rates and oil prices—for example, for every 1 cent per litre jump in the price of aviation fuel, PWA's bottom line changes by $2.2 million—conservative investors will sell their shares." Other investment houses claimed to have sufficient faith in Eyton and the good financial discipline that he had instilled in PWA.

Only the airline unions were nervous about the takeover. As a labour-intensive service industry, airlines are highly unionized, and duplication of services at airports used by both airlines could mean layoffs. Some of the unions wanted to know how, after holding firm on wage restraint, Eyton suddenly had $300 million with which to buy CPAL. PWA had 2,873 employees to CPAL's 8,578, and the International Association of Machinists, which represented 3,500 workers at CPAL and 600 at PWA, was "concerned at the harsh evidence that deregulation had brought about." So were the Brotherhood of Railway and

Airline Clerks and CALFAA, both of which knew that PWA had its share of labour problems.

The man who should have been most worried, wasn't. In a seemingly complacent mood, Max Ward commented, "One less airline doesn't really mean anything. There won't be a major change for Wardair. Air Canada will be competing with PWA/CPAL like mad, and so will we. We always have and we always will."

"I felt all along that it would be substantially better if we were merged," recalled Carty, who, according to the press, had brought Stinson and Eyton together. "You would have a carrier approaching the size of Air Canada, a true competitor." He predicted, "There won't be any massive changes, at least for a few months, as most of our routes in the West are profitable for both carriers." The weak link in the combined PWA/CPAL network was Ontario, where Air Canada had outmanoeuvred CPAL in purchasing Austin Airways as a "feeder." Carty did warn that "some rationalization" of routes and jobs might take place.

What was causing some "gnashing of teeth," he said, was the need to find a new name for CPAL, recently CP Air. Since it was no longer part of the Canadian Pacific conglomerate, PWA had until the next eighteen months to change it. "The new name will have to have international significance," said Eyton. Above all, it had to retain its national identity. Various combinations were tried; "Canadian International Airlines" (CIA) was rejected for obvious reasons. "Maybe we should run a contest, and knowing that eventually both airlines will come together, we have to have one name for both."

PWA waited until February 26, 1987, when the CTC had approved the purchase, to plan the unveiling of its new name and image. On March 24, after "an extensive testing process of image packaging" by OVE Design of Toronto, the theme "Our Spirit Takes Wing" was proclaimed. As dawn broke over Toronto's Pearson Airport, a PWA captain ceremoniously shook hands with a CPAL captain before a DC-10. In Halifax, Calgary, San Francisco and Montreal, employees of both airlines met to view videos of the name change. Beginning April 20, television commercials, billboards and newspaper advertisements announced the name "Canadian Airlines International Ltd. (CDN)" to the world. During the unveiling, Eyton explained that the new name and image of the airline "recognized its Canadian heritage and proud pasts of the founding airlines." On April 26, PWA ceased to exist, and

On March 24, 1987, the airline changed its name to Canadian Airlines International Ltd. The new logo symbolized "wings across five continents." The red wing, shown in April 1991 on the first A 320-211, denotes speed and motion; the pewter-coloured bars represent the five continents served. CAA (UNN)

all its flights were now designated "CP," with its 737s having "Canadi>n" painted onto them.

Stylized to include the logo, Canadian Airlines International Ltd. was now the same in both official languages, the logo symbolizing "wings across five continents," the red wing denoting speed and motion, while the pewter-coloured bars represented the five continents served by CDN. Eyton wrote in the special commemorative issue, "Our goal will be to strive for a debt/equity ratio of 1:1 by the end of the year and 1:3 by 1991." "It was a Herculean task to accomplish in just 45 days," remembered Albert Chappell, director of scheduling. "The cooperation and hard-paced work of the staff in payload control, maintenance and operations from both PWA and CDN was tremendous." On May 2, the inaugural flight to Rio de Janeiro and São Paulo was to carry the new image to Brazil.

Don Carty was not there to see it. Realizing that there could only be one chief, Carty quit the month before, along with four vice-presidents, and returned to American Airlines as senior vice-president. "Don isn't going away because he is mad at us," Eyton told the *Vancouver Sun* on March 3, 1987. "People reach crossroads and decide to move on." Carty also downplayed the controversy of moving the headquarters from Vancouver to Calgary. He was made chief financial officer of American Airlines' parent corporation, AMR, rising to executive vice-president in 1989. When Pierre Jeanniot left Air Canada abruptly in August 1991, its board turned to Carty for leadership. The recently privatized airline was demoralized with million-dollar losses and layoffs of 2,900 employees, and Claude Taylor hoped to parade the Canadian-born Carty before nervous investors at the next annual general meeting. On January 30, 1992, he gave them his answer: "I'm staying where I am."[4] Donald Carty's day would come on May 20, 1998, eleven years after leaving Vancouver, when he would have the job he had always wanted: succeeding Robert Crandall as CEO and president of AMR.

As for Air Canada, if truth be told, it was relieved. With one less player, the dogfight over Canada's skies would now be a lot easier. A bleeding CPAL might have been a dangerous adversary, and who knew if its parent CP Ltd. didn't want to throw the money from the Cominco sale into rebuilding it. The government airline began a media campaign to press Ottawa to take away CDN's overseas routes and form a single Canadian international carrier. But history wasn't on Air Canada's side. On August 28, 1987, Bill C-18, the National Transportation Act, was given royal assent, demonstrating the government's de facto recognition of deregulation in Canada. His airline facing privatization, Pierre Jeanniot had wanted to make it more seductive to investors by updating the fleet. He hoped that Ottawa would see the PWA/CPAL consolidation as a reason to absorb part of Air Canada's long-term debt, freeing up money for the purchase of fuel-efficient A-310s. When this showed no signs of occurring, he sold the Air Canada DC-9s to Citibank and then leased them back, raising enough capital to replace them and the less adaptable Boeing 727s with thirty-four Airbus A19s, which would begin delivery in 1989. There was no way CDN, with its narrow-bodied, non-hushkitted Boeing 737s, would compete against them.

In October the government redefined the international routes of both companies. Canadian Airlines was designated for Munich, Frankfurt, Denmark, Sweden, Norway, the USSR, Mexico, Central and South America (except Venezuela), Indonesia, Taiwan and Delhi or Calcutta. Air Canada was given Korea, Singapore, Malaysia and the Philippines, but shut out of Hong Kong and Tokyo, the cities it really desired.

With a buoyant world economy throughout 1987, domestic aircraft capacity in balance with traffic demand and a relatively weak Canadian dollar, the PWA Corporation achieved a healthy 8.4 percent operating margin. Income before the write-off of the airline integration costs was $58 million. This was helped in no small way when Air Canada ceased operations for nineteen days in December because of a machinists' strike. If its directors only knew that it would not see another profit for ten long years . . .

As was the case in the United States, the nation's skies had become a battleground in which the airlines, unlike other industries, were slow to reinvent themselves. In comparison with the airlines, domestic auto makers and telephone companies adapted to deregulation and incursions from foreign competitors with minimum difficulty. But North American air carriers continued on as they had since the 1950s, building up arsenals of new aircraft and expanding networks rather than economizing and paying down their debts. Commercial aviation, it has been said, is a testosterone-driven, ego-satisfying industry, and this was certainly so in the late 1980s. In response to the traffic growth in 1987–88, Air Canada, Wardair and PWA all ordered new fleets of aircraft. In April and May, CDN accepted delivery of the first of its four Boeing 767-300ER aircraft, and orders were placed for three Boeing 747-400s, and seventeen Airbus A320-200s, all to be delivered over the next six years.

Wardair was in the throes of transforming itself from a charter operator into a scheduled airline, with special focus on the domestic transcontinental market. It possessed the wrong equipment for domestic regional flying so in early 1987, Max Ward prepared to sell his DC-10s to British Caledonian and two of his Boeing 747s to GPA and to order twelve A-310s at a bargain $42 million each. His staff at first considered purchasing Boeing 767s and even put a down payment on them to reserve spots on the production line. After three days of hard bargaining in Toronto, Max sealed the deal for the Airbuses, seven of which were slated to be introduced on the "transcon" market the following summer. In October, he placed orders for twenty-four Fokker F100s and sixteen McDonnell Douglas MD-88s. Like steroids, the purchases were meant to make Wardair bigger and stronger. But also like steroids, they weakened the airline's long-term health. By December, Wardair's net loss was $21.6 million.

This additional capacity in competition with CDN and Air Canada

Maxwell W. Ward. In 1989 he acknowledged that "my era is gone but Rhys is the man to carry my legacy." Ward remained true to his bush pilot origins, naming his aircraft after Punch Dickins and Wop May. In turn, the first 747-400 that CDN bought was named for this great Canadian. CAA (UNN)

caused domestic yields to decline, resulting in a revenue loss for all three carriers. For CDN, the strong traffic demand in 1988 and the declining price of fuel resulted in costs $25 million lower than those of 1987. Yet, despite a capacity growth of 17.6 percent, unit costs (operating expenses per available seat mile) edged up toward 6.1 percent, a result of Wardair's entry into the domestic market and the onset of a recession. Faced with the prospect of an excess domestic capacity in 1989, CDN presold thousands of 1989 seats in November 1988 at yields well below those experienced in 1988. The airline reduced domestic capacity by turning over some of its routes to its Canadian Partners, but in the short term this had a negative effect because of the time lag in eliminating expenses associated with the removed capacity.

On December 31, 1988, with a net income of $30.3 million, CDN operated a fleet of four Boeing 767-300ERs, thirteen DC-10-30s and sixty-six Boeing 737-200 aircraft. It now had 13,985 full-time employees, of whom 83 percent were covered by labour agreements and had just been introduced to their new uniforms. Services to thirteen communities were transferred from Canadian Airlines to its partner "affiliates": Time Air, Calm Air, Ontario Express, Inter Canadien and Air Atlantic.

On the same date, Air Canada had sixty-seven aircraft that were unencumbered and had a market value of $1.4 billion. It also had $1.3

billion in unsecured, undrawn credit facilities and commitments for a further $400 million. The $1 billion in debt on its balance sheet was completely unsecured and at rates available only to a crown corporation.

It is said that while other industries go through good and bad times, airlines in Canada gyrate between bad and just plain awful, and this was the case as the new decade began. As the British airline entrepreneur Freddie Laker had discovered, the big boys play rough. In the summer of 1988, Max Ward got his comeuppance. What killed Wardair was the "hub and spoke" virus. All the revenue that Air Canada and CDN received from their "feeders" at the hubs was pure profit. A major carrier was already committed to fly between the two cities, so when connecting passengers filled the empty seats they provided added revenue at no extra cost. Wardair might offer individually cooked steaks served on Royal Doulton china and a flower on each passenger's tray, but it didn't have a Boeing 737 feeding passengers into it. Although Wardair may have won the fight for scheduled flights between the main cities in Canada, it couldn't offer its passengers connecting flights to their home towns. With their "alliance families," Air Canada and CDN controlled 90 percent of the domestic market and matched Wardair dollar for dollar with, as Ward ruefully noted, "the best gates" at the hubs. Losing $1 million a day, he approached Rhys Eyton. "Trust is the key," Ward later said. "I trust Rhys very much. My era is gone, but Rhys is the man to carry (my legacy) on." On January 19, 1989, Eyton offered Max Ward a generous $248 million for his airline. Ward had also received an offer from the AMR Corporation, which included three Canadian partners— a merchant bank, a Toronto pension investment group and a Newfoundland investment company—who were to own 75 percent of it. Ward later said that it wasn't an offer to be taken seriously.

"Maybe we should have waited six months until Wardair died," said an unnamed Canadian Airlines executive. The airline was $700 million in debt and perilously close to default. It could have been bought for a lot less, stated Frederick Larkin, an aviation analyst and investment dealer at Bunting Warburg. Peter Wallis at CDN said Ward and Eyton had struck a deal and that "PWA had carried through with its obligations."[5] Was this just an expensive way to eliminate a rival? There were rumours that to save Wardair from bankruptcy, Ottawa was about to designate it as the second carrier for Hong Kong, a move that would have crippled CDN.

One of the by-products of swallowing up the disparate cultures of

EPA, Nordair, Quebecair and Canadian Pacific over such a short time was the sinking of employee morale to new depths as many workers lost jobs or saw friends and family members lose theirs in an industry that had until then seemed secure. "The newly formed company soon became a cultural battlefield. Employees continued to identify themselves according to 'their' airline's colour—blue for PWA, orange for CP and yellow for Nordair. Coffee circles began to form around colour codes rather than work groups, and prospective transferees were asked what colour they came from. In some areas colour heritage decided a promotion, explained a botch-up, or was offered as the reason why things couldn't change."[6]

Customer service became incidental to the politics of colour coding. In 1989 passenger satisfaction on CDN's in-flight catering alone was rated at a dismal 6 percent. Long-serving CP Air cabin crew and experienced travellers had warned of PWA's low service orientation, especially when compared with the service offered on CP Air's long-distance flights; it was like expecting Cunard service on the crosstown bus. Wardair's own high standards were so prestigious as to be on another plane altogether, and that airline had built up a loyal following in the leisure market. Business executives were also abandoning CDN in droves. In 1987, when they asked to name an airline for business travel in Canada, only 58 percent recommended CDN. By purchasing Wardair, Eyton hoped to acquire a team of service-oriented staff that would raise the in-flight quality of meals and service on his own airline.

Besides strengthening CDN's position at the Toronto hub, where Terminal 3 was about to open, the most obvious reason for buying Wardair was that CDN finally had access to the prestigious markets of London and Paris. Even without Canadian's feeder services, Wardair had derived $50 million annually from these routes. Finally, some of Wardair's twelve new A-310s could be used to go up against Air Canada's latest Airbuses on the trunk routes within Canada.

The initial impact of the purchase on Wardair itself was the cancellation of the order for Fokker F100s and MD-88s, the closing down of its Winnipeg base and the laying off six hundred of its employees. As had occurred with the CPAL sale, the promise of keeping the two airlines separate was initially made: Wardair could fly internationally, and Canadian would take over its new domestic routes. Both airlines tried to operate independently of each other, code-sharing only on a selected number of

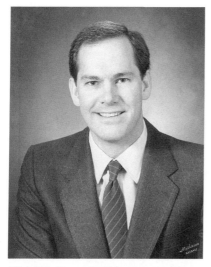

Hailed as a "boy wonder," the thirty-two-year-old Kevin J. Jenkins was appointed president and CEO of first Wardair and then Canadian in April 1991. He understood that the airline's survival depended on attracting an alliance partner and considered a merger with Air Canada. CAA (UNN)

transcontinental flights, but in September 1989 it was decided to trim costs by a full integration.

Max Ward joined the PWA Corporation Board, and Kevin J. Jenkins was appointed president and chief operating officer at Wardair. Born in Edmonton thirty-two years earlier, Jenkins studied law at the University of Alberta and after a short time practising in Edmonton entered the Harvard MBA program, completing his studies there in 1984. Intent on a career on Wall Street, Jenkins met Murray Sigler at a Harvard management seminar. As Bruce Samis had persuaded Eyton, so Sigler was able to talk Jenkins into joining PWA in June 1985. Jenkins led the PWA team in October 1986 that arranged the sale and lease-back of the sixteen Boeing 737s with GPA and halfway through the negotiations he was made group vice-president, finance. Drew Fitch, PWA's treasurer, said: "Kevin has a maturity beyond his years, but he has a way about him that does not intimidate. He is as sharp as a whip, very 'big-picture' oriented." After the CPAL takeover, Jenkins was made group vice-president and chief financial officer of both airlines. Although promises had been made to keep both airlines flying, by December 1989 economics ruled and CDN eliminated 1,900 jobs, of which 1,017 belonged to former Wardair employees.

In July 1990, CDN borrowed $300 million from Mitsui, a Japanese consortium, and planned to sell two of its 747-200s to American Airlines. The money was partially used to buy all of Time Air, for $28 million. PWA's working capital deficit was now $300 million and its shares valued at $9.50, compared with Air Canada's at $8.37. Then Iraq invaded Kuwait on August 2, causing oil prices to rocket from $18 to $30 per barrel. To keep up with rising costs, CDN instituted four separate fare increases. In a further effort to cut costs and penetrate deeper

The 747-400 required only two pilots; its computer technology was referred to as the "glass cockpit." The power of its four General Electric CF6-80 C2 engines—each capable of producing 62,000 pounds of thrust or 2,000 pounds more than all four engines on the DC-8—are demonstrated here to good effect. CAA (UNN)

into Europe, on September 27, CDN began services to Frankfurt but closed its operations in Amsterdam—the first European city it had flown to. It also sold its production delivery spots for six Boeing 767s, three DC-10s and one Boeing 747 and began selling its 50 percent stake in GPA Jetprop, netting it $6 million.

December 13 was an inauspicious date for the first of the three Boeing 747-400s to be delivered to the Vancouver base. A derivative of the original 747-200, the 400 had first flown on October 27, 1988. Capt. Rick Wiley recalled the difference between the two versions: "The first thing that pilots noticed was that the circular staircase to the upper deck had been replaced by a straight staircase, which was now at the rear of a much larger upper deck. The utilitarian light grey colour of the 747-200 flight deck was now a pleasant tan. Gone too was the second officer's station, its large instrument panel replaced by shelves containing the 'aircraft's library'—manuals on operations, maintenance and the aircraft's documents."

The forward instrument panels consisted of six cathode-ray or TV tube displays, two in front of each pilot, showing aircraft altitude, speed and navigation data; and two in the centre panel, showing engine data and systems displays. On the centre pedestal between the pilots, three control display units had small screens and keyboards. The left and right ones allowed the pilot to enter and retrieve data from the aircraft's two flight-management computers, and the centre one gave access to the aircraft's maintenance computer and the pilot's airborne datalink system. Also in the centre pedestal was the aircraft printer (not unlike an airborne fax machine), which printed out messages, including weather forecasts from the datalink system, to the pilots.

Since the 747-400 required only two pilots to operate it, everything possible was done to reduce the workload. Most of the aircraft's systems were automatic and backed up by standby systems, eliminating the need for pilots to monitor them. If a system failed, the aircraft's maintenance computer recorded the failure and displayed it to the pilot as a written message. There were 360 lights, gauges and switches on the flight deck, compared with over 1,000 on the 747-200. Pilots referred to the technology used in the 400 as the "Glass Cockpit" and to the older 747s as "Classics" (an analogy from a current Coca Cola ad).

The 747-400 was then the world's heaviest commercial aircraft. It had a gross takeoff weight of 870,000 pounds (100,000 pounds heavier than the 200 model) and was powered by four General Electric CF6-80 C2 engines, each capable of producing 62,000 pounds of thrust—2,000 pounds more than all four engines on the DC-8.

Boeing had designed the 747-400 to be a very long-range aircraft and despite its size, the 400's engines were 25 percent more fuel efficient than those on the 200. The wingspan on the 400 was increased by over 12 feet to 213 feet and a vertical 6-foot high winglet was added to each wing tip to reduce drag. Finally, additional fuel tanks were installed inside the horizontal stabilizer or tail fin, increasing the aircraft's fuel capacity by 22,000 pounds to a maximum fuel load of over 380,000 pounds—30,000 pounds more than the maximum gross takeoff weight of the DC-8. All these modifications give the 400 a still-air range of over 7,500 nautical miles or over fifteen hours in the air.

The 400 was also equipped with a full-time auto-throttle system and three autopilots. Only one autopilot was required in flight, but a pilot wanting the aircraft to land itself for any reason (usually poor visibility in fog) could engage all three autopilots on approach. The pilot then lowered the gear and the flaps at appropriate times and set the desired speeds on the control panel. Adjusting the autobrake system to the desired setting, the pilot simply monitored the aircraft as the auto-pilots flew it down the localizer and glide slope. The ground-proximity warning system had been enhanced to call out the required radio altimeter altitudes in a recorded voice. The auto-throttle system controlled the speed, including bringing the thrust back to idle on landing, and the autopilots landed the aircraft. The pilot then selected the thrust reversers on the engines, and the autopilots tracked the centre line of the runway as the autobrakes brought the aircraft to a stop.[7]

When the Persian Gulf war's "Operation Desert Storm" began on January 16, 1991, business travellers and vacationers everywhere were frightened to fly, and airlines all over the world staggered under the twin blows of exorbitant fuel prices and empty aircraft. CDN grounded seventeen of its aircraft and cancelled thirteen return flights to London, Milan, Rome and Frankfurt. Fear of terrorism caused a drastic decrease in bookings for February and March. The airline had temporarily reversed the flow of red ink in the third quarter of 1990 with earnings of $36.7 million, but because of the war, 1991 was shaping up to be a bad year. Worse, Kevin Jenkins couldn't generate any money by selling off Wardair's ten Airbuses for $900 million. Montreal-based International Airplane Co. Ltd. (IACO) had hoped to unload them onto Russia's Aeroflot, but was still stuck with the first two A-310s that it had accepted the previous year. The sale was the last chance to build up cash reserves before the effects of the war were felt. It died on February 7, when a deal that would have given CDN $710 million in exchange for the remaining eight aircraft collapsed. It was small consolation that the airline received a $52 million letter of credit from IACO for backing out and that the Department of Defence, looking to replace its antique Boeing 707 transports, considered buying three of the former Wardair A-310s.

Both Air Canada and CDN were unanimous in their drive to change the federal legislation preventing more than 25 percent of foreign ownership of airlines in Canada. Air Canada was contemplating an alliance with US Air. Jack Lawless told the *Financial Times* that CDN favoured a percentage increase to as high as 49 percent. In the midst of the turmoil on February 21, the airline celebrated the opening of its fortress hub in the futuristic Terminal 3 at Toronto's Pearson airport. But these were desperate times and even with CDN being run more efficiently its unit costs in 1991—feeding the passengers and fuelling the aircraft—were 14.6 cents an available seat mile, compared with Air Canada's 18.2

cents, but a long way to go before arriving at US Air's 13.3 cents—and American Airlines' 11 cents.

On March 1, Rhys Eyton gave up his position as president to Kevin Jenkins. With Dave Murphy, the senior vice-president (finance), Jenkins explained the company's serious financial position to its employees and asked for their help. The employees were given until June 1 to reply to a labour cost reduction of $120 million annually for the next three years. In April 1991, in response to pressure from both countries for unlimited access to each other's markets, bilateral air negotiations between the United States and Canada began. The first A-320 was delivered in May, just as the discouraging first-quarter results of a $67 million loss were made public. PWA arranged for $141.8 million financing to pay for the Boeing 747-400s. By the second quarter, losses were $35.7 million and the stock at $8.25. The company decided to sell two of the Boeing 767s that were on order as well as five of the A-310s to Polaris Leasing, netting it $150 million.

On September 11, Jenkins announced his plan to save $116 million annually by eliminating 1,600 jobs, planning fleet changes and, with the ratification of the unions, implementing wage freezes. Not unexpectedly, the unions were lukewarm to the last, and the savings made through the balance of his plan barely totalled $84 million. When the airline announced that it planned to lay off 1,300 employees, Jenkins toured all of its major bases across Canada between September 13 and October 2—surely the most unenviable task for a president—explaining the airline's situation to the employees. Reviewing options, he said "that the status quo was out" and that the company's survival depended on attracting an alliance partner. Significantly, he mentioned that this did not rule out a merger with Air Canada. The total loss for 1991 was $162 million.

Unable to entice Don Carty to the president's job, Air Canada's board selected Hollis Harris, former president and chief operating officer of Delta Airlines and later Continental Airlines, both of which had been to the brink of bankruptcy. When he reported for work February 20, 1992, Harris was told that part of his mandate was to put CDN out of business. With him came another Delta Airlines employee, Lamar Durrett, hired as Air Canada's executive vice-president and chief administrative officer. Harris thought that a merger with CDN was not only inevitable but immediate and felt personally betrayed when this did not happen. In subsequent discussions with Jenkins, Harris played the nationalistic card, saying that a marriage between the two airlines would produce a strong all-Canadian giant that could effectively compete on world markets. The stumbling block was for both sides to raise $1.5 billion in new money, and finding it demonstrated Harris's lack of understanding of the transportation industry in Canada. While CDN was prepared to go cap in hand to the federal and provincial governments for it, Harris, as an American, was vehemently opposed. "I was adamant that we would not get a penny in financing from the government," he said. That summer a strange drama was played out: the southerner, Harris, warned against Canada selling its aviation birthright, while Jenkins looked toward Fort Worth for his airline's salvation—where Canadian Don Carty was executive vice-president of American Airlines.

The very public fight over which airline was to survive carried over not only onto full-page newspaper ads but to airports across Canada, where employees of both companies eyed each other warily. Labour unions like the Canadian Union of Public Employees (CUPE), which now represented the flight attendants from both Air Canada and CDN, lost their effectiveness. The fear that CDN staff had of being swamped by Air Canada led Eyton to reassure them. "We have no plans to give up control. But we have not ruled out any alternatives, including a made-in-Canada solution."

Rhys Eyton had played a weak hand for as long as he could and with laudable finesse, but the pot had gotten too rich for PWA's blood. He once said that he would leave the airline business when it ground him down. "I have to say that I've done everything I can," he said in one of the last interviews given while in office, "but I've never seen a time when it's been so difficult to project any length of time ahead."

Eyton had spent his career creating an airline that could (almost) go head to head with Air Canada in the domestic market. His bold moves and obsession with the bottom line increased the company's share of internal travel from 30 percent in 1988 to 45 percent in 1992. The problem was that even without the Persian Gulf war, the world of aviation was shifting too rapidly for either Air Canada or CDN to carve out their own destinies. This was also one of the few instances in which an airline duopoly hadn't worked effectively, for anyone. Canada was next door to the most aviation-oriented nation on the planet. CDN might be opposed to "cabotage"—or the right to carry passengers from point to point within another country—but because of Canada's ribbon-like demography,

Hired in 1992 as Air Canada's president, Hollis Harris was told that part of his mandate was to put Canadian Airlines out of business. He turned Air Canada around with fuel-efficient A-319s and a corporate makeover.
ACA Slide 10

US carriers already served 90 percent of its population, whereas the scattered population centres in the United States meant that Canadian carriers served only 25 percent of the US market. Adding to the imbalance is the fact that the seven major carriers in the United States were *each* larger than the two Canadian airlines combined.

Television viewers had never seen a commercial such as the one aired during the 1992 Winter Olympics. Two Boeing 767 airliners were shown flying parallel through the clouds, wings close together, almost touching. One had red-and-white markings, the other the distinctive blue-and-red corporate logo of CDN. What happened next was straight out of Steven Spielberg by way of Walt Disney. Passengers left the red-and-white plane and transferred to the CDN aircraft via the wings. The voice-over said, in part, "Whatever the reason, more and more travellers are switching from one airline to the other." Produced by ad agency Chiat/Day/Mojo, it was the most costly, technically complex commercial ever made and had been based on CDN's claim that it pulled in the equivalent of one planeload a day from other airlines. Directed by Ousama Rawi, film footage of real aircraft was used along with twelve-foot models of aircraft filmed against a green background

317

The Chiat/Day/Mojo "Wingwalker" commercial aired during the 1992 Winter Olympics. The voice-over said: "Whatever the reason, more and more travellers are switching from one airline to the other." Its success can be measured by Air Canada going to court for a restraining order.
CAA (UNN)

(green and blue are the easiest colours to make disappear), allowing them to be "matted." The actors' hair was "mussed" by wind machines as they walked along a blue platform placed at the exact angle of the wings, and a computer named "Harry" simulated the effects of the airliners flying through the clouds. Although CDN claimed that the abandoned red-and-white aircraft was generic, an Air Canada spokesman vociferously claimed that only his airline had those markings in Canada and unsuccessfully applied to the Superior Court of Quebec for an order to restrain Canadian from airing its "Wingwalker" ads.[8]

On March 19, PWA and American Airlines announced "that they were undertaking detailed discussions regarding a comprehensive strategic alliance." The airlines signed a letter of intent on March 22 under which AMR would acquire a one-third equity interest in CDN and a 25 percent voting interest, the maximum allowed under Canadian regulations. Two weeks later, Hollis Harris revealed that PWA had rejected two proposals from Air Canada that would have alleviated its financial crisis:

a proposal for Air Canada to buy PWA's international routes, especially those to Tokyo and Hong Kong, and a proposal to merge the two carriers that would make Canadian Airlines a second-level domestic carrier. Harris filed a formal complaint with the National Transportation Agency over a proposed CDN/AMR deal. Eyton replied that "PWA had given full and careful consideration to Air Canada's proposals but that talks with American Airlines would continue to maximize shareholder value, minimize job losses and maintain competition."

The first quarter loss was $73.9 million just as Ontario Express, Time Air and Inter Canadien were merged into "Canadian Regional." On July 1, Air Canada saw its alliance with US Air scuttled when the latter chose British Airways instead. Inexplicably, the minister of defence, Marcel Masse, took this opportunity to announce that a way to assist Air Canada might be to "alter the agreed-upon sale of three ex-Wardair A-310s to DND." A week later, Canadian announced that it was no longer talking with American Airlines but that merger talks with Air Canada would begin. Plainly, it could not provide AMR with adequate guarantees for servicing its operating-cash-flow requirements or its long-term debts. At this point, the federal government decided to help Jenkins make up his mind. In a heavy-handed gesture, Glenn Shortliff, the clerk of the Privy Council, advised Jenkins on July 24 that if CDN wanted to sell the Wardair A-310s to DND it had to firmly break off talks with AMR. The turmoil did nothing for PWA stock, which valued at $3.95 on July 28, dropped to $2.08 in the following two days.

CDN now had little choice but to entertain offers from Air Canada. This was the signal for rank-and-file CDN employees to begin organizing rallies, petitions and letter-writing campaigns to prevent any merger with Air Canada. In Calgary, the employees even launched a plan to forego a percentage of their salary to revive the AMR deal—anything, it seemed, was better than falling into the clutches of its fellow Canadians. As a result of this show of support, PWA's board now considered two proposals, one from the Council of Canadian Airline Employees and the other from Air Canada.

CDN's second quarter loss was $34.4 million, and by now both airlines were losing $2 million a day. Even by industry standards, both Air Canada and CDN were highly leveraged—80 percent of their fixed-capital base was encumbered by debt, compared with a figure of 70 percent for an American airline.

On August 18 Air Canada announced that it had found another

partner, American Airlines' main rival—the largest air carrier in the world, United Airlines. The AMR talks having firmly ended, on August 25 the federal government bought the three A-310s for $150 million, one of which was fitted out for prime ministerial travels and became known as "The Flying Taj Mahal" or "Air Force 1½." Harris made another merger offer in September. Now alone and with its stock down to $3.00, PWA's board of directors accepted. On October 8 a committee from both airlines submitted a pre-merger agreement to their boards of directors. (Bay Street derisively referred to the single carrier as "Mapleflot.") Echoing the mood of public acceptance, the *Vancouver Province* editorialized: "Canadian is a victim of a thousand knives. Its merger with Air Canada was really inevitable . . . It's just impossible for small countries like Canada to have two international flag carriers."

The committee wanted a new holding company to be formed on the basis of a one-for-one swap of shares in which PWA shareholders would own 40 percent and Air Canada shareholders 60 percent. But when Kevin Jenkins attended a board meeting at Air Canada on October 27 to work out the merger of the two airlines, the situation had altered within three weeks. During the discussion, he recalled, "it became clear that the Air Canada board believed that the merger was not financeable and that in any event, Canadian Airlines would not survive past the first quarter of 1993, so why go through all of this when they could soon have it for nothing?" Later that same day, Claude Taylor stated that unless Ottawa provided more than $1 billion in 'rent-free capital' the merger was off."[9]

The boards of both companies rejected the holding company idea on October 28; one area of contention was how both airlines would function under the combined debt load of $7.7 billion. Financial analyst Ross Healy, at National Credit Rating Services, concluded that "because of PWA's classic first-order insolvency . . . merging the two airlines would have produced a desperately sick carrier that would have to sell about $2 billion in assets immediately if only to arrive at the state that Air Canada had been in before the merger." Clearly, the accountants at Dorval had done their sums as well.

Spurned by Air Canada, CDN returned to AMR on November 3, seeking ways to form an alliance. Once more the key was being able to present its intended with some form of dowry. Mike Harcourt, the premier of British Columbia, offered to commit $20 million in loan guarantees if the federal government also helped out. CDN asked

The Vancouver hub with airline staff on an atypical day. The airport's strategic role in the AMR relationship allowed Kevin Jenkins to buy more time to extricate the airline from the CRS entanglements.

Rick Sloboda/ Canadian Airlines

Ottawa for loan guarantees that totalled $190 million to secure $500 million in new equity from the airline's employees, American Airlines and a public share offering. It was incomprehensible, Harris said, that the taxpayers should pour money into a poorly run, sinking enterprise.[10] If a free market were operating in Canada, CDN would be allowed to go the way of Braniff, Eastern Airlines and Pan American. Air Canada staff agreed with this sentiment, and between November 14 and 19, they rallied across Canada against the federal government providing loan guarantees. "If the government would step back from supporting Canadian, there is a very good chance that it would fold," pontificated the *Montreal Gazette,* "but the Liberals don't want to be seen to be dumping it because of their fragile support in western Canada." Now it was Air Canada calling for fair play and Canadian Airlines seen to be favoured by Ottawa; the traditional roles for both airlines had become reversed.

As if PWA did not have enough on its plate, Gemini filed a $1.5 billion lawsuit against it on November 12, claiming a breach of fiduciary duty, and a $500 million lawsuit against both PWA and AMR. The airline's stock sank to $1.50.

On November 24, the federal government agreed to lend PWA $50 million, suggesting that it ask the provinces for help and its employees to make further sacrifices. Later it would be revealed that as part of the $50 million deal, Ottawa insisted that PWA stop paying creditors. The airline stopped payments to lease and loan creditors for five months but continued to meet its daily payments to tradespeople. It was now losing $600,000 a day—its stock at $1.15 in November, fell to 70 cents on

December 1, 58 cents December 2, and 54 cents the next day. Then, on December 18, the governments of British Columbia and Alberta threw it a tow line, promising loan guarantees of $20 million and $50 million, respectively.

"It was the employees of CDN that saved their airline," Harris later admitted. Cynical observers would disagree: the critical factor was the unwillingness of the Tories to approve a takeover that would have been unpopular in the West. Whoever was responsible, this was the sort of Christmas present that Jenkins had prayed for. Somewhat reassured, on December 29, 1992, Robert Crandall and Don Carty signed a tentative agreement with CDN for a $246 million infusion of cash to be channelled through a subsidiary, Aurora Investments Inc. AMR would get 25 percent of CDN's voting stock and two seats on the eight-member board of directors, and the American Airlines computer reservations system, SABRE, would be given a twenty-year contract. In return, PWA had to:

- persuade its creditors to exchange $900 million in debt for common shares;

- win the approval of its unionized employees for wage give-backs and a three-year wage freeze;

- divest itself of contractual obligations to the shared Gemini system, and transfer to SABRE;

- get regulatory approvals from the federal government for foreign investment in a Canadian airline.

All these conditions had to be met by December 31, 1993 (this deadline would later be extended). The agreement was portrayed as primarily a service deal in which AMR would provide services like pricing and yield management, operations planning, international base management, food and beverage support services, capacity planning, technical services and accounting. All of this would be done through the SABRE system, for which CDN would pay $100 million a year. "Because we underinvested in these areas," explained Jenkins, "AMR will take them over and do a better job more cheaply than we could." AMR would not be represented on any of CDN's operational boards, and Crandall and Carty would only have two of eleven votes in approving the airline's overall business plan. However, they could veto capital expenses of $50 million or more that were not included in the preapproved business plan. "It's nonsense to think that control of this airline is going to US hands," said Jenkins. "What it is, is an injection of US capital into the Canadian economy."

Both airlines were to be reoriented on a north–south basis in which American's passengers passed through Canadian's hubs in Toronto and Vancouver while Canadian's passengers would use US cities for their connections. Because of its limited access to other airlines' hub airports in San Francisco and Los Angeles, American would use Vancouver as its transpacific hub, where Canadian staff would be available to handle ticket sales and to unload, service and maintain its aircraft. Toronto's Pearson airport would take pressure off American's overcrowded hubs at New York's La Guardia and Washington's National airports.

Vancouver's new terminal, then being built, would give American Airlines greater access to the Orient, an area where the airline was weak. As Grant McConachie knew, Vancouver lay 1,200 miles closer to the Orient than were San Francisco or Los Angeles. Crandall had just paid $150 million (US) for the right to fly a single route from Seattle to Tokyo, so buying a $246 million interest in CDN with its established connections was a bargain. American's faith in Vancouver would be rewarded in 1997, when it became an "open skies" airport, and American codeshared all transborder flights into it—but this was far in the future, and the year 1992 lurched to an end disastrously for both airlines with Air Canada's loss at $454 million and CDN's at $543 million.

Trying to extricate itself from Gemini to join SABRE proved more difficult for CDN than it could have imagined. The Ontario courts denied PWA's bid to declare it insolvent (Gemini had failed to pay down loans to both airlines because it was installing expensive computer equipment), and the Competition Bureau didn't have the jurisdiction to let PWA out of the bid. Jenkins juggled as many balls as he could as on April 30, PWA resumed payments of $35 million to its creditors. But on May 20, Covia Technologies too sued PWA for a breach in fiduciary duty and civil conspiracy, bottoming out PWA stock to 46 cents. There was some good news when the National Transportation Agency unanimously agreed to the AMR deal, causing Air Canada unsuccessfully to pressure the cabinet to overturn the decision. Rhys Eyton proposed a negotiated settlement on May 31 for CDN to leave Gemini, and the Federal Court ruled that the

Competition Tribunal had the power to free it. Eyton chastised Ottawa for not intervening in "this ludicrous" war between the country's two largest airlines before CDN was driven to bankruptcy. "Surely there comes a time that [the government] would be interested in acting to see that the industry doesn't damage itself beyond repair." A management shuffle at CDN on June 15 saw many of its experienced managers, including Murray Sigler, leave.

Air Canada proposed its second merger on August 18, 1993. The airline was enjoying a $14 million profit in its second quarter (the first in two years) and began to upgrade with the purchase of forty-eight Canadair Regional jets for its cross-border market. Harris gave Air Canada a new look, took in maintenance work from Continental and Northwest Airlines and began phasing out the old Boeing 747 and DC-9 fleet. He raised $240 million through the sale of five A320s to six international banks (including the Canadian Imperial Bank of Commerce), and then leased them back. This time Air Canada offered $200 million in cash for CDN's international routes, particularly those to Japan, Hong Kong, Thailand and Taiwan, and an assumption of $800 million in debt and the leases on the five Boeing 767s and three Boeing 747s. In return, it wanted PWA to be its North American carrier, drop the AMR deal and stay in Gemini. If destroying CDN was the first part of his mandate, making Air Canada a global carrier was the second. If it wished, Harris told the press, CDN could remain alive as "a powerful domestic carrier." Air Canada also ran a nationwide newspaper campaign promoting the merger. Canadian Airlines gamely countered with rebuttal ads that began, "Air Canada, are you confused?"

The usually polite Eyton was visibly furious and called this "a diabolical plot to kill Canadian. Air Canada wants a monopoly. Because I can tell you, the sale of our international routes at the number they've suggested would end up killing Canadian Airlines in very short order." Later he added, "I've reached a point where I've just had enough of this. I've had enough of the arrogance of a former crown corporation and of the gunslinger who arrives from Georgia to tell us how to organize our airline structure."[11] The PWA board of directors rejected the Air Canada offer for the international routes and aircraft, and the airline's stock rose slightly to 62 cents. With the ensuing federal election, both prime ministerial contenders got into the act: Liberal candidate Jean Chrétien called for a federal mediator to act between the airlines, and Conservative Kim Campbell stated on September 20 that neither airline wanted a federal mediator. The one ray of hope was the Gemini ruling by the Supreme Court of Canada on October 14, when it allowed PWA to transfer to SABRE.

Along with seeing the inevitable party political broadcasts that fall, television viewers were being treated to a CDN advertisement that rivalled its highly successful "Wingwalkers." It featured a DC-10 autographed by the airline's employees, accompanied by the voice-over "If your good name were riding on something, wouldn't you do what it takes to make a difference?" Laurie Schild, manager of marketing communication, confirmed that there was a perception among the public that employee-owned companies provided better service and that this was the concept that Chiat/Day went for. All it needed was an employee and a 747-400. The agency auditioned more than ninety maintenance and engineering staff before selecting John Alvares from Vancouver. "He was a natural as the representative of the airline, he appeared hardworking and honest and he moved comfortably on camera." Auditioning for aircraft proved a lot more difficult. The first choice was a Boeing 747-400, but none were available for the necessary eight days of shooting. There were several 747-200s sitting out in the aircraft graveyard in the Arizona desert that could be used, but there wasn't time to construct the unique 400 winglet and repaint a whole aircraft in Canadian's colours. There were two A320s in Canadian livery at the Airbus Industries plant in Toulouse, but neither could be flown out of the country yet, and filming in France would have cost an additional $400,000. With the deadline to shoot only ten days away, someone discovered that a DC-10 owned by Potomac Capital Investment Corp. happened to be in Vancouver for maintenance. The company allowed the agency to use the aircraft but it was painted all white. Thanks to the overtime efforts of maintenance staff in Toronto and Vancouver a fourteen-day paint job took five days. (Since the plane was all white, the top didn't have to be painted.)

As for the autographs, the signatures were actually decals. Each letter in each name had to be carefully aligned. Employees' names were randomly selected, primarily from those nominated for the President's Award. What did the airlines' employees think of it? Some of the comments were "It really hits home—very emotional" and "It tells everyone that the company is still around and that the employees are investing in its future."[12]

When the Liberals won the federal election on October 25, 1993,

"If your good name were riding on something . . ." Unable to get a 747, the ad agency used a DC-10 instead. The signatures were actually decals, but many of the airline staff took the spirit of the commercial literally and began signing their names with felt pens to aircraft they had worked on, until ordered to stop. Rick Etkin/CAA ADV 0024

the new minister of transport, Doug Young, appointed former prime ministerial aide Stanley Hartt to the role of mediator. However, the battle was now far beyond federal mediation. Hollis Harris considered launching an antitrust lawsuit against American Airlines, a move that would have escalated the fight to international levels. It was all settled in a low-key, typically Canadian way. Gemini's head office happened to be in the Winnipeg constituency of the new human resources development minister, Lloyd Axworthy, where the loss of seven hundred jobs would have been a poor beginning for the Liberals, especially in the West. As in McConachie and Baker's day, Ottawa brokered a deal. If CDN could be freed from its contract, the federal government might see its way to granting Air Canada rights to Japan, Korea and Hong Kong. On December 10 PWA and Gemini arrived at an out-of-court settlement in regard to the $1 billion lawsuit, and CDN's Pegasus (its internal inventory system used by payload, reservations, airports and cargo) was removed to the SABRE multi-host database in Tulsa, Oklahoma.

Air Canada made a third offer on December 16 to buy CDN's golden egg, its international routes and eight aircraft, for $300 million and to assume $800 million in debt, but this offer was rejected on December 22. By end of 1993, CDN's loss stood at $291.8 million and Air Canada's at $326 million.

Early in the new year, former premier of Alberta and PWA board member Peter Lougheed warned that the Liberals were considering giving Air Canada the rights to China and Japan. Air Canada announced that if Gemini were dissolved, it intended to join Galileo, the largest computerized reservation system (CRS) in the world. On January 27, 1994, it discontinued litigation that would preclude the completion of the AMR/CDN deal. As a reward, Young designated Air Canada the second carrier to Japan (but not to Hong Kong) and only to Osaka, not Tokyo's Narita Airport. On January 29, PWA's stock closed at $1.40, and by March 28, the airline had completed all arrangements for transfer of the CRS from Gemini to SABRE.

Rhys Eyton resigned his position on April 15 but agreed to remain as non-executive chairman. "The key to this business," he often said, "is operating from a strong balance sheet. If you don't have that, you won't make it." After eighteen years, it wasn't that he had lost sight of the bottom line, but it must have looked to him more and more like a contest to see who could destroy his airline first: the unions, the federal government or Air Canada.

CDN employees ratified the financial extension plan and new collective agreements were signed with four of its six labour unions that would result in annual labour costs being reduced by an average of 17 percent for three of the groups and 24 percent for the fourth over the next three years, saving $93 million per year. The employees scraped together $200 million of their own salaries to buy an equity stake of 25 percent in the airline, the share price being $16.00. It had been a gruelling rescue campaign for Kevin Jenkins, who felt that he had become "a lightning rod" for employee grievances that dated back to the EPA takeover.

The signing of the "open skies" pact between Canada and the United States on February 24, 1995, gave Canadian carriers immediate access to any point in the United States from any point in Canada, while phasing in US carrier access to the largest home markets, Montreal, Toronto and Vancouver. This came with a sporting-chance clause that allowed both Air Canada and CDN to establish some sort of foothold before the market was totally opened to the US airlines. Both airlines multiplied their cross-border flights from 102 in 1995 to 461 in 1996, but Air Canada won the first round, increasing its market share from 25 to 30 percent in a year and predicted that it would go to 50 percent by 1997. It had ended the year with $775 million in cash, giving it flexibility for the purchase of more Canadair Regional Jets and two Airbus A340s for long-range flights to Osaka. In December, because of a new government policy that allowed a second carrier to a destination where the primary Canadian carrier handled more than 300,000 passengers per year, Air Canada was finally permitted service into Hong Kong, where its name was translated into "Maple Leaf Airlines" in Cantonese. This wasn't the daily service that it would have preferred but, as CPA veterans recalled about their airline's first transcontinental flights, the monopoly had been breached.

CDN's answer to the Regional Jets were its venerable Boeing 737-200s that had soldiered on in various fleet colours and configurations, some having been with CP Air since 1968. They would be upgraded in January 1996 as the price of the airline's "cost of entry" into the transborder market and launched on March 4 on two battlefields, the two domestic triangles: Ottawa–Montreal–Toronto and Calgary–Edmonton–Vancouver—and the United States. CDN's in-flight catering had already been upgraded (hot meals, healthier pasta replacing meat, instead of Air Canada's sandwiches) and gone were the cramped, three-abreast seating for twelve wider seats in first class and the eighty-eight in economy spread wider, giving each passenger an extra 2.5 centimetres of legroom. Tickets issued by either American Airlines or Canadian Airlines could be used interchangeably, and the service and seating would be consistent in both airlines.

The Montreal–Toronto shuttle was increased to forty-one times daily and Toronto–Ottawa to thirty-three daily. Aviation analysts thought CDN's entry into the rich Eastern Triangle, where Air Canada's Rapidair had traditionally cornered 70 percent of the business, a desperate ploy. Going up against Air Canada, which even had flights from the Toronto Island airport, was said to be "like trying to kill a bear with buckshot."

Neither was CDN's Western Triangle safe anymore. Calgary-based WestJet Airlines and Winnipeg-based Greyhound Air were eroding both major carriers' markets, capitalizing on a version of CP Air's old "Skybus" and PWA's "Airbus" idea—low fares and basic service. But CDN was looking beyond its shuttle, hoping the frequent business fliers would stay with the airline and continue on to London, Taipei or Bangkok. The airline's infatuation with executives dated from the 1970s, and with 5 percent of business travellers making up 37 percent of all airline revenues in 1996, the strategy remained constant. It provided gateside writing desks, "dedicated check-in," fax machines, Club Empress lounges that had automated massage-chairs, a concierge at the gate, AT&T phones on board . . . anything it seemed, to woo the businessman.

American Airlines, which already had the largest codesharing agreement in the world with CDN, wanted to expand its business alliance so that, in Carty's words, "it would provide consumers with a truly seamless travel experience across the full breadth of our route systems." Ads showed two jets nuzzling nose to nose, and the caption read, "One continent. One call. One ticket. One check-in. One team."

Buoyed by all of this, in June 1995 Kevin Jenkins predicted that Canadian was "back on track" and that by the year's end it would even make a profit of $50 million. It was bad enough when this figure became a projected *loss* of $35 million, but worse was to come. On December 20, when news of a huge adjustment to the company's figures was made public, its shares on the Toronto Stock Exchange dropped by 50 cents within a single day, to $4.50. The stock price, especially at Christmas, was a sore point with the employees because most of them had taken stock in lieu of higher salaries. Jenkins knew that the rank-and-file were at the boardroom door, baying for his blood. It seemed that the recently hired chief financial officer, Kevin Benson, had reassessed the accounts in his own way. "He didn't just house-clean," explained company spokeswoman Laurie Schild, "he vacuumed." Because of this, the total deficit had plummeted to between $185 and $195 million.

The differing figures resulted from the different techniques of Benson and his predecessor, Drew Fitch. Fitch, then on a sabbatical after five stressful years, used to account for the cost of one-time charges in the period in which the bills would be paid. Benson, on the other hand, put all charges into the year the decisions were made—in this case, 1995. "Mr. Fitch wasn't right or wrong," commented Schild. "He was just different." But this meant that Benson transferred $50 million of the restructuring charges into 1995, whereas Fitch would have spread it out over several years.

It also touched on the shady practices of travel agent kickbacks and rebates. In some Asian countries several months elapse before travel agents report their monthly sales of airline tickets, so a company can only project its sales for that period. CDN sold over a million tickets a month in Asia, and the small discrepancy multiplied to large figures.

Added to this was another $40 million to $50 million of costs caused by higher fuel prices and the change in the value of the Japanese yen, items that Fitch could not have foreseen.

Kevin Jenkins was understandably furious. "It looks like we don't know what we're doing." His credibility with the financial markets was gone, he told his senior staff, along with what little reputation he had among the employees. A massacre at senior levels ensued, and a number of vice-presidents were removed, including Terry Francis, the senior vice-president for marketing and sales, whose $100,000 bonus for his role in the PWA–AMR deal had come under scrutiny at the May AGM. Also out were Douglas Ball, Ian Bootle, Gilles Dagenais, Chris Nassenstein, Ted Ranson and Laura Safran. They were replaced by four internal appointments—Rod Dewar, Keith Pope, Barry Rempel and Mark Williams—but more pointedly, there were three imports from American Airlines: Barbara Amster, Richard Haddock and George Mueller.

In Amster's hands rested much of the fate of the airline. At the age of forty-nine, she had been vice-president of yield management and distribution planning at Fort Worth and was made responsible for setting price and determining the profit for every seat on every flight. "It was a company in shock," she recalled when she arrived in Calgary. "I did find something of the battered child syndrome. People were just trying to hang on."

Jenkins, meanwhile, had been correct on both accounts. In early January 1996, members of the International Association of Machinists and Aerospace Workers (IAM&AW) signed a petition demanding that the board dismiss Jenkins immediately. The union was still smarting from the loss of 208 jobs that Canadian's consolidation of maintenance from Calgary to Vancouver had cost. This was nothing compared with the fury of all 16,500 employees. A group describing itself as "The Canadian Airlines Shareholders' Action Committee," many of whom were employees, took out newspaper ads denouncing Jenkins: "We believe that it is high time to address this company's continuing losses, depressed stock prices and the credibility of the top management." Union leaders approached Harry Steele, the company chairman, asking him to review the airline's leadership. The spectacle of the embattled Jenkins standing up at a business luncheon, then turning dramatically around and saying, "I want to show you that there are no arrows in my back," received wide coverage. But the board gave Jenkins a vote of confidence: "We believe that this criticism is ill-advised, ill-informed

President on June 29, 1996, the reserved, soft-spoken Kevin Benson earned his campaign ribbons by returning Trizec to profitability. He made it clear from the outset that nothing could be accomplished in Canadian Airlines without the help of the unions. CAA (UNN)

and inappropriate." Pointing out that the memo was signed by the two union representatives on the board, Sid Fattedad and Judy Korbin, Jenkins put on his best face possible, saying that the airline "had caught its second wind." It was not behind in its debt repayment, which had dropped from $200 million in 1995 to $120 million the following year. Jenkins also voluntarily reduced his own salary by 30 percent, accepting $367,500 in 1995 from $447,500 the previous year. Canadian's stock inched up to close at $2.70.

In contrast with CDN's nosedive into trauma, in February 1996 Air Canada announced a profit for 1995 of $129 million. Hollis Harris said his goal in 1996 was to raise shareholder value in the airline by paying down debt, aiming to cut $1 billion off the $2.4 billion long-term debt by the year 2000. The airline was doing so well (and in keeping with Harris's streak of independence), that he refused to take advantage of Ottawa's and Quebec's offers of grants to upgrade the Dorval maintenance base. At the Air Canada annual general meeting of shareholders on May 14 Harris, satisfied that he had completed his mandate, resigned his position as chairman and CEO to return to Atlanta—from where he had been commuting—to be succeeded by Lamar Durrett. He moved quickly to block any forms of closer co-operation between CDN and AMR by asking the US Department of Transport to check the arrangements for the breaking of antitrust laws. In this it was disappointed. On May 28, the USA granted anti-trust immunity to the AMR–CDN deal, something that Air Canada now hoped it would do for its own partnership with United Airlines.[13]

On May 29, 1996, the Canadian Transportation Act received royal assent and replaced the 1987 National Transportation Act. It created the Canadian Transportation Agency (CTA), which, much like the NTA, was a quasi-judicial tribunal with the powers and rights of a superior court. But that was where the similarity ended. Now a Canadian air carrier licensee could operate with whatever aircraft it chose, when it chose to do so and as often as it wished—subject in some cases to certain financial requirements. With this the latest recognition of the de facto breakdown of national airlines and the emergence of truly global skies, Ottawa's relationship with transportation had evolved dramatically in three decades. In the cities of Winnipeg, Vancouver and Calgary, the former headquarters of Canadian Airways, Canadian Pacific Airlines, CP Air and CDN, it must have seemed too little and too late.

Kevin Jenkins resigned as president and CEO of CDN on June 28. "This is the most difficult decision I have ever made," he said in a prepared statement. "However, I believe that Canadian Airlines' best interest will be served by a change in leadership." Vice-president Tony Johnston added, "Kevin told me it's time for a change. He has had to lead this company through the most difficult period in aviation history . . . not just in Canada but around the world." A former Canadian Airlines flight attendant, Joan Vanstone, who had campaigned for three years to have Jenkins ousted, could barely control her excitement: "I'm absolutely delighted—I don't know why it's taken so long. Kevin Jenkins didn't have his heart in the company." Jenkins later said that for him the last straw had been the IAM&AW rejecting the new contract by a vote of 62 percent—for the second time. (Had his resignation not been so rash, he would have discovered that two rejections without a strike vote under union rules meant acceptance, and the contract was passed by default.)

Deeply religious, Jenkins bid farewell to the airline by assembling his senior staff for the last time and reading from Philippians 4: "Whatever is true, whatever is noble, whatever is right, pure and lovely, admirable, if anything is excellent or praiseworthy, think about such things." In September he was recruited by the Westaim Corporation as president and CEO. The company had a portfolio of surface-engineered products, electroluminescent flat-panel displays and biomedical coatings, and the closest it came to aviation was in manufacturing aerospace materials.

His place was taken by the chief financial officer, Kevin Benson, and the predictable statement was issued: "I look forward to building on our first-quarter results and our better than expected cash position in the months ahead." Vanstone feared that Benson, too, didn't have the passion for the airline industry necessary for success, but sympathized: "it's a bit like being given captaincy of the *Titanic*."

Benson's own successor was Doug Carty, the brother of Donald Carty, who also became senior vice-president. Carty told *The Financial Post* that CDN had made great strides in improving asset utilization and that revenue per available seat mile had declined only by 0.4 percent to 10.26 cents. The disposal of property and equipment, including the sale and leaseback of eleven Boeing 737s in April, had generated $93.7 million, compared with $1.1 million the year earlier. Operating costs per available seat mile had declined by 2.1 percent to 11.81 cents, despite a rise in fuel prices. A contributing factor in Jenkins' resignation had been the first-quarter loss of $110.9 million, and the airline continued to lose $23.8 million in the second, but at a reduced rate. Its cash in hand at the end of the first half of 1996 was $111.9 million, significantly higher than the $19.8 million for the same period in 1995. Revenue passenger miles had increased by 12 percent in the first quarter and 8.5 percent in the second. The sale of "Canadian Holidays" in July did help, as did the elimination of money-losing routes such as Frankfurt and Paris. The airline lost its codesharing agreement with Lufthansa to Air Canada but replaced this on June 15 with a major agreement with British Airways. The day before Benson took over as CEO, the IAM&AW ratified a thirty-eight-month contract that would save the airline $34 million annually. Only the Canadian Union of Public Employees (CUPE), which represented 2,600 flight attendants, still held out on the 17.1 percent cost-reduction agreement.

The Boeing Company had waited six years after the initial flight of Aerospatiale's Airbus to propose the development of a family of advanced commercial aircraft with the designations 757, 767 and 777, and by then it had lost 30 percent of its longstanding market. The 757 was a short- to medium-range airliner and would be bought in large numbers by the big three: Delta, United Airlines and American. In Canada, seven were sold to Canada 3000, the charter airline. In 1983, following an order for thirty from United Airlines, Boeing announced that it would build the longer-range model 767-200. Air Canada bought twenty-one of the 767-200 and 200 ER (extended range) and Canadian Airlines eleven of the 767-300 ER, a 254/290 passenger derivative of the original Boeing 767-200 model. The 300 was distinguished from its predecessor by a fuselage that was twenty feet longer, increasing seating

Boeing 767-375 C-FCAE was the second of eleven delivered to Canadian Airlines and put on the ten-hour Toronto–São Paulo run. With its roomy cockpit originally intended for a three-man crew and mix of older-generation instruments and CRTs, it is said to be a "pilot's dream." CAA (UNN)

capacity by 22 percent and cargo volume by 31 percent. The ER version allowed greater range as well as long, overwater extended twin operations flights because of an additional auxiliary power unit with high-altitude capability, plus a fourth hydraulic motor-driven generator, increased cargo compartment fire-suppression capability and cooling sensors for the electronic flight instruments. Canadian Airlines put its nine Boeing 767s on the European routes, on the ten-hour Toronto–São Paulo route, as well as to Honolulu and over the North Pacific to Japan and China. Because of the 767's two-aisle, wide-body cabin, passengers rated it equal to the larger, wide-body 747 for in-flight comfort.

First Officer Donald Muir reckoned that from a pilot's point of view,

> . . . the B767-300ER is a dream. It is one of the easier and safest aircraft to operate. Because of its automation and design, there is less "hand flying." Although it is possible to ignore the computers and fly it manually, the automation is designed to ease the pilot's work-

load, increase safety and decrease the aircraft's operating cost. Unlike on other aircraft, the pilot wastes no time troubleshooting faults but in most cases merely turns a system on or off. The cockpit is very big (the original layout was designed for a three man crew) and quiet, making long flights pleasant. The instrumentation is a mix of older-generation instruments as well as the newer-technology cathode ray tubes (CRT), and this allows for an easier transition for a pilot from the older-generation Boeing 737 or DC-10, than if he were to go directly to a fully automated Airbus or Boeing 747-400. In fact, the 767 is known as the "swansong" aircraft for many of the older pilots as they approach retirement, pleased that it is the last of the commercial jets they will fly.

Canadian 767s were outfitted with Club Empress seats that had power outlets for personal computers. Now busy executives could use their flying time usefully—tapping away on their laptops—provided they switched them off while taking off and landing, because the laptops interfered with the instruments on the flight deck. Working at 35,000 feet had become increasingly important to "frequent fliers," which was why Canadian focused on improving seating, installing lumbar support and increasing seat pitch for more room. (Seat pitch is the difference between the front edges of two upright seats, one behind the other; the average seat measures twenty inches from front to rear, and to be comfortable a seat pitch of thirty-six inches is the norm. In business and first class, CDN (like Air Canada and British Airways) installed a seat pitch of forty-five inches.

In July 1996, as part of its expansion program at the Vancouver hub, CDN hired between 500 and 700 staff, the first major expansion since the layoffs. Among the 3,000 applicants, 150 men and women were chosen as flight attendants and put through the eight-week training program, the first since 1988. Under the collective agreement, they started at $22.42 an hour, with a minimum pay credit of seventy hours per month and an average of about seventy-five hours in the air.[14] Canadian Airlines supervisors knew that about 10 percent who made it through the screening and training would quit before graduation or in their first year. Half of the training was taken up by safety and emergency drill, aircraft systems, emergency equipment usage, evacuation techniques using cabin simulators and door mock-ups and gaining a first-aid certificate at the end. It was impressed

In July 1996, as part of the expansion of the Vancouver hub to Asia, Canadian Airlines chose 150 men and women to train as flight attendants. After six weeks of classroom work, some of them celebrate their graduation before serving on scheduled flights. Rick Sloboda/Canadian Airlines

upon trainees that they couldn't dial "911" at 27,000 feet. Because Canadian Airlines was competing along the Pacific Rim for customers with such heavyweights as Singapore Airlines, Cathay Pacific and Japan Air Lines, multilingual attendants (especially those speaking Mandarin, Cantonese, and Japanese) were in demand.

Besides the safety training, a fortnight was devoted to more traditional subjects, such as service etiquette, computer training, cultural awareness and the selling of in-flight products. Once the six weeks of classroom work had ended, the trainees served on scheduled flights and were matched with an experienced attendant. The ability to work as a team was the chief guideline, said Francis Fiorillo, vice-president of in-flight service, but as one graduate said, "You also had to be a nurse, teacher, mother and entertainer."

The strong performance on the long-haul routes to Asia was offset by CDN's lower yields in Canada as a result of competition from the new low-cost carriers. "That's the dilemma," Doug Carty said. "The Asia–Pacific offers huge growth, but we're fiscally challenged to take advantage of it." The airline was flying at 64 percent capacity, meaning a third of the seats on each flight were empty. The airline's net income for the third quarter was $85.7 million, compared with the previous year's $98.9 million. November 1 would ever after be known as "Black Friday," the day that Canadian Airlines announced that it was on the brink of collapse, and it entered into a month that every airline employee and many other Canadians would remember for a very long time.

"Our continuing operating losses are threatening the company's viability," said Kevin Benson. "Thus we are planning decisive action to return the company to profitability in the face of increased competition and falling ticket prices." In his operational restructuring plan, Benson outlined the cost-based restructuring of the airline's operations, which targeted four years of annual improvements of $200 million through:

- $60 million from more effective use of the fleet to take advantage of the antitrust immunity and "open skies." Domestic capacity would be reduced by 11 percent, the transborder sector increased by 72 percent and the Asia–Pacific by 21 percent, catering to the "high yield" business traveler and last-minute flier. Domestic capacity would be decreased by 7 percent.

- $70 million from the reduction of overhead and other payments and the renegotiation of the AMR service fee of $140 million by $48 million annually. The federal government and two provinces would also be asked to defer their payments.

- $70 million from a 10 percent wage reduction for employees. Wages were the airline's single biggest cost, at 26 percent of total operating costs, and Benson wanted the six unions to hold a referendum-style vote by November 30 to decide whether to accept the wage cut. The success of the plan would depend, the president warned, in large part on the strength and determination of the employees.

In total these changes would eliminate no more than 250 positions out of the total work force of 16,400.

When Benson took the helm of Canadian Airlines in July, its employees had two questions about their new chief executive: who is Kevin Benson, and why is he running our airline? "I don't know anything about him because he's new to the industry," said one. "That may be just what the company needs."

Buying time to restructure a company was not new to the forty-nine-year-old accountant. He had led the Calgary-based property company Trizec Corp. through the same exercise in the early 1990s. The father of three, he was born in Durban, South Africa, and trained as a chartered accountant in 1971. He emigrated to Canada in 1977 as the protégé of another South African–born accountant, Jack Cockwell, of the Bronfman group of companies. Trizec, their real estate division, was in trouble, and Benson steered it carefully through heavy damage-control operations, displaying what his colleagues called "an innate sense of fiscal caution." He is supposed to have promised himself that he would never get involved in such a painful, complex battle again. Benson left Trizec in 1994, when it was sold to Horsham Corp., and was working for his own consulting firm when the headhunters called. Canadian Airlines needed someone who could raise money, and the Trizec experience showed that Benson had the necessary contacts in the financial community, particularly among foreign bankers.

Benson's weakness was auto racing; he owned part of Payton-Coyne Racing, based in Illinois. Perhaps it was the excitement of being responsible for even faster aircraft that led him to Canadian. When he was recruited to replace Drew Fitch in the fall of 1995 he is supposed to have said, "If it doesn't have wings or wheels, forget it." The job initially looked easy: a plan was already in place, and Benson's duties were simply to follow Fitch's methods through—until he discovered the $50 million shortfall.

Reserved, soft-spoken—in the heat of negotiations with union leaders or answering reporters, he had been called the essence of "cool"— the only time Benson ever seemed uncomfortable was when a group of employees surrounded him at a suburban Vancouver hotel and praised him for saving the company. Like Grant McConachie he had a pilot's licence, and like Russ Baker he did nothing by halves. But, unlike Kevin Jenkins, he won praise almost immediately from the employees for dealing directly with them, not their union leaders, on the possible reopening of their collective agreements. He also made it clear that the airline needed two years to restructure itself and emphasized from the outset that nothing could be accomplished without the help of its six unions.

CDN made its required interest payment on October 31 but still owed $30 million on the $120 million loan made to it during the 1992 restructuring. Ottawa and the provinces of Alberta and British Columbia agreed to defer the $10 million quarterly principal loan payment for thirty days. Doug Carty said that they would probably ask for a break when the next deadline came around. In a classic understatement, he admitted, "We're in pretty deep trouble."

The other alternative was for AMR to buy more of the company than the 25 percent limit currently allowed. In return, it would get Canadian's coveted Asian routes. But this time even if Ottawa eased the foreign-ownership rules, there was no guarantee that American Airlines was still interested. It stated in a press release, "at this point we do not contemplate a further investment in Canadian. If an additional cash infusion would have ensured Canadian's profitability, we would have given such an investment strong consideration." Fort Worth had seen the value of its $246 million investment dwindle down to $30 million. Nor was CDN the only access to the Orient any more. If it failed, analysts predicted that American would "do a deal with Air Canada." The doomsayers also predicted that the value of CDN's Vancouver hub was hardly eternal. The next generation of jets, like the Boeing 777, would put Tokyo within reach of New York or Chicago.

The crowd of employees listening to Benson in the freezing Calgary airline hangar on November 2 knew where they had heard this before. Asking for concessions from governments and major shareholders, the deferred payments and loans, the cutting of domestic routes, the pay cuts—they were all there. Their new president might have been using Jenkins' notes from previous meetings. The promise that when the $200 million was saved the employees would be repaid in profit-sharing was greeted with hollow laughs. Everyone present knew that Canadian hadn't made a cent in seven years. Their union advisers told them that the $70 million Benson wanted from them wouldn't keep the airline in operation for more than a week; time and again they had bought their jobs with pay cuts. One employee even pinned her hydro bill to the bulletin board, saying she was struggling to pay that now.

On November 9, CDN issued a terse ultimatum to its workers to give up 10 percent of their wages by November 30 or start looking for new jobs. "There is no Plan B," warned Benson. Spokeswoman Diana Ward was even more blunt: "If all employees, including the unions, do not accept this, then the next step is an orderly and dignified shutdown of the airline. We will begin recalling our aircraft from overseas." The airline's financial crisis only served to make the public skittish about their tickets and frequent-flier investments, and travel agents spent much of their time soothing frantic passengers.

Buzz Hargrove, president of the Canadian Auto Workers (CAW) union, representing reservation and ticket agents and a quarter of the airline's work force, called Benson's plea for his members to accept a 10 percent wage cut a charade. "We've been there, done that, and look where we are today," he told a meeting of union representatives. Instead he proposed that the money-losing airline be helped by a bail-out package from Ottawa, the provincial governments and the banks;

Their faces reflecting the despair within the airline, on November 15, 1996, the entire board of directors resigned to avoid personal liability.
Rick Sloboda/Canadian Airlines

regulation of the airline industry, to stop CDN from ever again expanding on money-losing routes and, finally, allowing AMR to buy more of the airline.

The minister of transport, David Anderson, the member of Parliament for Victoria, had already stated that the Liberal government would not think about any form of assistance, but that the 25 percent limit on foreign ownership was not sacrosanct. After all, what was so Canadian about owning a collection of aircraft built in the United States and Europe and already leased from Japanese, Israeli and Dutch banks? Anderson warned that it was too late for the unions to want to "put the genie of deregulation back into the bottle." And, as to discouraging CDN's expansion, his department gave the airline access to routes to Chile, Spain and Russia.

CDN's plight in the midst of Prime Minister Chrétien's parsimony caused westerners to dredge up old grievances. There might be little appetite for a financial bailout either federally or provincially, but the perception that Air Canada was currently prospering because it had been subsidized for so long drew cries of regional favouritism. On November 15, on the advice of their lawyers, CDN's entire board of directors quit to avoid personal liability in the growing crisis. This would prevent creditors from going after the board's private assets, including homes and savings, to settle the company's accounts. The chief concern was wages. Directors of federally regulated companies can be held responsible for up to six months' wages for employees. The CDN Directors included Kevin Benson; Harry Steele, chief executive, Newfoundland Capital Corp.; John Cassaday, president of CTV; Sid Fattedad, chief financial officer, BC Workers' Compensation Board; Ross Fitzpatrick, chief executive officer, Viceroy Resource Corp.; Peter Lougheed, lawyer for Bennett, Jones, Verchere and former premier of Alberta; Art Mauro, lawyer; and Ron Southern, chair and chief executive, Satco Ltd. and Canadian Utilities and, from Aurora Airline Investments Inc., Don Carty, president, AMR Airlines Group; and Robert Crandall, president and chief executive, AMR Corp.

The two AMR appointees seen to be extricating themselves from direct involvement caused the newspapers to speculate whether this was a precursor to complete separation. "This is not a good sign, given the company's liability," one wrote. "Things are getting critical." AMR spokesman Al Becker attempted to quash such rumours by stating that "no one is running away from Canadian Airlines. We have a huge investment and we continue to feed thousands of passengers to them every year." Benson urged the employees to be patient and ignore "the many rumours that seem to gather legs of their own accord." He criss-crossed the country for a series of secret talks, playing his cards close to his chest, talking with former Ontario premier Bob Rae, who had helped restructure Algoma Steel in Sault Ste. Marie and Spruce Falls Power and Paper Co. in Kapuskasing. The IAM&AW, the most adamant union to refuse the cuts, had hired him as a consultant.

The employees themselves put on a brave face, especially at Vancouver, where a shutdown of the airline and the accompanying loss of 6,400 jobs would severely damage not only the airport's but the province's economy. The loss to British Columbia was compared with General Motors closing down in Detroit. It was estimated that 70,400 people would be unemployed nationally if the airline folded—16,400 employees and 54,000 in the food, travel and fuel business. It would cost $1.5 billion to pay the UIC benefits and $21 million in GST rebates that they would be entitled to while unemployed. In total, the *Vancouver Sun* guessed that the national economy would take a $3 billion hit.

On December 13, Benson met with five of the six unions—only the

CANADIAN AIRLINES' MAJOR CREDITORS

Lienholder, lessor or trustee (partial list)	Collateral
Royal Bank of Canada (major lender)	Not available
General Electric Capital Corp., of Stamford, Conn., With GPA Group, Ireland (GE Capital); Banque Paribas; Natwest Aerospace Trust, London; Hambrose Trust, London; Gecas Saskatoon Inc., Barbados (GE Capital)	2 Airbus A320s 7 Boeing 737s Boeing 747
Montreal Trust Co. of Canada, Calgary (trustee)	Boeing 747 spare jet engines and parts 4 flight simulators all future receivables
Bank Hapoalim, Tel Aviv	17 Boeing 737s
PLM Worldwide Leasing Corp., San Francisco, with First Security Bank, Salt Lake City and Christiania Bank, New York	14 Boeing 737s
USL Capital Corp., San Francisco	1 Airbus A320 4DC10s
Kreditanstalt fur Wiederaufbau, Frankfurt	Boeing 747
Nissho Iwai Corp., Toyko	Boeing 747
Castle Management, Oslo	2 Airbus A320s
International Lease Finance Corp., Los Angeles	Airbus A320
Aircraft Lease Portfolio Securitization, Jersey (UK)	Airbus A320
Sanwa Business Corp., Chicago	Airbus A320
Greyhound Financial Services, London (UK)	2 Boeing 737s
LTCB International Leasing; International Leasing Corp., Ryoshin Leasing Corp; Showa Leasing; YTB Leasing Co., NTT Leasing Co., all of Tokyo	Boeing 767
Fuyo General Lease Co., Yasuda Lease Co., Tokyo Marubeni Corp., NK Lease Co., Kobota Lease Corp., Hitachi, all of Tokyo	Boeing 767 Boeing 767
Algemene Lease Co., Amsterdam	Boeing 767
JJHD Aircraft Inc., Wilmington	Boeing 767
Potomac Capital Investment Corp., Washington	2 DC 10s
Diamond Lease Co. Ltd, Tokyo, and three others Tokyo Lease; Sumisho Lease; Central Leasing, all of Nagoya, Japan	2 DC 10s DC 10
JL Rocky Lease Co., Tokyo CFM International, Cincinnati IBM Canada, Toronto Xerox Canada, Toronto	DC 10 4 jet engines computer equipment office equipment

Source: Ontario Ministry of Consumer and Corporate Relations; Alberta Personal Property Registry

Canadian Airline Simulator Technicians (which represented twenty-two people) were not at the Richmond hotel meeting. While the talks plodded on, protesters called on Prime Minister Chrétien for federal aid. Ottawa, after all, had just given thriving Bombardier Inc. of Montreal an interest-free $87 million loan. But the prime minister had said that a government handout wouldn't help the airline gets its finances in order. The first breakthrough occurred on December 21, when the IAM&AW agreed to the wage reduction, provided employees that earned less than $30,000 a year were not affected. The Canadian Air Line Pilots Association, whose members had the most to lose, also supported the 10 percent reduction. Then the Canadian Union of Public Employees (CUPE) reached a tentative thirty-eight-month deal on December 23 and agreed to join the discussions on restructuring.

Other offers were also forthcoming. Premier Ralph Klein said that Alberta would give CDN an eighteen-month extension on the 1992 loan and speed up a planned reduction in aviation fuel to save the airline $8 million annually. On November 27, Ottawa decided to take part in a taxpayer-funded rescue by agreeing to refund jet-fuel taxes to any airline that had posted large losses over a number of years. The condition was that all six unions agree to Benson's restructuring plan.

Of the holdout unions, Hargrove's CAW had the highest media profile. A commonly heard complaint was that mismanagement, not workers' wages, was responsible for CDN's problems. Members argued that the company had come to them as many as five times in this decade for concessions, only to throw the savings into new planes and money-losing routes. The management presented their latest pay-reduction proposal, which would see those earning up to $22,500 annually take no cut, while those making $30,000 took a 1 percent cut. BC premier Glen Clark flew to Fort Worth to discuss with AMR the role it would play in any survival plan. CDN wanted to pay AMR $48 million a year less than the $110 million annual management fees it then paid. Robert Crandall wrote to transport minister Anderson that despite the drop in value of AMR's investment, "we are considering Canadian's request for an adjustment in our service contract." Benson now had four unions, two provincial governments and Ottawa agreeing to take the $35 million in wage cuts and $35 million in fuel tax cuts and appealed to CUPE and CAW leaders to allow their membership to vote on the restructuring package.

In Toronto on December 2, Benson announced that he was about to suspend payments to the seventy banks, trading and leasing companies of about $170 million of the $265 million due over the next six months. Benson and Carty gave the group of creditors a short rundown of the company's financial picture and explained the debt referral plan, describing it as a "short-term bridge." The time bought would carry the airline through the money-losing fourth and first quarters. Then Benson took his show on the road to the money markets in San Francisco, London and Tokyo.

As Benson scrambled to put together a rescue package, CDN put up its regional airlines in central Canada for sale. Canadian Regional Airlines (CRA) operated fifty-two aircraft: ten ATR 42s, eighteen Fokker F.28s and twenty-four Dash 8s, which included fourteen of the larger fifty-passenger Series 300. On the block were Inter-Canadian (1991) formerly Quebecair and Ontario Express Ltd. Both wholly owned subsidiaries lost money, but they did feed thousands of passengers into the Toronto and Montreal hubs and would continue to do so, whoever they were sold to. The 45 percent interest in Calm Air ensured that it collected traffic from Manitoba, the Northwest Territories and northern Ontario into the system. The Maritime carrier, Air Atlantic, was an independent regional carrier. Fortunately, there were no buyers, and the regionals that accounted for 13.7 percent of the corporation's passenger revenue and 25.1 percent of its total domestic revenue in 1996 were able to show improvement in the following year with 16.8 percent and 30.8 percent respectively. The increasing use of the refurbished Fokker F.28 jets now with full business class service on the high-density Eastern and Western Triangles (and an Air Canada connector strike) were the cause of a 25 percent jump in revenue the following year. Duncan Fischer was replaced as the president of CRA by Mary Jordan, who reported directly to Benson. Although she had come to CRA via the senior-management ranks at American Airlines, Jordan was familiar with the problems of regionals and—Canada. Born in Saint John, New Brunswick, after earning an MBA she joined American Airlines in 1983 because her parents had moved to the southwestern United States. Although she had cut her teeth on the huge American Eagle family, Jordan was under no illusions about the differences between US and Canadian markets. "US regionals operate from multidirectional hubs with strong hub-and-spoke patterns. Within CRA, we see considerable point-to-point traffic over a much more linear network, with fewer dominant hubs." Under Jordan, CRA would take a more active role in

Canadian-born Mary Jordan came to the presidency of Canadian Regional Airlines via the senior ranks at American Eagle. In April 1998, Jordan assumed the office of senior vice-president. CAA AA92-95-3/a

supporting the parent company, particularly on the business triangles. Like the Asian market (now with a fourth Boeing 747-400), the regional airlines had the potential for profit, but only if the restructuring program could be successfully implemented.

AMR said it was prepared to lower its fees by $48 million annually over the next four years. In Ottawa, finance minister Paul Martin and transport minister David Anderson had been careful to characterize the fuel tax rebate offer as not a bailout but as being open to all other airlines and insisted that CDN surrender the potential tax-loss benefits it had accumulated, which would be close to $120 million.

In early December, Hargrove, Anderson and Benson met in Ottawa to try to settle their differences, but the meeting went sour and broke up after twenty minutes. Industry minister John Manley said that Hargrove would have to shoulder the blame if Canadian Airlines went down. "It's one thing to lead your people over a cliff; it's another thing to push them from behind." The members of the CAW felt that they had taken the struggle as far as they could, and the committee was unanimous that its members should back the deal. Using a broadly worded section of the Canada Labour Code, labour minister Alfonso Gagliano threatened a forced vote, and the CAW agreed to a deal. "We are recommending to our members," said Hargrove, "that they vote in favour of this restructuring agreement, including the wage cut. It's a major victory for all Canadian Airlines workers." After five weeks of hard bargaining, the breakthrough came when Anderson softened his earlier opposition to government involvement in the industry and told Hargrove that he would put this in writing. The letter went back and forth several times, the union and the minister's office both making

Praised for his direct approach to the employees, Kevin Benson negotiated a rescue package with six unions, two provincial governments and Ottawa for his four-year restructuring plan. Rick Sloboda/ Canadian Airlines

changes to the wording. Finally at 1 a.m. on December 7, the minister accepted the union's version, with Anderson noting that "while he did not share Hargrove's views on the industry, he did share the concerns." Hargrove announced that his people had "given and given and given and that the wage cuts were not necessary or just, but that the union had fought and fought and fought and were now tired." Anne Davidson, a Vancouver-based CDN employee who had led the bargaining committee, warned Benson, "Don't ever come back again and try to pretend that cuts are the answer. You think you've had a fight this time, it's nothing compared with the fight you'll get." The month had been "just pure hell for the employees—the stress of worrying is tremendous."

"Now we can go out and tell people who have flown with us to come back," said Benson in a telephone interview with the *Toronto Star*. "They don't have to worry about their points." The agreement meant that of the airline's 16,400 members, not just the CAW members, earning less than $25,000 annually would not have to take the pay cut, but reductions on a sliding scale would increase to about 8 percent for pilots. The average cut for members of the CAW would be about 3.7 percent and in effect for four years. Salaries would decrease in 1997 by

$47 million, or a 5.8 percent reduction from 1996. The operational restructuring plan would result in a reduction to expenses of $31.4 million in 1997.

Its financial difficulties notwithstanding, in 1996 Canadian Airlines carried 11.9 million passengers to over 240 destinations in nineteen countries and on five continents. In Canada alone, it flew to 108 places. Its domestic passenger revenue was $1,053.5 million, accounting for 47.6 percent of its scheduled passenger revenue. International passenger revenue accounted for 52. 5 percent or $1,164.2 million. There was a net loss of $187.1 million, compared with $194.7 million in 1995, but then the airline was moving into the first year of Kevin Benson's operational restructuring plan.

Nordair, Eastern Provincial Airways, Wardair and Canadian Pacific Air Lines had all disappeared into the graveyard of aviation history, and in November 1996 Canadian Airlines nearly joined them. Twenty years had come and gone since Don Watson had resigned and Rhys Eyton's Mexican vacation had been interrupted. Few airlines have transformed themselves so completely. Canadian Airlines had changed direction, and with care and caution—and renewed enthusiasm—it had stepped back from the abyss.

Even in July 1997, seven months after the crisis, when the company had shown an encouraging profit of $2.6 million in the second quarter, Kevin Benson refused to speculate on its future. "If a driver thinks of winning in the sixth lap," he told reporters, "he rarely makes it past the third. We are where we hoped to be at this point—no more, no less." His restructuring plan had three years to run, and whatever the fiscal wizardry, the soothing of bruised employee relations or seizing of Asian opportunities, the fundamental economics of running an airline in Canada had not altered since the days of James Richardson, Grant McConachie or Rhys Eyton. He knew that government tourniquets and employee wage concessions can never fundamentally correct the geographic and economic disparities that have plagued this country since the Canadian Pacific Railway laid two bands of steel across it more than a century earlier.

CONCLUSION

Final Approach

That Canadian Airlines, having lost $1.53 billion between 1989 and 1996, was still in business in 1997 was something of a miracle. That the company not only pulled out of its decade-long debt spiral that year but also earned a modest profit of $5.4 million was even more of a miracle. In his company's annual report for 1997 Kevin Benson would write: "The first year of our four-year Operational Restructuring Plan focussed on cost reductions and network realignment. The next three years will focus on the subsequent phases of the Plan: revenue growth, balance sheet improvement and fleet renewal." This survival plan, however, hinged on four pegs over which Benson had no control: the economic health of the Far East, the negative exchange rate in Canadian/US dollars, the vindictiveness of Air Canada, and the support of American Airlines.

In 1997 traffic from core Asian markets, Canadian's traditional mainstays, had increased by more than 40 percent over the previous year. While there were some troubling signs on the horizon—Hong Kong had just been returned to China, and Japan was experiencing a decline in the yen—traffic from Taipei and Beijing more than compensated for this. Canadian was confidently forecasting that its Asian traffic to North America would rise 10 percent in each of the next two years. The International Air Transport Association (IATA) was more cautious—it predicted a mere 4.4-percent traffic growth in the Asia-Pacific region.

The dollar exchange rate was a growing problem. Except for its

349

operations to the United States, the airline earned its revenue primarily in Canadian dollars, though also in Japanese yen, the Taiwan dollar, the British pound sterling, and the Hong Kong dollar. Unfortunately, while the yen was weak compared with the American dollar, the Canadian dollar was weaker still, and all Canadian's aircraft finances, equipment rentals, fuel bills, and AMR service fees were paid out in aviation's international currency: US dollars. Thus, the company's operating expenses were deeply affected by fluctuations in the US/Canada rate of exchange. In 1997 every penny of change in that rate made an impact on Canadian's expenses of approximately $9.5 million. The obvious solution was for the airline to expand as far and as fast as it could into the United States and earn more US dollars.

Then there was Air Canada. Not counting its regional carriers, in 1997 Air Canada had an overall yield (the average revenue generated from each mile travelled by a paying customer) of 17.5 cents, while Canadian's yield had fallen to 13.94 cents. Worse still from Canadian's point of view was that in the fourth quarter of the year Air Canada's yield had risen to 19 cents while Canadian's had fallen to 13.87 cents. Taking full advantage of the 1995 "Open Skies" treaty with the United States, Air Canada had not only captured more than 59 percent of the high-paying business travellers' market, but its trans-border routes had generated 25 percent of its revenue and had accounted for almost 40 percent of its profits. Ironically, at the same time Air Canada remained as unpopular with the general public as ever. When author Peter Newman said that "if God had meant us to fly, he would never have invented Air Canada,"[1] many of his countrymen would have agreed with him. But none of this seemed to poison Air Canada's bottom line, and both its domestic and international flights were doing well. Now run by Hollis Harris, its second American CEO, Air Canada had re-equipped with a fleet of the latest Airbuses and ended 1997 with a war chest of $650 million. However, having been shut out by Ottawa from much of the Far East, Air Canada's management complained that Canadian was too protected by the federal government, even to the point that the Liberal Party always used a Canadian Airlines Boeing 737 when campaigning. Most people at Air Canada believed that in the "real world" Canadian would have gone bankrupt years earlier.

As to AMR, American Airlines' parent corporation which had bailed Canadian out and now owned 33 percent of it (with a 25-percent voting interest), that Dallas-based company had already been unfaithful to its northern partner. While still funnelling its Asia-bound passengers onto Canadian's aircraft through the Vancouver hub, American was taking advantage of a new bilateral air transport treaty between Japan and the United States that allowed 90 round-trip flights a week for US carriers—including AMR. Linked with this was a code-sharing agreement between Japan Airlines and American Airlines that would divert some of American's feed away from Canadian. American also made it clear that it planned to strike more code-sharing deals with other Asian carriers, further diminishing its need for Canadian.

All these problems barely showed on Canadian's radar screen at the end of 1997. It had, in fact, been a boom year for the whole airline industry and Canadian's aircraft had been filled to 75-percent capacity. Passenger revenue was up 2.3 percent from 1996, net income was a credible $5.4 million, compared with a net loss of $187 million the year before. Money-losing routes to Europe had been dropped, and the Fokker F.28s with their lower operating cost had taken the eastern shuttle into the black. Online services such as "inter@ctive" had proved popular, two provinces had given fuel tax rebates, creditors had once more agreed to defer payments, jet fuel prices had fallen, and discount carrier "Greyhound Air," a potential rival in the West, had bitten the dust. The introduction of Canadian's "Executive Platinum" reservations desks, which concentrated on the high-yield business market, had opened more trans-border routes and led to closer code-sharing with American Airlines and British Airways.[2] Stock in parent Canadian Airlines Corp. had climbed from $1.50 in mid-January to a high of $4.65 in October. The future for Canadian looked secure at least as far ahead as December 31, 2000. But, characteristically cautious, Benson was not about to break out the champagne just yet. "There are no big celebrations going on anywhere in the company at these results," he said. "They [the results] are as targetted, no more or no less, and we remain very focussed on the remaining three years of the plan."

But the evil star that Canadian had laboured under for so many years hadn't finished with the company. Six days into the new year an ice storm caused a four-day closure of both the Montreal and Ottawa airports, postponing the Team Canada trade mission that Canadian was flying to Latin America. Although the mission was only delayed by a single day and Canadian got the contract to bring in troops and generators from Edmonton to Mirabel Airport, the storm would cost the company $11 million of its first-quarter earnings. The airline attempted to recoup some of this by extending the New Year seat sale to January 20.

Canadian hadn't tapped the capital market since 1991, but it now needed to raise equity to replace its fleet. On April 7, 1998, Canadian Airlines Corp. (CAC), the holding company which held a 67-percent equity interest in Canadian Airlines Inc., issued notes worth $175 million US of high-yield debt in secured notes due in 2005. "It's a first step in showing the markets and showing the public that we're serious about raising capital," said company spokesman Jeff Angel. But the money was going to be used to repay indebtedness, he said, not to buy much-needed new aircraft. Analysts knew that Canadian, even with its debt of more than $800 million, would have no trouble selling the issue. On the strength of the US high-yield "junk" bond market, Moody's Investors Service assigned a B3 rating to the issue while Standard & Poors gave it a B-minus rating and the company's credit a triple C-plus. As a result, small investors, betting on a successful turnaround, were prompted to put their money into Canadian stock, hopeful that its 1997 profit was a foretaste of things to come. By spring Canadian's shares—at $1.40 the year before—had reached $5. That April Canadian even had some spare change available to repay the Alberta government $12 million on the loan guarantee package stemming from its 1992 rescue.

Then in May 1998 the US Federal Aviation Association grounded all early model Boeing 737s in order to inspect their Teflon-coated wiring systems. By that time the nineteen 737-200s in the Canadian fleet averaged eighteen years of service. Since these old models had to be brought in for inspection, the company took this opportunity to hush-kit them as well, then proceeded to hush-kit the other twenty-three 737s in its fleet. This left passengers scrambling to make alternative arrangements, and consumer confidence was lost—especially on those routes where Air Canada was competing with its new Airbus 319s. That August, a three-year agreement to overhaul the DC10-30s and 737s was signed with Hawker Pacific Aerospace of Sun Valley, California. This was an expensive exercise—financially and public relations-wise—because it lowered the second quarter's profits to $4 million and highlighted the need for fleet renewal.

The next blow came from American Airlines which, caving in to the demands of its pilots' union, revised its code-sharing arrangement with Canadian and withdrew fourteen of the airline's nineteen trans-border "designators," the most significant being Vancouver–Dallas, Toronto–

Dallas and Vancouver–Chicago. American's pilots were able to force the issue because their airline hadn't provided them with enough trans-border flight hours the previous year to satisfy their union contract, mainly because Canadian had been capturing an increasing share of that traffic to the tune of $1.1 billion in 1997 or 55 percent of the company's scheduled passenger revenue. Angel played down the loss, saying that the codes would be reintroduced at the start of next year and that passengers wouldn't see any change other than a different flight number on their tickets. Most important, the alliance with American, vital to Canadian because it provided passengers for its Asian routes, was unaffected.

In the meantime, the airline increased cash flow by selling assets and cutting costs. It gave up its expensive downtown Calgary premises and leased 100,000 square feet in a tower at Calgary Airport. The company's regional carriers were the next on the block. While they had done well as a whole, delivering $500 million the previous year, some of them posed potential problems. Canadian North, connecting Edmonton with Yellowknife, Inuvik, Resolute Bay and Iqualuit, was sold in June to Norterra Inc., an aboriginal-owned company. Montreal-based Inter-Canadien (1991) was sold off in August to a group of Quebec investors. This sale allowed Canadian to break its commercial agreement with Air Atlantic, but as Inter-Canadien (1991) would remain a partner of the parent airline, Canadian could maintain the feed traffic from the Maritimes without the economic hardship. The biggest sale took place in November with the transfer of ownership of the company's engine shop with its 200 employees to MTU Munich, a Daimler Chrysler Aerospace affiliate. All of these cost-saving measures, coupled with programs such as MAPS (Minimum Acceptable Performance Standards), which were designed to raise the efficiency of staff, allowed Canadian to report a happy second-quarter profit of $4 million—the best second quarter since 1990.

When Air Canada pilots went on strike on September 2, Canadian's management recognized it as a fresh opportunity to win over the hearts and minds of the public. For the two weeks that it had a national monopoly on most routes, Canadian filled its planes to capacity, added more flights, worked its staff overtime and launched a media campaign thanking everyone for their patience. But when the strike ended, the hard-won customers, lured by Air Canada's "kiss and make up" seat sale and anxious to protect their Air Canada frequent flyer points, deserted

Canadian, and the company was left paying for heavy overtime and maintenance fees—with no permanent gain to show for it.

On the brighter side, on September 17 the company took possession of the first of two Boeing 767-300ERs from Asiana Airlines—its first new aircraft since 1995. Two brand new 767-300ERs were also leased from GE Capital Aviation Services to replace two of the DC10s. At Vancouver Airport a new crew room was opened on Level Two and the Operations Centre spruced up with new carpeting and signage. But the biggest boost for Canadian came on September 24 when Kevin Benson, addressing the Empire Club in Toronto, broke the news that Canadian had been invited to join "oneworld." In the late 1990s airlines had been coming together to form global alliances to co-ordinate ticket coding, frequent flyer programs and schedules. The two biggest blocs were Star Alliance (Air Canada, Thai Airways, United Airlines, Lufthansa, Scandinavian Airlines System and Varig) and oneworld (American Airlines, British Airways, QANTAS and Cathay Pacific). Oneworld was the older one and well placed on the crucial New York–London route. In 1997 its 220,000 employees had flown 174 million passengers to 632 cities in 138 countries. Being a member would allow Canadian to offer its passengers more destinations, common passenger and baggage handling and access to 223 airport lounges worldwide. Grinning and cracking jokes at the expense of Air Canada, Benson said this was the best thing to happen to his airline.

However, at the end of summer, traditionally the busiest time in air travel, Canadian reported a third-quarter profit of only $93.6 million—down from the $106 million of the same period in 1997. Management knew they had to generate more revenue quickly. But with Air Canada's "kiss and make up" seat sale, the continuing loss of its American Airlines code on some of the trans-border routes, the weak Canadian dollar, and crippling NAV CANADA fees, it would take real imagination to increase company profits. Their first move was to initiate a fresh campaign to capture a larger share of the business travellers' market. By 1998 the airline held a 32-percent share of that market, but while corporate travellers accounted for only one in every 200 passengers, they represented 40 percent of Canadian's revenue. In the past, industry studies had shown that while business travellers valued price, schedule and connections very highly, they also wanted creature comforts such as opulent airport lounges, larger seat size, and good food. Canadian's campaign, therefore, included new "Millennium J" seats on its A320s,

Acting on survey results that showed business travellers placed a high premium on creature comforts, Canadian spent $38 million on new and renovated executive lounges at airports. Canadian Airlines

power seats on the 767s, and the introduction of the "Chef's Conclave," a series of signature menus from a dozen of Canada's best chefs. (One passenger paid the airline the ultimate compliment: "This meal is so nice that I forgot I was on an aircraft.") At major airports the airline also opened new "Canadian Club" lounges, the gathering place for pampered road warriors, and redesigned existing ones. The first to get the treatment was the Empress Lounge at Vancouver where the massage chairs, cyber café and eighteen-hole putting machine were a hit. The Empress Lounge at Pearson's Terminal 3 was redesigned to feature a business centre with fourteen work stations, Internet and music-listening stations, shiatsu massage chairs, freshly prepared food, a cappuccino machine and an array of single malt Scotches.

A snowstorm on January 2, 1999, closed runways at Pearson International Airport, and the country's two major air carriers were forced to cancel flights to and from Toronto, backlogging national service into the following week. Air Canada, with 45 percent of the traffic into Pearson, failed to communicate the cancellations to its customers in time, and hundreds of fuming would-be passengers continued to show up at the airport. Although the company insisted that the fault lay with Mother Nature and not with the staff cutbacks instituted since the summer strike, its customers were not consoled. "It wasn't Air Canada's finest hour," said one travel agent. Calgary-based Canadian Airlines, however, was able to bring in aircraft quickly and, as its customer service staff levels had not fluctuated, they were able to fall back on the contingency plan they had used during the Air Canada strike. Customer Service Manager Laurie Schmitz recalled that "the only staff who could make it to work were those who had spouses and four-by-four trucks," but once at the airport they worked round the clock, grabbing naps when they could. Staff waded through the masses of disgruntled, bleary-eyed passengers to get to their desks, set up standby counters behind the regular check-ins, and climbed on top of baggage counters to shout out the names of fortunate passengers who were given the precious few seats available.

Then, with this emergency behind them, on January 13 Canadian launched its new "Proud Wings" logo—a bright blue stylized version of the old Canada Goose in flight—when the newly repainted and redecorated CP30 (a 747-400) arrived at Vancouver airport. Work on the aircraft had been done in the utmost secrecy in Beijing so that the unveiling would have the maximum impact for shareholders, employees, media and customers. Test painting had taken place at an aircraft graveyard in Arizona, then the plane and the paint masks sent to China, far from prying eyes. While the exterior was being painted, the seats had been reupholstered and the entire cabin refurbished. At the same time employees were issued new dark blue uniforms, which were variously described as "elegant," "crisply tailored" and "very executive." The new uniforms, the redesign of cabin interiors, and the repainting of everything from planes to hangars to trolleys would cost the company $38 million over two years, but it was hoped that this new image would mark the turning point for the carrier.

To ensure that the travelling public got the Proud Wings message, Canadian also launched a new advertising campaign. Ad agency Gee Jeffery and Partners set out to show business travellers that Canadian was a caring, "home team" airline that understood the stresses and strains of their busy lives. Since it was well known that businessmen do not have to worry about the price of airfares, the five TV spots they created aimed at the heart rather than the wallet, extolling serenity and wholesomeness. Newspaper ads targeted the business traveller with "Menu by Design" meals, more leg room, and lap-top plug-ins. Sadly,

In January 1999 Canadian Airlines International unveiled its new "Proud Wings" livery, a stylized version of the old CP Air Canada goose. There would be only one more logo change—the addition of Air Canada's maple leaf. Canadian Airlines

the ads' "caring" message missed the mark because just at that point the industry was becoming more price-driven. (As a footnote, it is interesting that the airline shared the prize for the most innovative advertising spots that year with the Eaton's department store chain—which was also foundering.)

But while all systems seemed to be "go" for Canadian in early 1999, there were already rumours that Canadian's losses for the year just ended could reach as high as $130 million. Barbara Amster confirmed that it would be significant. "I think we have said that to us minus $50 [million] to plus $50 [million] is break-even. This is not break-even." Rumour became fact on February 10 when the company posted its disastrous fourth-quarter loss for 1998—$150 million. Capital expenditures for the year had totalled $81.4 million, most of which had been spent on hush-kitting the 737s and on investment in the new company image. A net loss of $137.6 million for the whole year was recorded— a considerable difference from the profit of the previous year. On the plus side was the fact that on December 31, 1998, the airline had more cash on hand than it had ever had—$307.4 million compared with $190 million a year earlier. Labour costs were under control: Canadian's 14,123 full-time employees were covered by labour agreements that didn't expire until December 31, 2000.[3] In 1991 employee payrolls had been 31.2 percent of the operating revenue; by 1996 they were 26.4 per cent and still dropping. And as a member of oneworld, the airline now had powerful friends. Furthermore, in February 1999 American Airlines returned use of the codes it had cancelled, and a new code-sharing agreement was reached with rival Alaska Airlines and its sister carrier Horizon Air. These developments, however, were somewhat offset by the news that Air Canada had been given permission by Ottawa for up to three flights weekly into Taipei, a market that Canadian had developed.

The consensus in the financial markets that Canadian's immediate future still looked secure continued even after the company posted first-quarter losses for 1999 of $102 million operating and $107 million net. "Can we continue to lose $100 million a quarter?" Benson said at a media conference held in a Toronto hotel. "Absolutely not." Standard & Poor analysts reacted by giving Canadian's CEO credit for being a "straight shooter" and said that it wasn't a case of a badly run opera-tion. Benson had made all the right moves with good service and cost cuts but had never got an even break from the markets or the Canadian dollar (currency swings had cost the airline $70 million in 1998) or from the American Airlines pilots' union. Its Asian core markets had finally recovered, and this fact together with increases in trans-border and British traffic had pushed the first-quarter load factor to 69.4 per-cent and 40 percent of the business market. Membership in the oneworld alliance was paying off, particularly with increased United Kingdom flights.

At the same time there were hints of a deepening crisis for the air-line. The annual general meeting scheduled for Vancouver in May was postponed until the end of June. The reason given was that Benson "was talking to the financial team about a lot of different options for recapitalizing the company" and that delaying the meeting would allow shareholders to better assess the expected gains. Management, Benson said, was also busy on a "short haul" plan, shifting the focus from the business traveller to the discount "no-frills" airline concept to take advantage of the coming vacation season. But revamping to capture a share of the short haul market that WestJet had exploited would mean scaling back the fleet and breaking leases on the gas-guzzlers in favour of fuel-efficient 737-700s. And while these changes might have been possible, Canadian's long-suffering unions would not allow the company to reduce the unionized workforce in order to take on the non-unionized staff required to make a no-frills service cost-effective. The short haul plan was dropped.

After the CEOs from the seven airlines of oneworld held their quarterly meeting in Vancouver on May 14 to welcome Lan Chile and Finnair into the fold, it was suggested that one of them might come to Canadian's rescue. Benson agreed that a cash windfall from a oneworld partner such as American or British Airways would have suited him but, he said, "You don't go to friends first." In any case, AMR certainly wasn't keen on investing any more in Canadian—not only because of the 25-percent limit but also because its powerful pilots' union opposed further investment in the carrier. In fact, it would have been a very foolish white knight who would ride to Canadian's rescue at this point, putting money into a chronically money-losing company whose stock had dropped more than 90 percent in five years. In addition, Canadian was not only nearly $1 billion in debt, but any rescue bid was effectively booby-trapped by AMR's 1994 $246 million purchase of 827,000 Canadian Airlines preferred shares. Since AMR had the option of converting them into common shares and getting as much as $430 million for them—many times the value that the stock market was then giving Canadian's shares—any purchaser of Canadian Airlines would have to buy back AMR's stock at that inflated price.

To raise money while management looked for a way out, Canadian sold off its interest in Equant, the telecommunications company, for $23.5 million and gained another $45 million by selling two of its DC-10-30s. It also deferred its order for 10 new Airbus A320s that were scheduled for delivery in 2001 and 2002. As well, to recoup some of its NAV CANADA fees, which in the previous year had risen from $70 million to $130 million, the airline introduced a navigation surcharge on all tickets. The airline followed this up on June 1 with an issue of 1.7 million non-voting shares to holders of non-interest-bearing convertible notes (due in April 2002) who had elected to convert them into shares. There were now $33.4 million in notes outstanding, and every shareholder was aware that if Canadian went into liquidation they would not get a penny. With $21 million of junk bond interest eating away at any profits, the airline now needed an estimated $200 million to just break even, and industry watchers expected either a cash infusion or a retrenchment—or both. David Bell, the airline's treasurer, assured the media on June 9 that a deal was imminent. "Canadian expects a major transaction in 1999 and American Airlines is going to be a big part of this." As to the bond issues of the previous year, Bell admitted, "In hindsight, that wasn't the right decision. But who was to know the way markets would go?" Then to everyone's surprise, on June 11 Canadian quietly celebrated an increase of 5.6 percent in traffic for the fifth consecutive month, while Air Canada reported a dip of 2.2 percent. Canadian also hired 25 new flight attendants to cope with the summer schedule.

On June 29 the company held its long-awaited AGM in Calgary. Rumours swirled about the packed house that a deal had been crafted with AMR and British Airways that would include the participation of Canadian investors to keep the foreign ownership within legal limits. Instead, Benson announced that, despite failed attempts in the past, he wanted to seek some kind of route-swapping deal with Air Canada once he had firmed up the $300 to $500 million equity infusion he was looking for. He explained that he had opened discussions with Air Canada's crown prince, Robert Milton. The 39-year-old wunderkind Milton was scheduled to inherit the office of president and CEO from Lamar Durrett, who would officially resign on September 1. Unknown outside Air Canada, he would be the third successive American to run the airline. This wasn't the first hint that the two were talking. In March, when Jeff Angel had confirmed that Canadian was involved in talks designed to help Ottawa create a new policy for designating international routes, members of the media had asked if "other parties" were involved. He had responded that "private discussions were not for comment." No one really believed that talks now between the two airlines would lead to a merger, yet Benson's remarks at the AGM caused some speculation. Had he given up the struggle with Air Canada? Canadian's media office later said that the CEO had just been "musing." Air Canada spokesperson Priscille Leblanc responded when asked for a comment that her company was always open to business proposals and that it did have ongoing discussions with Canadian on any number of industry issues but that Air Canada was well aware of the Competition Act.

Under Commissioner Konrad von Finckenstein, Canada's Competition Bureau was set up as an independent law enforcement agency responsible for merger reviews and the conduct of business as defined by the 1985 Competition Act. Its rules forbade competitors from talking to each other on prices, routes or capacity or taking any action that either smacked of collusion or was likely to lessen competition. Although its powers were to be limited to identifying competition issues and indicating to the government means of mitigating them, it was expected that the bureau would be decisively involved in approving any merger. The year before, Finance Minister Paul Martin had taken its advice regarding the prospective merger of big banks in Canada, and on the strength of the bureau's report had disallowed it. Having reviewed previous attempts at airline mergers—Canadian and Wardair (1989), the proposed Air Canada and Canadian merger (1992), the alliances between Canadian and American Airlines (1994) and between Air Canada and United Airlines (1996)—the bureau had expressed concern that competition was fast becoming extinct in the airline industry. A combined Air Canada and Canadian Airlines—or, alternatively, Canadian's bankruptcy—would leave Air Canada the dominant carrier. But after a lengthy investigation, on July 15 the bureau issued a statement saying that it had looked into the allegations of complicity between the two airlines but could find no evidence of such.

The precariousness of Canadian's plight became clear on July 28 when the company released its second-quarter financial report, showing a net loss of $17.8 million—compared with a net profit of $4 million in the same period the year before. Calls for a federal lifeline for Canadian now came from an unexpected quarter: Buzz Hargrove, leader of the Canadian Auto Workers (CAW), which represented the 3,159 passenger service agents who made up 22 percent of the company's workforce. Ottawa, Hargrove said, should inject $100 million annually to buy a stake in the airline and get rid of the 25-percent cap on foreign ownership restrictions, thereby allowing AMR to increase its investment. The federal government's response came on July 30 when Defence Minister Art Eggleton announced that Canadian had been awarded a $96 million, three-year contract beginning April 1, 2000, with an additional two-year option to continue to provide domestic charter services for the military. Canadian had won its original five-year DND contract, valued at over $233 million, in 1995, operating 14 regular and charter flights weekly to places like Petawawa, Ontario, Cold Lake and Wainwright, Alberta, and Val-Cartier, Quebec. The new contract was not, of course, enough to save the company, and everyone waited to see what else would be offered.

The wait ended on Friday, August 13. The markets had closed and Parliament had been in recess for weeks when Transport Canada called a press briefing for 4 p.m. Industry Minister John Manley and Transport Minister David Collenette announced that the government would invoke Section 47 of the Canada Transportation Act, suspending for ninety days all competition in the airline business to allow Air Canada and Canadian Airlines to discuss a restructuring of the industry with a view to co-operation in some areas. The request to invoke Section 47 had come from Canadian's CEO that morning, Collenette assured the media, adding, "We believe that Canadian came to the view that unless some measure of this nature was used, then they would start to perhaps not discharge some of their service obligations. It was no secret that Canadian had made a valiant effort to meet their restructuring plan and they had come to the conclusion that it would be in their best interests and the domestic industry to have this restructuring." Manley (whose department oversaw the Competition Bureau) threw in, "I'm committed to ensuring the outcome will be as pro-competitive as possible for the benefit of Canadian consumers." Unfortunately, no one asked him how suspending anti-trust legislation so that both airlines could consolidate fares and divide up routes was "pro-competitive."[4]

The announcement by Manley and Collenette on August 13 was a historic one as it spelled the end of the federal government's two-of-everything policy, which dated back to Confederation. The public's perception was that Canadian's protective shield had been taken away and that Air Canada had been given the green light to move in on its weaker rival, a perception that the transport minister had wanted to avoid. Government intervention into a free market economy was inflammable enough without considering the effect that abandoning the airline would have on its shareholders, consumers and employees. What the government did have in its favour was that Canadian had initiated the request for the suspension by letter on August 9. Had the request for suspension come from Air Canada, the Liberals might have been under serious attack in the West by the Reform Party. However, given the Liberal government's track record, there was, even on that Friday afternoon, a whiff of misgiving as to what was behind the press conference. *Globe and Mail* columnist Hugh Winsor was closer to the truth than he realized when he wrote that "Mr Collenette and Mr. Manley are hanging their hopes on some outside (that is, not Air Canada) investor or investors who can be persuaded to invest in some restructured entity which would be more attractive than either of the two airlines in their current form."[5]

In fact, behind the Collenette/Manley announcement was a drama that had begun months earlier. In January 1999, when Canadian's board of directors had met to discuss the state of the airline and okay the launch of the Proud Wings campaign, they had concluded that the status quo in Canada's airline industry was obviously no longer working. Benson was given the authority to begin discussions in February with Air Canada about merging the two companies under the oneworld banner. As it was part of the rival Star Alliance, Air Canada at first rejected this idea but came back in April with a counter-proposal. The merged company would switch its alliance to oneworld and out-source its reservations to oneworld's SABRE system, but in return Air Canada wanted a $1 billion payment from Canadian to compensate for the potential erosion of its balance sheet. Canadian refused.

Meanwhile, back in March, with Canadian approaching bankruptcy and in need of friends in high places, Kevin Benson had heard rumours of a Cabinet shuffle and had approached Jean Pelletier, a senior aide to the prime minister, to find out who would be heading up the transport ministry. Pelletier introduced him to David Collenette, who was slated for the job. He was the one man in the federal government who, according to *The Globe and Mail*, could make or break Canadian and its employees "with the swish of an ordinary pen."[6] First elected in July 1974, the MP for Toronto's Don Valley East had worked through two previous portfolios—multiculturalism and defence—side-stepping the landmines in both. His misstep had come instead in October 1996, when he wrote to the Immigration and Refugee Board on behalf of a constituent regarding a visa application. The uproar led to his resignation as defence minister, but within nine months the prime minister returned him to Cabinet, this time as transport minister. Collenette found himself called upon to decide the fate of the very cornerstone of the country's commercial aviation industry—the two-airline policy, something that both Liberal and Conservative governments had affirmed since 1947. If Canadian was to be at last allowed to sink or if the two airlines were to consider a merger, everyone from passengers, labour unions, mayors of small communities, travel agents and members of Parliament would want a voice in deciding the outcome—and Collenette would be stuck in the middle.

A sign of the government's increasing fear that Canadian was about to fold came in June when Collenette convened a secret cabinet committee, made up of Justice Minister Anne McLellan, Environment Minister David Anderson (representing British Columbia), former Treasury Minister Marcel Masse and Public Works Minister Alfonso Gagliano.[7] The committee concluded that Canada was about to suffer the fate of Germany, France and Australia—that is, of having one major international airline fed by a number of regional airlines. They already knew that Benson was not going to ask Ottawa for more taxpayer dollars for Canadian. A government rescue, Benson had candidly observed, might please his 17,000 employees but would anger 26 million other Canadians. Still, Collenette's committee knew that those 17,000 jobs translated into $500 million in tax revenue and 35,000 votes, most of them in the Liberal-weak West. And there was a federal election coming up.

While the cabinet committee was stewing, Air Canada had come back to Canadian on June 23 with a surprise offer. It would give Canadian $400 million in cash and an additional $125 million for aircraft spares, as well as assume all Canadian's debt and lease obligations for the aircraft used on its international routes. There would be no layoffs and a mere 1,400 job losses through attrition. In return, Canadian's international routes, slots and rights would be transferred to Air Canada. In essence, Air Canada would then be the international carrier with Canadian limited to code-sharing on domestic and worldwide flights. Canadian had to agree not to compete internationally for ten years, and the company was not to try to find capital, sell or re-finance assets until the end of September. Benson's rejection stated that the amounts being offered for the routes were far less than they were worth and that such a deal would result in layoffs for his company. Then he countered by asking that the offer be extended past Air Canada's July 23 deadline while his officers examined Air Canada's business and financial models on code-sharing. Air Canada's response was to suggest that the sharing of such sensitive information would make it necessary to bring in the Competition Bureau.

What happened next depends on who tells the story. According to Air Canada's spokespeople, "As Canadian refused to enter into a confidentiality agreement preventing communication of the information to a third party, discussions were terminated." Canadian's Jeff Angel said, "We have studied the proposal and Canadian has come to the conclusion it's just not acceptable. Canadian's international routes are the most profitable part of our operation." Air Canada's Priscille Leblanc rejoined with, "We have no reaction. As far as we know, we'll still be meeting next week. The proposal is on the table. It will be next week."

But a week is a long time in politics . . . and airlines. As WestJet's chief executive officer, Clive Beddoe, has said, "Long-term planning in the airline business is about ten days." While Air Canada and Canadian had been exchanging proposals for a merger, in Ottawa Collenette had realized that he would soon have to take action. Although he had promised to seek the Competition Bureau's advice on the airline issue, he would have to go over the bureau's collective head and suspend the competition rules in order to prevent the complete collapse of Canadian Airlines. He knew, however, that any changes in legislation to favour Canadian Airlines would require an Order-in-Council and full Cabinet approval—the only way to implement policy without parliamentary consultation. Taking such a bill before Parliament would ensure a fight from MPs with a heavy concentration of Air Canada workers in their ridings. It was said that Collenette personally walked his special Order-in-Council[8] around to get the necessary signatures from other ministers.

With Cabinet scheduled to meet on July 28, all that remained was for the Privy Council Office to be briefed by Transport Canada and the matter placed on Cabinet's agenda as an "MC" or Memorandum to Cabinet. MCs are presented by the minister responsible for the issue and are intended to brief colleagues on policy changes that are about to take place and make sure that they form a united front. On the day that Collenette's MC was introduced, whether by coincidence or design, Canadian released the figures on its second-quarter losses, underlining the need for action. On August 4, 1999, four days before Canadian Airlines officially asked for it by letter and nine days before calling a press conference to announce it, the minister's office had abruptly informed the bureau that it was going to suspend the competition rules.

On Monday, August 16, the first business day after Manley and Collenette announced that suspension, share prices for both airlines rose, reflecting investor optimism that something good could come out of the government's move. Air Canada gained 60 cents to $7.60 and Canadian rose 14 cents to $1.90. But there were doubts. Analysts knew that a merger of Canadian with Air Canada was not feasible because of Canadian's massive debt load, the incompatible fleets, and the intricacies of combining two workforces with different levels of seniority. And they knew that Air Canada had little incentive to sit down with its rival—except to steal its Asian routes. As for AMR, Canadian was worth more to it dead than alive. The American company still held the veto over all vital decisions, and if Canadian did ground its aircraft and shut down, American's equity stake was convertible into debt, which would give it a greater precedence over all other creditors. The national consensus, therefore, was that Collenette had opened a Pandora's Box and that, whatever happened now, the status quo in the airline business had ended and Canadian could never be the same.

Kevin Benson told the media that Ottawa's decision had been a sensible move that would allow the troubled industry to at least talk about restructuring. "This is just one more step—it doesn't mean anything and it may not lead to anything. But no one can turn around and say that there was no forum for discussion."[9] And he emphasized that it did not mean a merger with Air Canada. "If you limited every date to one that would lead to a fifty-year marriage, you wouldn't have been on too many dates in your life." He refused to say exactly when Canadian would run out of cash, though it was speculated that the company would hit the wall in the spring of 2000. His dating analogy was not too far from what most people suspected Air Canada had on its mind, and industry analysts warned that Air Canada pretended to want to talk, but all it was looking for was to take advantage of Canadian's vulnerability and get at its books. Besides, Air Canada's new American president, Robert Milton, was said to be a "take-no-prisoners kind of guy."

But Collenette's suspension of the competition rules had opened the door for a third party to enter, and on August 24 Gerry Schwartz made public his own proposal to merge the two airlines. Known as the "leveraged buyout king" by admirers and as a corporate vulture by critics, the 57-year-old Winnipeg native commanded unlimited credit— both financially and politically. His Onex Corporation, founded in 1984, had a 1998 annual revenue of $8.8 billion, assets of $6.8 billion and 53,000 employees. Among the companies that it owned or operated at that time were Sky Chefs, Celestica Inc., Dura Automotive Systems Inc., ClientLogic, American Buildings Co., J.L. French Automotive Casting, Phoenix Pictures and Lantic (the sugar company that he had beaten BC entrepreneur Jimmy Pattison out of). Four years before, Schwartz had failed in a $2.3 billion hostile takeover bid for another national icon— John Labatt Ltd. and its Toronto Blue Jays baseball team. All the chutzpah in the world couldn't prevent him from being outbid by $4.50 per share by the Belgian company Interbrew. This time he wasn't going to make the same mistake. Schwartz and his wife, Heather Reisman of the Indigo bookstore chain, were well connected with the Liberals, having raised millions of dollars for the party. Many saw a conflict of interest in this,[10] but the Onex CEO shrugged off the questions about his political influence with, "I'm a Liberal. I always have been. I've been a fundraiser for the Liberal Party in the past. So what? I haven't asked for any favours. It would be a shame to somehow think [the proposed merger] was inappropriate because I, like many Canadians, take part in the political process."[11]

As far back as 1996 Canadian Airlines had talked to Schwartz and his vice-president Tony Melman about an infusion of capital. The talks had ended when it looked as if the AMR solution would work, but they had been restarted in early 1999 at about the same time that Benson began talking to Air Canada about a merger. It was obvious that the airline industry in Canada was crying out for restructuring, and Onex was experienced in doing just that. Melman approached Tony Hine, a First Marathon investment banker and expert at takeovers. Then in early March he met with Peter Buzzi, a merger and acquisition specialist with RBC Dominion Securities Inc. Representing Canadian Airlines, Buzzi confirmed what Melman and everyone else in the industry already knew: Canadian would need up to $500 million to keep flying. In late March Melman approached Benson and Doug Carty with an alternative to investing in Canadian. What about merging both airlines? Benson promised to take the suggestion to AMR; his "musings" at the AGM on June 29 about co-operation with Air Canada had actually been "a trial balloon" to test public opinion. The idea of an Air Canada/Canadian Airlines merger under Onex management appealed to American Airlines, which saw it as an opportunity to recoup some of its $246 million investment. So in May AMR managing director John Boettcher flew to Toronto to meet with Onex and Canadian at the Inn on the Park. As federal law prohibited a foreign takeover of Air Canada, AMR would have to stay hidden while the kill took place. The whole operation depended on it being presented to the federal government as an Onex-led initiative. Called "Project Tornado," it was to be packaged as a wholly Bay Street affair, a deal that a cabinet minister from Toronto could field in Parliament.

But "Project Tornado" had to be carried out quickly, and everyone on Bay Street knew that the Competition Bureau would delight in frustrating the process by studying the proposal with glacial speed, submitting it to nameless committees for further reflection, by which time a bankrupt Canadian would be parking its aircraft in the Arizona desert. The task of circumventing the Competition Bureau was therefore given to Nigel Wright, a one-time speechwriter for former prime minister Brian Mulroney and now a corporate lawyer at Davies, Ward & Beck—Onex's lawyers. Invoking Section 47 was crucial to his company's bid. "If we didn't know what the rules were for dealing with all the issues that would come up," Wright later said, "we had very serious reservations about proceeding without a map. Section 47 was enough for us to go on." Therefore, in June Benson was told that if Onex was to get involved he must convince the government to suspend the competition rules.

Ever since March Kevin Benson had been dutifully making the pilgrimage to Transport Canada in Ottawa. What was not publicly known until Schwartz told the Senate transportation committee that November was that Benson had actually registered as a lobbyist for Onex at that time, and on June 27 he had showed up at Collenette's office with the good news that a "private sector solution" was in the wings (no pun intended), but only if the competition rules could be suspended. Only three days earlier Robert Milton had also come to see the minister to outline Air Canada's proposal to inject cash into Canadian in return for its Asian routes. With either option—Canadian's private sector solution or Air Canada's cash injection into its rival—not a penny of taxpayer money would be needed, no foreign airlines would be involved, and transferring the routes from one airline to the other was well within Collenette's jurisdiction.

Although Collenette had stated in his announcement on August 13 that the government was suspending the rules so the two major airlines could "talk and compare notes," Onex documents filed later in Quebec Superior Court contain two references to "Bidco" (the code name for Onex) being assured that the government would invoke Section 47. The documents, in which Air Canada is code-named "Moco" and Canadian Airlines dubbed "Caco," also indicate that the sole reason for invoking Section 47 was to take out the Competition Board as an obstacle to immediate government approval of the Onex proposal. They also show that Manley had to be told this fact as he and Collenette would have to act in concert.

While this was going on, Onex was buying Air Canada stock

through its numbered companies and RBC Dominion, its investment bankers. This is common enough practice in a takeover bid, as buying enough stock would entitle the predator company to request a shareholders vote if it loses. By July 15 Onex had spent more than $43 million on Air Canada common and non-voting shares, and rumours began circulating on Bay Street that a takeover of that airline was immediate and that an announcement could be expected after Cabinet's July 28 meeting. Wright's response was to order RBC Dominion on July 23 to stop buying. When nothing public came out of the Cabinet meeting, on August 9 RBC Dominion resumed buying. By August 20 Onex owned 3,689,800 Air Canada common shares, representing 3.1 percent of those outstanding, and 4,468,230 Air Canada Class A shares, approximately 6.6 per cent of those outstanding.

It took a two-hour conference call between Schwartz and Canadian Airlines' board of directors on August 23 to get their approval for the proposed Onex deal. Afterwards Schwartz leaned back in his chair and blew loudly into an imaginary trumpet. That same day he registered a new company called AirCo (Airline Industry Revitalisation Co.) to use as a vehicle to purchase a controlling interest in Canadian Airlines and Air Canada. The next day, August 24, Schwartz unveiled his "private sector solution" before a news conference in Montreal, Air Canada's home base. "We all know that the Canadian airline industry cannot maintain the status quo," he said. "This is the time for a bold step." Once AirCo had completed purchase of the two airlines, the new company would merge them to create a strong national airline. It would operate under the name "Air Canada," be based in Montreal, have an annual revenue of $9 billion and rank as the 15th largest in the world (based on revenue). The deal was valued at $5.7 billion, based on Air Canada shareholder value ($1.5 billion), Canadian Airlines equity ($54 million), Canadian Airlines debt ($930 million), Air Canada debt ($3 billion) and AirCo debt ($250 million). AirCo would invest $1 billion with AMR, Onex, and related entities. AMR had agreed to invest $625 million in AirCo equity and own 14.9 percent of the airline. In return it would take over the merged airline's reservations and accounting operations but sell off its stake over time. Onex would invest $250 million and own 31 percent but would also sell off its stake over time. The remaining stock would be widely held by Canadians. Air Canada would be taken from the Star Alliance and put into oneworld. Holders of Canadian Airlines common and non-voting shares would be entitled to receive either $2 per share in cash or .24 common shares of AirCo. Holders of Air Canada common and Class A non-voting shares would receive $8.25 per share in cash or one common AirCo share. The offer, which would expire on or around November 4, would be conditional upon at least 66.6 per cent of shareholders tendering their shares. The only sour note in Schwartz's proposal was that an estimated 5,000 jobs would be cut because of duplication. "This is an opportunity to stop the bloodbath" said Schwartz, noting that the two airlines had lost a combined $2 billion since 1990. "As the government has acknowledged, the status quo is clearly not sustainable. We feel strongly that this is the right private sector solution—comprehensive, long term, sustainable and requiring no government or taxpayer's money."

According to *Maclean's* magazine, Air Canada was reportedly "blindsided" by Schwartz's announcement.[12] Until that moment, the company's management had thought they had the upper hand in their scorched-earth, waiting-for-Canadian-to-die campaign. Now, late night comics cracked that Milton was out looking for shark repellent, and the airline actually fielded calls from sympathizers already mourning its demise. "Hunkered down in their headquarters," said the *National Post*, "Air Canada's US-led management team were livid at yesterday's turn of events."[13] Schwartz's surprise attack had been conducted in the best Pearl Harbor manner, something that Air Canada's present and former American CEOs must have recognized. What's more, from the Montreal airline's view, this pre-emptive strike had been mounted by the Dallas–Toronto–Ottawa axis. If there was some paranoia in all of this, it was because the former government airline knew that it had few friends in Ottawa, and after its mishandling of the January snowstorm, even fewer elsewhere in the country.

The airline's initial public reaction to Schwartz's plan was a pair of stunned press releases that said, "It is difficult to see in the Onex proposal any benefit for Air Canada or its employees." Should Schwartz succeed, it warned, control of Air Canada would be with AMR and Onex, and it characterized the $8.25 offer for its shares as an "unsolicited, below-market offer." After all those months of secret liaisons, Air Canada was not a little hurt by Canadian's rejection in favour of a foreign suitor. "We felt that our proposal was a win-win-win situation for the issues facing American Airlines, Canadian Airlines and the federal government." Milton was also disturbed by Canadian's summary dismissal of Air Canada's last offer, made just three days before Schwartz's bombshell. "The rejection . . . and the announcement of the Onex-American Airlines bid within days of the rejection make it clear that our offer was not seriously considered . . . Canadian was not prepared to engage in any serious discussions with us despite the Section 47 order. Based on our information, American Airlines would have had a veto over such a transaction—and presumably exercised it." Schwartz countered that he had tried to arrange a meeting with Air Canada executives but they had declined. Benson was surprised at Air Canada's lack of enthusiasm to accept Schwartz's proposal. "It's being offered to them under their name, in their town, with them as CEO. Why are they turning it down?"

Former Canadian Airlines CEO Rhys Eyton was disappointed at how things had turned out for his old airline but was resigned to the change. He said, "You're always a little sad to see something you spent so many years with change, but I've been around long enough that certainly I've been prepared to accept change." Among the members of Canadian's present board of directors, Peter Lougheed, Alberta's former premier, had the most difficulty justifying his acceptance of the Onex offer. A champion of the two-airline policy, he had brought PWA to Calgary in the first place. In an October 1999 *Globe and Mail* article titled, "Why I changed my mind,"[14] he explained that Air Canada had been given significant competitive advantages when the federal government privatized it. The playing field had always been "unlevelled" in its favour—a "sweetheart" balance sheet and priority landing slots at the best airports, which it used to full advantage once the Open Skies Agreement had been negotiated. Canadian had survived only on the profitability of its Asian routes, and once the economic crisis of 1997 took the impact of those away, it had been a losing battle to keep flying. When the board received the Onex proposal that August, they realized the advantages of a merged airline, and Lougheed decided that they had a fiduciary obligation to shareholders, creditors and employees to accept. Everyone would win, he said, including the consumer.

Reaction to Schwartz's plan from Canadian's anguished employees ranged from fatigue to cautious optimism. Those who had been with CPA, CP Air, EPA, Nordair and PWA had seen it all before—restructuring plans, corporate mergers and management turmoil. To keep the planes flying in whatever the current colour and logo, workers had endured belt-tightening pay cuts and not protested. They had seen colleagues laid off and held their tongues, praying that they would not be next. They had been required to buy company shares that were now a fraction of their original value. What they liked about this plan was that this time no one was looking to them to bail the airline out. If Schwartz succeeded, it would end the gloom and anxiety that they had been under since 1986.

"A smart move" and "just like the old Gerry—the master of the leveraged buyout," applauded portfolio managers. Investors reacted to the Onex proposal with guarded optimism: Canadian Airlines shares went up 3.0 cents to $1.86 and Air Canada's rose 70 cents to $8.70. Schwartz was only offering 45 cents less, but long-suffering Air Canada employees recalled that when the airline went public a decade before, its shares had been $8. In the same period Onex shares had increased nine-fold on the Toronto Stock Exchange (TSE), and on August 24 they rose another 75 cents to $26.75. Fund managers who had made millions of dollars investing with Onex thought the deal very positive and recommended its acceptance to Air Canada's shareholders. With Onex's business plan and AMR's support, shares in the new Air Canada would triple in value.

The prime minister, relieved that a private sector solution for Canadian's perennial problems had been made public, endorsed it—or rather he endorsed Gerry Schwartz. "We're not talking about somebody starting a new company here," he said. "Onex has been on the stock exchange for a long time and . . . he's not a guy who runs a nickel and dime store." But he was more cautious on August 25 when he told reporters, "The proposition has to be solid to be acceptable. We have to see if the proposition makes sense economically and financially." Opposition Leader Preston Manning mourned the demise of Canadian Airlines as "a severe blow, not only to Western Canada, where most of the job losses will occur, but to the entire nation, which will soon feel the impact of having only one transcontinental airline." Alberta's Premier Ralph Klein judged the proposed deal to be the lesser of two evils. "I don't feel good about it," he admitted. "It's about the only thing that was available because everyone has come to the conclusion that the two airlines were not sustainable, and it's a lot better, in my mind, than Air Canada taking out completely Canadian Airlines."

But before the amazing month of August ended, there was another twist in the saga. Everyone assumed that Air Canada, busy preparing a counter-offer to the Onex bid, was waiting to get the backing of United Airlines, its heavyweight partner in the Star Alliance. But since that could be a long time in coming, Air Canada prepared in the meantime to adopt the "poison pill" strategy which had been developed during the merger and acquisition madness of the 1980s when Trans World Airlines was taken over and asset-stripped by Carl Icahn, an experience from which it never recovered. To make future takeovers of this kind unpalatable, companies made it impractical for pirates to convince their shareholders to sell them their shares. On August 30, Air Canada swallowed such a pill, calling it the "Shareholder Rights Plan." In essence it said that lock-up agreements and voting arrangements in connection with a takeover bid over the 10-percent threshold would automatically trigger the "pill." This bought its board time to mount a credible defence, that is, negotiate with other parties.[15] Onex knew that the poison pill had to be neutralized, that is, the courts would have to force the Air Canada board of directors to remove it.

Air Canada also set a special meeting of shareholders for January 7, 2000, ostensibly to consider the Onex offer, but this was two months beyond the date that Onex wanted and well past the Section 47 "window" that the federal government had opened. AirCo, therefore, filed an application in the Ontario Superior Court on September 7, arguing that its August 24 offer would fall victim to "serious prejudice" if the shareholders' meeting was not held by November 8. Air Canada promised to "vigorously contest" it, and they hired Calin Rovinescu, a takeover specialist in the law firm of Stikeman, Elliott, as chief counsel.

One of the things that all the players understood was how power in Ottawa worked, and no one understood better than Schwartz that final approval for his proposal rested with Cabinet. Consequently, Onex signed up the most powerful lobbyists—or "government relations consultants," as they like to be known—in the capital. Both Earnscliffe Strategy and the Capital Hill Group could count on a lot of connections in both the Chrétien and Martin camps. Even poverty-stricken Canadian Airlines scraped up enough change to hire two lobbyists, Global Public Affairs and former Liberal cabinet minister David Dingwall's firm. Against this, Air Canada lined up the country's biggest lobby firm, Government Policy Consultants Canada, and brought in reinforcements in the person of 71-year-old Hugh Riopelle, its former "in-house House of Commons lobbyist." In mid-October, Schwartz countered by employing his friend Paul Pellegrini, president of Toronto's Sussex Strategy Group. The extra hired guns had been added, Nigel Wright explained, because, "Unlike our friends at Air

Canada, we don't have a large public affairs or media relations group. So for us, it is an out-sourcing issue."

Schwartz also stumped the country like a politician, his let's-make-a-deal manner and affable demeanour at odds with Robert Milton's. He even appeared on the Canoe current affairs website for an on-line chat. This gained him wide media attention as he explained his vision for a revitalized air industry. On September 14, standing before the Haida Gwaii sculpture at Vancouver Airport, he spoke to employees of both airlines and later, on a popular radio talk show, took calls from listeners, most of whom were appreciative Canadian employees. After all they had been through, one Canadian staff member said, Schwartz looked like "the best bet."

Although their unions could not seem to save their airline, Canadian's employees were so well protected that their labour representatives even had two seats on the board, and now both Gerry Schwartz and Kevin Benson were careful to consult with Buzz Hargrove of the CAW, Judy Darcy of the Canadian Union of Public Employees (CUPE) and Dave Ritchie of the International Association of Machinists and Aerospace Workers (IAMAW). In a meeting with Hargrove on August 30, Schwartz told him that "Jobs are probably the most important issue and we are committed to making sure the vast, vast majority of any job losses can be taken care of through attrition." Hargrove now professed to be more comfortable with the deal, especially as Schwartz assured him that "hundreds of millions of dollars had been set aside in the proposal for one-time charges like the downsizing effort."

The unions were in an unenviable position because some represented employees in both airlines. CUPE, for example, defended 3,998 Air Canada and 2,565 Canadian flight attendants. Only the Air Canada pilots had their own exclusive union, the Air Canada Pilots Association (ACPA). Now Air Canada employees wore anti-Onex stickers, and Canadian employees held meetings at which they yelled, "Better dead than red"—red being their rival's colour. Rather than waiting for the government to act, Dave Ritchie, vice-president of the IAMAW, met with his lodge representatives on September 3 at the Constellation Hotel near Pearson Airport. They represented workers in both airlines—a total of 19,000 baggage handlers, aircraft refuellers, caterers (including Onex-owned Skychefs) and mechanics. After the meeting Ritchie issued an ultimatum: Ottawa had 30 days to get directly involved and guarantee its members' jobs or they would "lay down their tools for as long as it takes." And he struck a chord in the aviation world when he said, "I don't want to see what happened to the Avro Arrow happen here. I don't want the industry to be lost. Shareholders can decide what's good for them but we elected a government to ensure that our rights would be protected." While the CAW declined to support the ultimatum, Gerry Schwartz met with Ritchie for an hour. "But," said Ritchie, "he wasn't able to indicate to me that he could give me any guarantees."[16] Collenette, who was still denying that his Liberal government favoured the Onex bid, urged the IAMAW to reconsider. The strike was called off when the minister assured Ritchie that his members' jobs would be protected.

Milton finally met with Schwartz on September 8 and formally received the Onex proposal for Air Canada. The minutes of that meeting remain confidential; all that was admitted was that no agreement was reached, but shortly after that Air Canada's shareholders were mailed all the information necessary for voting on the deal. By now Bay Street wisdom was that by holding Onex off and postponing its shareholders meeting until next year Air Canada had made a good move that would get the best deal for the company's shareholders. And the same analysts who were advising Canadian Airlines shareholders (who would get $2 a share under the Onex proposal) to take the deal, now told Air Canada shareholders that they could afford to play hard-to-get with the undervalued $8.25 they were being offered.

Frenzied trading began in Air Canada shares with a new player emerging—the *Caisse de dépôt et placement du Québec*. Managing the pensions and savings of Quebecers, the *Caisse* had never been bashful in using its $68-billion clout. It already owned 19.9 percent of Air Canada's class A non-voting shares, but between September 8 and 10, it spent as much as $80 million to buy 909,450 common shares, giving it a common share stake of 5.6 percent. Its mandate was to keep jobs in Quebec, and if the Onex proposal meant the loss of 5,000 jobs and many of those were at the Air Canada base in Montreal, the *Caisse* could be counted on to block it.[17]

Air Canada fired its opening broadside on Monday, September 13—not at the bargaining table, but in the Federal Court of Canada, challenging Ottawa's decision to suspend the Competition Act. "The Section 47 order . . . was not issued to exempt unsolicited takeover bids or any other merger transactions," said Doug Port. His company wanted the bid subjected to a merger review by the Competition Bureau. In any case, he said, Section 47 was now useless as Canadian Airlines was "locked up," that is, it could no longer share information with Air Canada because under its agreement with Onex it could only talk to other "oneworld" airlines. Air Canada also attacked in two provincial courts: proceedings were begun in the Quebec Superior Court challenging the legality of the Onex offer, and an affidavit was filed in Ontario's Superior Court by Paul Brotto, Air Canada's senior vice-president of business development, to the effect that numbered companies linked to Onex had spent $43 million to buy Air Canada stock on June 15, one month before the federal government lifted the competition rules. Onex fired back that it was routine to quietly acquire shares in a targeted company so as not to upset the market and that Onex had never tried to hide the acquisitions. Meanwhile, Onex and Collenette's spokesmen kept up a Greek chorus of "there was no insider status, inside track or privileged information."

"What part of 'Yes' do they not understand?" Schwartz wondered. "Under our proposal the headquarters of the new airline is in Montreal, the culture will remain fluently bilingual, the name will remain Air Canada, and the senior management of the new airline will most likely come out of Air Canada." In Montreal on September 17 to meet with investors (including the *Caisse*) Schwartz pointed out that the Onex proposal was already responsible for increasing Air Canada shareholder value to $12, almost double what it was before August 24. But on that same day, to prove that it was doing just fine without Schwartz, Air Canada released its interim results for July and August. Not surprisingly everything was up: operating income by 11 percent, passenger revenue by 3 percent, passenger yield by 5 percent. The company followed this on September 21 with a director's circular that recommended unanimous rejection of the AirCo offer on the grounds that it violated the "Air Canada Public Participation Act," which limited individual holding of Air Canada common shares to 10 percent. On the same day it made public the details of the $2 billion offer it had made to Canadian on June 23.

Meanwhile, revelation of AMR's "lock-up" of Canadian had drawn adverse media reaction, and on September 22 Don Carty of AMR was forced to hold a press conference in Toronto. "We at AMR believe that the Onex proposal represents the opportunity to truly create a great Canadian airline . . . The one weak part of our network is Asia and [we] intend to work with the new Air Canada to build Vancouver as the premier North American hub . . . there's nobody in Dallas interested in running a Canadian airline. We have 800 airplanes of our own to figure out what to do with everyday." When asked for a comment, Collenette would only say that the matter was now before the courts, but his officials admitted that while the minister knew before August 13 that the Onex bid was in the pipeline, he had been unaware of the "lock-up" clause with AMR.[18]

While its fate was being decided in the media, stock markets, law courts, and union and government offices, Canadian carried on. The company had already done five "Team Canada" flights, three similar missions for the Quebec government and two for Alberta. Now, after a long tendering process and charging only a break-even price for the prestige and honour involved, Canadian had once more won the privilege of flying Team Canada. Between September 11 and 18, the company flew the prime minister, eight provincial premiers and 200 businessmen to Tokyo and Osaka. The aircraft used was the 747 C-FGHZ, "Rhys T. Eyton," stocked before the trip with the newest ovens and trolleys, desks, photocopiers and fax machines, swept by a security team prior to departure, and closely guarded by the RCMP. Despite the uncertainty of their airline's future in the air, "It was apparent from the moment delegates and media boarded the aircraft," said Scott Bradley of Canadian's government affairs department, "of the pride and dedication our crew and employees took in ensuring the success of the mission."

In a management circular issued on October 8 Air Canada's directors unanimously recommended that the shareholders reject the Onex-AMR resolutions and forward their proxies for the vote to be held at a special meeting on November 8. They warned that under the AirCo offer, Onex and AMR would own 38.1 percent of the new Air Canada. The publicity mills at Air Canada now went into full gear. In the week following their public rejection of the Onex offer the airline launched a "Spread Your Wings" seat sale with reductions of 40 percent, offered special relief fares to earthquake-stricken Taiwan, and was the first to introduce the Airbus A330 in North America. At the same time the Star Alliance was showing new strength, signing up powerful new members—Singapore Airlines and All Nippon Airways (ANA)—while all that oneworld could attract were Iberia and Finnair, which were hardly in the same league.

On October 8, the same day that Air Canada publicly rejected his offer, Schwartz held a news conference outlining his iron-clad commitments to the industry and the country, as part of his company's broad lobbying efforts to win support in the lead-up to the critical Montreal meeting of Air Canada shareholders—now exactly a month away. Onex urged Air Canada shareholders to agree to change their voting stock into restricted voting shares because this would allow them to skirt the 10-percent cap and tender to Onex's offer. The company could then go to the Liberal government and ask it to approve Onex's ownership of more than 10 percent of Air Canada. Schwartz said his company was committed to Canadian control, jobs ("There will be no layoffs until at least one round of fair, voluntary offers has been made to the affected employees"), published airfares, service to smaller communities ("Every Canadian community that receives scheduled service from Air Canada, Canadian Airlines or both—including their owned affiliates— will continue to receive scheduled air service for at least five years"), competition, quality of service, bilingual service, seat sales and frequent flyer points. "Upon completion of the merger, the shareholder rights that AMR negotiated with Canadian in 1994 will terminate. Apart from a right to nominate two of the 13 directors, neither American nor its parent AMR will enjoy any special rights, control or influence over the new Air Canada."[19] In addition, Onex promised to appoint an ombudsman reporting to the new airline's independent directors to ensure the commitments are followed. "The new Air Canada will stand behind all its commitments for at least five years, which is well beyond the expected two-year integration period for the merger," Schwartz said. AirCo sent out its pink proxy forms to Air Canada shareholders on October 13 to vote their shares against the "poison pill." The accompanying letter said: "This is a unique time in the history of Air Canada and for you as an Air Canada shareholder . . . We are committed to creating shareholder value. Who is going to look after shareholder interests—Onex or the same group that has been in charge for the last eleven years? Look at results and history, not just what they, or we, promise . . . We invite you to join in the creation of a new Air Canada."[20]

Despite Schwartz's canvassing, however, Onex was seen as the least trustworthy of all the players. In an Angus Reid poll taken between September 7 and 12, 54 percent of Canadians polled thought Schwartz's plan to merge both airlines "unacceptable." A majority thought the federal government, Air Canada and even the cowboys at AMR more scrupulous. What came as no surprise was that only 33 percent of those polled felt that Collenette was managing the issue well and that most Canadians believed the whole affair should be left in private hands. But what did stun pollsters was that, though there was a lack of support in all provinces for a bailout of Canadian Airlines, support was lowest in Alberta where only 29 percent favoured federal aid.

By October 12 the House was again in session, and the Bloc Québécois was questioning whether Onex had known about the government's August 13 decision to suspend the Competition Act before that company acquired its $40 million worth of Air Canada shares. Collenette found himself having to reassure MPs, especially those in rural ridings who would lose their air connections, that his government and not the private sector still had control over the nation's airline policy. "We want a Canadian solution. The market alone will not decide what is in the best interests of all Canadians." His boss was more blunt. When asked on September 27 about the estimated 5,000 layoffs in Onex's proposal, Chrétien sniffed, "We have never guaranteed everybody jobs. Even in the public service we cut 20 percent of them."[21]

The Ontario Securities Commission (OSC) was also nosing around. The timing in the Onex/AirCo purchase of Air Canada shares for its 3.1-percent stake was too coincidental. If there was any evidence that Schwartz knew about Ottawa's plan to suspend the competition rules, an OSC spokesman stated, the commission would launch an investigation. In response, federal officials continued to insist that Onex was not getting favoured treatment. They said the decision to suspend parts of the Canada Transportation Act until November 10 was based on many factors, one of which was Air Canada's August 20 proposal to buy Canadian's international routes.

On October 15, just as Gerry Schwartz was meeting with Nesbitt Burns Inc., Air Canada's financial advisors, to discuss his offer, Robert Milton's allies in the Star Alliance rode to the rescue. It wasn't unexpected. The Alliance CEOs had been meeting in Tokyo that month, and something more substantial than moral support for Air Canada had been expected of them. Now the two senior partners of the Star Alliance, United Airlines and Deutsche Lufthansa AG, announced that they were willing to put up $730 million, to consist of a share purchase in Air Canada of $230 million, the lease of three Airbus A330s for $190 million, and the provision of a ten-year guarantee to Air Canada of $310 million. In addition, backed by the Canadian Imperial Bank of Commerce (CIBC), which would contribute $200 million, the Alliance would buy back up to 35 percent of Air Canada shares at $12 each to a total of $800 million. Finally, the Alliance CEOs had put together their own "poison pill" in the form of a departure tax to prevent any of their players bolting to oneworld.

Four days later Milton announced a $930-million bid to overhaul the whole airline industry in Canada and to launch a discount airline from Hamilton, Ontario, to compete with WestJet. Air Canada, he said, would buy out Canadian Airlines at $2 a share for a total of $92 million, then revive it as a distinct entity with a smaller head office in Calgary. Canadian would be financially restructured, get out from under AMR's control, code-share with Air Canada and continue to operate its trans-border and international routes, where it would code-share with all members of the Star Alliance and Delta Air Lines. Each of the companies would have a separate board of directors and management team. Keeping Canadian separate would not only protect Air Canada from Canadian's debts but also keep its employees' wage gains and seniority levels intact. Under Milton's plan, no shareholder would own more than 10 percent of the Air Canada voting shares. "We have taken the time to craft a plan that is practical and built on the untapped potential of Canada's airlines," Milton said. "Canadian Airlines is an airline with great people . . . Its true potential has not been realized in recent years. Our solution will unleash that potential." Job cuts as the result of duplication would be half those of Onex's plan, but the downside was that Canadian Airlines workers would bear the brunt of them. Onex spokesman Nigel Wright thought that Milton's plan would be rejected by shareholders. "At first glance it looks like the 'Spruce Goose'—good on paper but it won't fly."

In the background, hearings before the House of Commons Transport Committee were allowing the Competition Bureau, airline CEOs, and the federal government to thrash out review and restructuring schemes. All of this excited the wrath of Robert Milton, who suspected that Ottawa, with its pro-Onex stance, was "pushing his airline into a corner . . . Air Canada is fine the way we are."[22] But Air Canada's legal manoeuvres were beginning to bear fruit, and on October 19, in response to its Federal Court application, the company got what it wanted from the Attorney General: when the Section 47 suspension ran out in 90 days (on November 13), so would the advantages it accorded the Onex proposal. Any merger transaction after that would face the full weight of the Competition Bureau's review powers.

CAW-Canada president Buzz Hargrove endorsed the Onex bid for control of Canada's skies, believing it offered the most security for his membership. Air Canada employees, however, accused him of selling them out. CAW

AirCo upped the ante on October 28 by offering $13 in cash per Air Canada share. "We urge you to vote for the Onex plan and to tender your shares to our offer," said the company circular. It also promised full national ownership and control of the new Air Canada as AMR's rights would terminate on completion of the merger, and the merged airline would also have the option of choosing AMR's "SABRE" as its information technology supplier. The circular derided Air Canada's offer "as totally and clearly unworkable" because "how [would] they finance over $1 billion in payments and costs that would result from terminating Canadian's current relationship with American Airlines?" But in a press release two days later Air Canada reminded the media of the AMR "lock-up" and the fact that the choice of SABRE came with a penalty of $525 million. How many more "secret deals" were there between AirCo and AMR? the press release asked. Air Canada now applied to the Securities Commissions in Alberta, British Columbia, Ontario and Quebec to force more disclosures.

Buzz Hargrove did not believe that Air Canada would meet the commitments that his union had been given by Onex. He told a news conference in Toronto on November 1, "I am endorsing the Onex proposal to restructure the airline industry as the proposal that is in the best interests of the members of our union." He said that his union had given Milton "every opportunity" to come up with wage and job commitments similar to those provided by Onex and he hadn't come through. While all the unions were angry that Milton was making plans to create a low-cost airline—which to succeed would need non-unionized labour—what now specifically angered Air Canada's union members was their leader coming out in support of the Onex bid. As a result, at a televised meeting in Toronto on November 4 Hargrove was roundly berated.

By now Canadian was bleeding heavily. Although on October 27 it was able to report third-quarter earnings of $71.3 million, it had racked up a nine-month loss of $54 million. It carried on with business as usual, however. The end of summer meant reducing capacity on its network—taking four Dash 8s out of service, down-gauging from DC-10s to 767s on the London, Honolulu and Japan routes and eliminating money-losing routes like Hong Kong–Manila and Mexico City–Monterey. The almost new Canadian Airlines office at Manila was closed on October 31 and traffic directed to Cathay Pacific. During 1999 Canadian had raised $50 million by selling and leasing back eight aircraft. Now four of these DC-10s were returned to their lessors, the Wells Fargo Bank of Salt Lake City and Finova Capital Corp. of Scottsdale, Arizona, to raise financing to buy two brand new 767-300ERs and lease one used A320. Two of the old "Douglas Diesels" had celebrity status and there was no point in repainting them in the Proud Wings image. One of them was the "Spirit of Canadian" (911), which had flown the world covered with 600 employee signatures. Nothing had demonstrated more that Canadian was an employee-driven airline.

What had begun on a Friday afternoon in August ended abruptly on a Friday afternoon in November. Aware that the crucial shareholders meeting was to take place on Monday, both sides knew that the weekend was the eleventh hour. On November 2, as expected, the Montreal airline, now flush with Star Alliance cash, raised the stakes to $16 per share. On November 5 Onex countered with a bid of $17.50 per share. However, on that same day in Montreal, Justice André Wery delivered the ruling of the Quebec Superior Court. Simply stated, it said that under the 10-percent ownership provisions of the Air Canada Public Participation Act the Onex-AMR Offer was illegal because Onex would hold 27.6 percent of the shares.

It was expected that Onex would either appeal the decision or try to influence Ottawa to change the 10-percent rule, but officially Onex said that it had abandoned its merger offer because of the Quebec court's decision. However, Schwartz did appear to be keeping the door partially open. "We hope both Air Canada and Parliament will now act to remove this restriction [of the 10-percent rule]," he said. Onex was expected to break even. Having spent $28 million at $5.77 on average per share to acquire its toehold stake in Air Canada, it could now sell its shares at the current price of $9.90, for a net gain of $18 million. But the legal and lobbyist fees during the 11-week battle would eat into that. Nigel Wright concluded, "It's probably a wash. The gain is spent money." Privately, advisors conceded that though the Quebec court's decision could easily have been appealed, by then Gerry Schwartz had lost the will to continue a bitter and expensive fight that threatened to drag on for months. "Gerry goes into these situations with the intellectual desire to win but [with] a number he won't go over if a bidding war erupts. As he gets closer and closer to that number, he loses interest," a former advisor said. The winner of the Onex/Air Canada war was said to be the *Caisse de dépôt et placement*. Not only had it picked the winning side but in keeping with its program of Quebec nationalism, it had kept Canada's airline industry unequivocally in Montreal. Canadian's share price fell 18 cents to $1.44, back to its July level.

On hearing that Schwartz had bailed out, Collenette must have suffered some palpitations. He spoke in Toronto on November 8, making it clear that he wanted to put the messy airline war behind him. Abandoned at the altar by Onex, Canadian's only hope now was Air Canada. Would Milton walk away too and let the Calgary carrier go into a death spiral, then pick up its routes and customers when the dust had settled? The Air Canada CEO, though initially reluctant to enter the very public fray, had gained in stature during it. He had played a cautious waiting game and won, but having taken on and defeated Ottawa, AMR and Onex, he was phlegmatic about it all. He told a Montreal press conference, "I guess today I'm supposed to be happy and nice, so I'll try." He called off the special shareholders' meeting scheduled for November 8, since Air Canada shareholders were no longer required to vote, but it was November 15—a full ten days later—before Air Canada made its formal offer to purchase Canadian.

First, Air Canada announced that it would buy back 36.4 percent of its shares at $16 a share. Then in a lengthy circular sent to Canadian shareholders on November 17, Air Canada said that it was not only prepared to step in with emergency financing during the restructuring of Canadian's debt but outlined that it was going to buy the airline for $92 million, or $2 a share, through a numbered Alberta company, 853350 Alberta Ltd., that was 90-percent owned by former CIBC debt-restructuring specialist Paul Farrar,[23] who had been hired by Air Canada, and

Don Carty, the former CEO of Canadian Airlines who now held that position at American Airlines, would go only so far to save his old company. "Is it a critical piece? No." oneworld

10-percent owned by Air Canada. The offer for Canadian's 46 million common shares, at a cost of about $102 million with fees and expenses, would expire on December 7.

In Air Canada's proposal only 2,500 employees from the two airlines would be let go through attrition, early retirement and voluntary severance packages over a two-year period. Canadian would operate as a separate subsidiary, but both carriers' routes and schedules would be streamlined to address surplus capacity. In addition, Air Canada promised to be fair in negotiations to establish a level playing field between the unionized employees at the two airlines. Existing service to all communities served by Canadian, Air Canada or their subsidiaries would be preserved. Capitalizing on their strengths—the combined airline would be the 10th largest in the world—the two carriers would launch daily Toronto–Tokyo and Toronto–Hong Kong flights and seek permission for additional flights between Vancouver–Toronto and Mexico. The two airlines would also seek to activate "dormant authorities" for routes including Vancouver–Shanghai, Vancouver–Sydney, Montreal–Milan/Rome, Toronto–Madrid and Toronto–Amsterdam.

There was speculation that Canadian's major shareholder, American Airlines, would refuse to sell its 25-percent control to Air Canada. But when American's Carty did suggest that severing his company's ties with Canadian would cost Air Canada $1 billion, Milton said, "Canadian Airlines is losing $500,000 a day. The figure of $1 billion is pretty far-fetched, and the sooner we can get on with sensible business-like discussions the better." From New York, Carty responded that American wouldn't hang onto its stake in Canadian at any price. "Is [Canadian] a critical piece? No. No single piece is. This is not an issue that is critical to us, our alliance or anyone else's alliance."

Meanwhile, Canadian Airlines said that it could meet its financial obligations through to the end of the following year, thanks to AMR's deferral of Canadian's $17 million per month fee payments. Analysts decided this claim was only part of a strategy to stall Air Canada for as long as possible, but after Canadian's announcement Air Canada shares dipped 50 cents to close at $10.55 while Canadian's gained 12 cents to close at $1.73.

Despite the uncertainty and poor publicity that their parent company had suffered through the year, the fortunes of Canadian Regional Airlines (CRA) had remained unaffected, and Air Ontario and AirBC both reported that business was normal. Based in a Calgary Airport hangar and run by Bob Reding, CRA employed 2,100 staff across the country and operated a fleet of 29 Fokker F.28s, 14 Dash 8-300s and 10 Dash 8-100s. It accounted for 32 percent or $451 million of Canadian's domestic passenger revenue in 1998. But Inter-Canadien, the eastern Canada "feeder" that Canadian had sold, was no longer part of the family. When it shut down on November 27 to "restructure," its president blamed his airline's financial woes on the bitter takeover battle between Canadian and Air Canada. "To the public we are Canadian Airlines in eastern Canada, yet we are not part of the solution. We are not included in the merger." In reality, his airline was experiencing the Air Canada/Canadian conflict by proxy: Inter-Canadien's ATR-42s were no match for the jets of Air Nova, the Air Canada feeder. As Inter-Canadien's planes were seized by creditors on December 5–6, the company's 400 workers demonstrated outside Transport Minister David Collenette's office, demanding that they be included in any takeover of Canadian. Collenette's officials met with them, the minister once more managing to sidestep the crisis by ensuring that he was not in Ottawa. It was a good thing too. The Senate Transport Committee had just released its report in unusually strong language. "The committee is not convinced the use of Section 47 was appropriate. Nor is it convinced that a disruption was imminent." Senator Michael Forr added, "We couldn't find anyone who could justify the minister using section 47." An aide to Collenette said the government disagreed with the committee's finding.

But it wasn't over yet. The end of November saw a last-ditch offensive mounted by Canadian's board, which twice refused to recommend that its shareholders either accept or reject Air Canada's offer. In a 17-page directors' circular mailed to company shareholders they asked for clarifications before they could complete their assessment of Air Canada's $92 million takeover. Was it a stalling tactic until AMR Corp and oneworld could do their calculations and come to Canadian's rescue? Meeting in New York at that moment, oneworld's CEOs piously claimed that Canadian was an important part of their network. "If Canadian isn't part of that mix there's literally a missing link," a British Airways spokesperson said, but no one provided specifics on what they planned to do. With time running out, Benson flew to New York to appeal to the executives of oneworld and especially to AMR to convince them to try to better Air Canada's takeover bid. Sadly, he returned empty-handed. Canadian had been abandoned to its fate.

Early on Saturday morning December 4, Canadian's board members linked up by conference call to hear from Benson. His one-man campaign to save the airline had failed. The only offer on the table was Air Canada's. The eleven men, some worried about their own personal liability if Canadian was declared bankrupt, issued a terse four-paragraph news release recommending that Air Canada's $92 million offer be accepted. Shareholders had now only three days left to tender their shares.

James Richardson, Grant McConachie, Don Watson and Rhys Eyton had all fought valiantly in their own time to keep this airline flying. But none had fought harder than Kevin Benson. He had been the Horatio at the bridge, holding the enemy off as long as he could. When his ammunition ran out, all he said was, "This is a momentous decision for Canadian Airlines."

The board's delay in recommending acceptance of the deal meant it was also going to be a close-run thing. Canadian employees were left with only one working day to make their decision, and as a result, thousands of them had not yet converted their "share entitlements"—which accounted for 20 percent of the airline's shares—to the common shares that Air Canada needed. Milton, therefore, extended the purchase offer to December 23. When all the shares were counted on December 8, more than 50 percent of all Canadian's shares had been tendered, but Milton decided to keep to the December 23 deadline.

The very next day Milton reached agreement with AMR. Now that oneworld had decided not to help AMR buy Canadian, Carty settled for payment of between $55 and $60 million and some non-monetary issues. The potential payoff for AMR was an understanding that it would continue to supply some airline management services to Canadian, with Air Canada promising $83 million in windup and unamortized start-up costs in the event that it cancelled these services. As a means of providing Canadian with interim funding, Air Canada bought its access to Narita Airport, Tokyo, on December 14 ($25 million), purchased and leased back the A320 simulator that Canadian had just purchased on September 15 ($7.5 million), purchased three of its Dash 8 aircraft for CRA ($13.5 million) and increased Canadian's line of credit at the bank by $50 million. Meanwhile, uncertainty about their jobs did not prevent Canadian's corporate headquarters staff from moving out of their downtown Calgary offices on December 16 and moving into the new premises at the airport.

On December 21, after two weeks of negotiations, Air Canada and Ottawa announced that they had found common ground in the airline's bid to acquire Canadian. The Competition Bureau too came on board. Determining that the merger was more pro-competitive than a liquidation through bankruptcy proceedings would have been, Commissioner Konrad von Finckenstein announced that he would not oppose the deal, provided that Air Canada sold off Canadian Regional Airlines (CRA), surrendered a number of slots at Pearson and refrained from establishing its own discount airline until September 30, 2001. In addition, for a period of three years the company would continue domestic service to all communities currently being served by Air Canada and Canadian and not lay off or relocate unionized staff until March 2002. Seat sales and frequent flyer points would continue to be honoured.

Milton had achieved what his predecessors, Gordon McGregor, Claude Taylor and Hollis Harris, had craved for years—control of arch-rival Canadian Airlines. But he didn't crow over this gift. "It's Christmas," he said at his news conference, but he did allow himself a rare smile. "There's no point dwelling on it. We just get on with the future." Perhaps Milton's low-key response was because he realized that he had only won a battle in a war that would continue on two fronts: Canadian's debt restructuring and employee integration.

Canadian Airlines ended 1999 with a loss of $222.3 million (compared with a loss of $137.6 million in 1998), although passenger revenue had increased 2.5 percent from the previous year. Its system yield had risen 3.1 percent since 1998 to 13.96 cents. Reduced operations and out-sourcing had barely decreased staff levels to 16,269 full-time employees. Losses on the eastern and western shuttles were offset by a 20-percent revenue growth on the trans-border routes, as well as the Far East and London—but all of this had come too late to save the company.

As the world celebrated the arrival of the new millennium, Canadian's Y2K project team sighed with the satisfaction of a job well done. All the aircraft, SABRE and air traffic control systems had operated smoothly. "The measure of our Y2K success was in the size of the non-event—and it was a huge non-event," said R. Bentkowski, Canadian's Y2K project director. On January 4, 2000, Canadian's board resigned, and an interim one took over. Besides Paul Farrar, its members included Kevin Benson, Robert Milton and Calin Rovinescu from Stikeman, Elliott. This was also the date that 853350 Alberta Ltd. acquired 82 percent of Canadian's outstanding common shares and non-voting shares at a cost of $79 million. The next day it also purchased the convertible preferred shares from AMR for $59 million.

Canadian's senior vice-president, Mary Jordan, and Air Canada's Rupert Duchesne headed the joint integration "QuickWins" team for streamlining the operations of both airlines. Some changes were immediate. Shifting capacity on some routes, especially where the airlines had gone head-to-head, was an obvious priority: Canadian continued to fly the Toronto–Ottawa–Montreal and Vancouver–Calgary–Edmonton triangles but, with the unpopularity of its F.28s, reduced the number of flights in favour of Air Canada's A319s. Though Air Canada and Canadian operated similar fleets, the subtle differences in the layout of cockpit controls meant that the pilots had to be retrained before flying the other company's aircraft. Similarly, the mechanics of both airlines required refresher courses in the other carrier's technology, but most symbolically, on January 30 Air Canada made space available in its Vancouver hangar for repairs on a Canadian 737. A transitional paint scheme, devised for the exteriors of Canadian's fleet, included the

Proud Wings goose and red Canadian name on the white forward fuselage and Air Canada's red maple leaf on the green tail. Chris Marshall,
project leader of the AC/CDN transitional livery project, said, "This is
monumental when you consider the long-standing rivalry. You cannot
find two finer symbols of Canada than the maple leaf and the goose.
Both certainly symbolize Canada on the world stage."

On January 19, 48-year-old Paul Brotto, senior vice-president of Air
Canada, was appointed chief operating officer of Canadian and on
February 29 succeeded Kevin Benson as CEO. In his farewell address
Benson said, "There is no room for two captains in the cockpit." He
would later write to his employees: "The past few years are ones that I
would never have wanted to miss, but I sincerely hope I will never have
to repeat. I leave with some very powerful memories, not of the sweaty
days and nights when we tried to find the right answer." On June 1 the
former CEO moved to Vancouver to become president of the Jim
Pattison Group, the country's third largest private company, run by the
entrepreneur who had once been touted as the white knight who would
rescue Canadian. It was July before many of the other senior Canadian
executives learned whether they were to be employed by Air Canada or
let go.

In February Canadian imposed a moratorium on all payments to
certain creditors in order to retain enough liquidity to sustain operations through the traditionally poor winter period. Then on March 24,
under the Companies Creditors Arrangement Act and Section 34 of the
United States Bankruptcy Code, the Court of Queen's Bench of Alberta
granted the company an order to start proceedings, thus requiring it to
file a viable debt restructuring plan. Presented on April 25, the "Plan of
Compromise" in effect "compromised" the company's liabilities and
reorganized its share capital. Under the plan, the debts of secured note
holders were reduced from $175 million US to $162.75 million US with
a modified interest rate in exchange for a guarantee from Air Canada of
the company's debt obligations. Certain creditors received payment
equal to 12 percent of the value of their claims, and the Corporation
agreed to relinquish its only material asset—its shareholdings in
Canadian Airlines—to 853350 Alberta Ltd. and cease to own it.

Canadian's secured creditors voted unanimously on May 28 to support the carrier's plan to restructure about $3.5 billion in debt, thereby
giving them the equivalent of approximately 98 cents on each dollar of
$250 million in long-term debt. This approval paved the way for

Canadian to be fully taken over by Air Canada. "I'm ecstatic," Paul Brotto said after hearing the result of the creditors' vote. "We were certainly hopeful of winning, but to win this overwhelmingly should give everybody confidence that this thing should succeed." However, unsecured debt holders, owed about $750 million, only voted 65 percent in favour of the offer because it was worth a mere 14 cents on the dollar to them. Among the creditors were Pacific Coastal Airlines of Richmond, whose contract had been unilaterally ended on April 1, 210 days short of its expiry date; Bear, Stearns Securities Corp., which was owed $84 million; and Cede & Co. of New York, owed $190 million. San Francisco-based Pegasus Aviation had leased eight 737 aircraft to Canadian, each worth an estimated $5 million. Even Canada Customs and Revenue was owed over $22.8 million. But as the restructuring plan had to be accepted by the Alberta Court of Queen's Bench before Air Canada could act on the merger, there was still time and opportunity for the dissident unsecured creditors to scuttle the restructuring plan. It so happened that a US "vulture" fund, Resurgence Asset Management LLC, now held some 60 percent of the $100 US unsecured bonds, and it set in motion an attempt to sabotage the deals between Canadian Airlines and its aircraft lessors. This fund's renowned resident "vulture," James Rubin,[24] claimed to see in the hiring of Paul Farrar to head up 853350 Alberta Ltd. an Air Canada plot to leave creditors with "an empty shell." At the hearings that followed, Robert Peterson, Air Canada's chief financial officer, testified that the cost of acquiring Canadian had now risen to $3 billion, two-thirds of this being in obligations that his company had reluctantly assumed so Canadian could refinance its fleet at more favourable terms. The remaining $1 billion was broken down into:

- $300 million to honour tickets for future travel already paid,

- $100 million for separation packages for redundant Canadian employees,

- $270 million to pay off secured note holders,

- $55 million to pay off unsecured creditors,

- $80 million in overdue bills owed to American Airlines,

■ $80 million to pay for 82 percent of Canadian's common shares,

■ $60 million for American's preferred shares,

■ $75 million to get out of a long-term services agreement with American.

Petersen explained that Canadian had been losing $2 million a day before Air Canada bought it and would have shut down by December 14 if it had not been merged. To keep flying, the airline had mortgaged everything from its telephones to the nuts and bolts used in aircraft maintenance. Canadian's trademarks had been used as collateral, with the Royal Bank claiming the airline's Canada goose. Dinner plates, napkins, and paintings owned by Canadian were also being used as collateral, the documents said. Underlining the airline's precarious situation was the release of first-quarter figures: not withstanding the integration team's reduction of head-to-head routes and "quick-win" cost savings, first-quarter operating losses were $153 million with a net loss of $255 million and a load factor of 66 percent.

After securing court approval of its debt restructuring plan, the most vital element in guaranteeing the success of Air Canada's takeover of Canadian was the co-operation of the labour unions. On March 27, the CAW became the first to reach a deal with Air Canada; it covered all 4,200 union members at Air Canada and 4,100 at Canadian. The Canadian Airlines pilots' union, ALPA, ratified a new contract on May 31, obtaining the same work rules and contractual language as their fellow pilots at Air Canada—who ironically were involved in a labour dispute with the airline at the time. The airline's 3,500 flight attendants, represented by CUPE, reached an agreement soon after. Admittedly, there were Canadian staff who left and sometimes en masse. In June, two flights out of Saskatoon had to be cancelled after 21 Canadian Regional Airlines employees resigned within three weeks.

On June 2 "the mother of all moves" took place as Air Canada and Canadian Airlines operations were merged at Pearson Airport. "This is one of the most ambitious airline projects ever undertaken," said Rob Ramage, Air Canada's senior director of customer service and development and the key architect behind the $80 million move. "But on June 3, I guarantee the confusion will be eliminated." Canadian bid farewell to its Terminal 3 home and moved its international operations into

Terminal 1 along with some of its domestic services. The remaining domestic and trans-border services were transferred to Terminal 2, sharing with United Airlines. All other carriers moved to Terminal 3. Canadian's computer systems were moved from SABRE to Air Canada's Res III, and on June 1 Canadian Airlines left oneworld to begin a code-sharing and frequent flyer partnership with Star Alliance, a move that also affected employee interline passes. On the domestic front, Canadian's Ontario Regional was replaced by the Air Canada connector, Air Ontario, while most Pacific coastal routes were replaced by CRA and Air BC. The merging of downtown ticket offices (Air Canada called them City Sales Offices) went smoothly, with Vancouver becoming the first where AC/CDN agents began working side by side.

The Toronto Stock Exchange advised on June 7 that, subject to the Alberta court's approval of Canadian's debt restructuring plan, it would stop trading and delist Canadian Airlines Corporation shares. This approval came on June 27, with the court ruling that the restructuring was fair, reasonable and equitable because it saved over 16,000 jobs at Canadian, protected consumers and preserved the integrity of the national transportation system. "With this ruling, we can implement the financial restructuring and close the purchase of Canadian Airlines by Air Canada," said Robert Milton. Although the Alberta court's judge, Marina Paperny, still had to hear motions appealing her decision, Paul Brotto knew that he was within sight of the end of his job. He revealed that 2,500 voluntary early retirement packages had been offered to employees in both airlines at a cost of $100 million. The next priority, he said, was to replace Canadian's aging fleet of Boeing 737-200s and Fokker F.28s. "When I came in," he said, "Canadian operationally was running exceptionally well: good service, good safety, good on-time performance. But financially Canadian was broke. It did not have enough money to meet the Christmas payroll."[25]

On June 30, as ordered by the Competition Bureau, Air Canada put CRA up for sale, but the process was really only Milton's lip service to the bureau. The CRA fleet of 56 aircraft—particularly the Fokker F.28s—were old and noisy, and the regional airline's unions, determined to pre-serve their seniority with Air Canada employees, were vowing to fight any new owner. "These were never created as stand-alone airlines," said potential buyer Barry Lapointe, the CEO of Kelowna Flightcraft. When, after 60 days, there were no buyers, the Competition Bureau allowed Air Canada to integrate CRA into its system.

By July, several of Canadian's senior officers—Doug Carty, David Bell and Brock Friesen—had moved on. Vice-president Rod Dewar was appointed chief operating officer of the British Columbia Automobile Association. Barbara Amster, who had come from AMR to be senior vice-president for sales, had long since left to become vice-president of Realtor.com, the Internet site for the National Association of Realtors. Mary Jordan, Steve Markey and Frances Fiorillo found positions with Air Canada. On a more junior level, employees such as Lisa Le, Juan Franky and Torben Bentzen moved from Calgary to work for Air Canada in Montreal. Bentzen, a 27-year veteran reflected, "I've been met with open arms. As far as my work experience is concerned, Air Canada and Canadian Airlines are very similar."

On August 30, Brotto could breathe easier when he was informed that Judge Paperny had dismissed all appeal applications; Air Canada no longer needed to operate Canadian as a separate entity—a promise that it had made during the Onex war. The Canadian brand would become extinct in a year, airline spokesperson Laura Cooke announced on October 5. "The airline changed course when it realized just how bad the financial situation was at Canadian when it acquired it in January. We developed the blueprint without having access to Canadian's true financial position . . . It quickly became clear that Canadian was not viable as a stand-alone company."[26]

Canadian's fusion with Air Canada continued through 2000, from little details like replacing Starbucks coffee on their flights with Air Canada's brew, "Second Cup," and moving its online services from www.cdnair.air to Air Canada's www.aircanada.ca, followed by the employee Intranet "HUB" being blended into Air Canada's employee website "aeronet." All three million members of the "Canadian Plus" frequent flyer program, which had been initiated two decades earlier, were sent a letter that summer advising them that the program was merging with Air Canada's "Aeroplan" on December 31. "We opened the doors on July 12, 1984," remembered Hilda Wilson, one of the first managers. "In those days membership files were not computerized. Every member's account was on microfiche that we had printed weekly with new updates. Our phone system consisted of eight lines on one phone. So the agent would take a call, ask the member to hold while they put the microfiche into the machine, then go back on the line."

Cargo integration begun March 6 in Miami was completed at all Air Canada/Canadian cargo facilities by October 1. Canadian's System

Operations Control (SOC) moved from Calgary to Toronto on October 21. A team of Air Canada and Canadian employees headed by Doug Scott (AC) and Jacek Romanowski (CDN) had been at work developing the SOC integration plan since January. The main task was switching Canadian's fleet from SABRE's Flight Operations System (FOS) to Air Canada's Flight Management and Integration System (FMIS). But Canadian's flight dispatchers, already working on Air Canada's flight dispatch technology, continued to work in Calgary until an agreement with their union could be signed.

The integration of cabin personnel, which was examined in detail by labour mediator Kevin M. Burkett, illustrated how complex some issues were. The best flying blocks, the most preferred vacation times, immunity from layoffs, seniority earned—whether by time or responsibility—was currency that employees at both airlines hoarded jealously. While both seniority lists were based on date of hire, the two "in-charge" lists were based upon different concepts of seniority: Canadian Airlines' list was based on date of hire while Air Canada's list was based on date of last appointment to the "in-charge" position. The hurdles in resolving such issues took months to surmount.

Some of the changes were more personal. For Canadian's employees, 20 or 35 years of service meant that they had joined when the airline was CP Air, Wardair, Transair or PWA. Now they were informed that as of January 1, 2001, the Air Canada service recognition program (pins, plaques and a letter from the president) would replace Canadian's. Employees reaching their service anniversaries prior to January 1 had until March 31 to select a commemorative gift from the Canadian Airlines Service Award program. After that they would receive an Air Canada remembrance. And finally in December 2000 the last issue of the Canadian employee newsletter, *Canadian Flyer*, appeared. It was sure to be missed—especially for its cartoons by Jake Visser. Starting in February employees would receive the Air Canada publication, *Horizons*. But it wasn't until September 25, 2002 that the Canadian Industrial Relations Board declared Air Canada and Canadian Airlines to be a single employer with respect to all bargaining units, allowing for a complete intermingling of their workforces.

When did Canadian finally disappear? Like the smile on the Cheshire cat, outside the airlines it faded out gradually. On May 31, 2002, the ID/travel cards of former Canadian Airlines employees and retirees expired, to be replaced by Air Canada ID/travel cards. On the

Staff of Canadian's employee newsletter, *Canadian Flyer*, put out their last edition in December 2000. Most of them now work for Air Canada.
Canadian Flyer

weekend of October 21–22, the computer reservations systems were—with a few hiccups—combined, and after that date all tickets issued bore only the name Air Canada. The following day, passengers looking at airport information monitors no longer saw listings of Canadian Airlines flights; the airline now belonged to history. Paul Brotto, the last CEO of Canadian Airlines, summed it up by saying that the new Air Canada was a child that would inherit the very best qualities of its parents. "Canadian's strengths as perceived by customers are in the human aspects, the service aspects, the warmth, the friendliness. Those are the aspects more associated with a mother. Air Canada's attributes are financial, get-things-done, operational excellence, which are more the hard attributes associated with a father. What we want is an offspring that has the best characteristics of both. It doesn't always work out that way, but that's what we'd like to be." Then he added with a smile, "At the very least, we control the gene pool."[27]

Why had Canadian Airlines died? Auto racing devotee Kevin Benson used to say that races were often lost, though never won, in the early laps. If this is true, Canadian had lost the race almost from the start. When Pacific Western Airlines entered the two-airline fray, it did so hobbled by more than $1 billion in debt from buying Canadian Pacific Airlines and Wardair Canada. The downturn in world aviation caused by the Gulf War hadn't helped either. By the time Benson took over the cockpit and implemented his four-year plan, it was too late. All it did was postpone the inevitable. In 1997 when airlines could not help but make money, Canadian barely made $5.4 million, the only annual profit in its last 11 years. A year later Benson's plan, already on shaky ground, was mortally wounded when the economies of Asia went into depression. This was especially true of Japan, which had accounted for 40 percent of the company's Asian traffic. The next year, when Canadian lost $137.6 million, it gamely attempted to stem the tide with various initiatives like its Vancouver hub and joining the oneworld alliance. Then in early 1999, in a desperate attempt to slough off its "Job" persona, the company even changed its image, bringing back the goose logo and spending money that it didn't have on luxury amenities to attract more high-paying business customers. Through it all, Air Canada competed relentlessly, adding to the problem of over-capacity and making it impossible for Canadian to cut routes or raise fares. As well, discount carrier WestJet Airlines, which began flying in 1996, had ravaged Canadian's western base. Operating costs rose 7.1 percent in 1998 while passenger revenue grew only 2.9 percent. The situation worsened through the summer of 1999 as fuel prices soared and new airport and air navigation fees were imposed. One former bond holder remembered, "The bottom line is that Canadian never had enough liquidity. They were under-capitalized and they were not able to use the capital efficiently."

Airlines rise and fall almost daily in these times, it seems. Indeed, even as Canadian was undergoing its death throes, Swissair (mistakenly thought to be the most carefully run airline in the world) and Trans World Airlines (the eighth largest US carrier) were also filing for bankruptcy. But Canadian had a mystique that far transcended its size. Its heyday was in the 1940s and '50s—Grant McConachie's era—when the CPA image was earned. It became a patriotic legend, an ambassador for its country, standing for such values as friendliness, audacity and luxury. But as competition grew tougher, its fleet was trebled, its route network expanded unwisely, and signs of the sickness that would eventually kill it became apparent. There was too little realization that the world had changed, that outstanding past performance and a gilt-edged name were insufficient to stay in the ring against Cathay Pacific, Japan Air Lines and Air Canada.

As for Air Canada, its victory in the takeover war may have been pyrrhic. On April 1, 2003, the country's remaining national carrier—hobbled by $12 billion in debt and two years of losses totalling $1.6 billion—filed for bankruptcy protection under the Companies Creditors Arrangement Act.

Similar financial hemorrhages afflicted many of the world's large airlines. The decline was blamed on a host of factors, including the war in Iraq and the outbreak of Severe Acute Respiratory Syndrome, but especially on the September 11, 2001, terrorist attacks. The truth was that Air Canada and other major carriers such as American Airlines, Continental, United and British Airways had become dinosaurs in a world that was increasingly suited to small, furry mammals that scurried about and stole their eggs. The airline industry had undergone a structural shift caused by rising competition from low-cost carriers like Southwest Airlines and in Canada, WestJet. The ease of shopping for cheap tickets on the Internet, the price of the new security requirements and the federal government's reluctance to pour large sums of money into the airline were all contributing factors.

The overriding problem, however, was that Air Canada was still a full-service carrier spreading itself too thin in a world more attuned to niche markets and discount airlines. It was, said one airline analyst, an Eaton's in a Wal-Mart world.

Air Canada did enter the industry downturn with substantial assets. It had a captive base at home with 70 percent of the domestic market, a strong global presence as the world's 11th-largest airline and, in Aeroplan, a profitable frequent-flyer program. Unfortunately, these assets weren't enough to stave off the Canadian version of Chapter 11 in the spring of 2003. In what must have been a familiar refrain to former Canadian Airlines CEOs, Milton went after the airline's greatest expense, labour—about 35 percent of pretax costs—and sought to cull rank and file. As Canadian had once done, Air Canada put up all of its unencumbered assets, including Aeroplan, as collateral for restructuring financing.

Whoever takes the blame, history will record that Air Canada was brought down by its inflexibility and lack of capacity to respond to a changed business climate. Perhaps, like the Anglo-French Concorde that Air Canada had once ordered, the airline belonged to another era.

N O T E S

Chapter One

1. Branckner lived and died by his ideals: he was killed when he embarked on the maiden flight of the airship R101 on its way to India.

2. Remaining true to the tradition that English gentlemen do not engage in trade, the British relegated commercial aviation to the Scots, two of whom would dominate Imperial Airways in its early years. George Woods-Humphrey and John Reith (later Lord Reith) were Glasgow classmates—the former playing a role in Canadian aviation and the latter moving on to run Britain's radio broadcasting corporation, the BBC.

3. The French government subsidized the salary of each commercial pilot up to 1,000 francs annually. The fact that one of the airlines in France, known only as "The Line," was begun by the arms dealer Pierre LaCoteire as a means of laundering money did not hinder the subsidization.

4. With an eye on the future, Luft Hansa courted all political parties, even giving complimentary air passes to the little-known National Socialists when most considered them street brawlers. The prestige of Nazi leaders arriving by air helped the struggling party enormously.

5. Commercial aviation was considered by the public to be so frightening that ticket agents were required to bring their hats and empty suitcases to work with them, so that they could board a plane as supposed passengers. The rationale was that the few genuine customers would think that the plane was full.

6. One of the first promoters of air freight was Henry Ford, who began his own airline (and built his own aircraft—the Ford Trimotor, or "Tin Goose") to deliver auto parts between his factories.

7. Since this assignment involved flying quantities of dynamite—the first time this had happened—no one knew what to expect, and other pilots shied away from the offer.

8. Before becoming famous as a polar aviator with Commander Richard Byrd, Bernt Balchen had trained as a pilot in the White Army of Finland, and before coming to Canada, had flown with Roald Amundsen to the North Pole. Immediately after leaving WCA, he signed on as co-pilot with Byrd to cross the Atlantic on July 1, 1927. They did fly from New York to France (two months after Lindbergh) but got lost in fog and turned back to ditch in the ocean. In 1946 Balchen pioneered the polar air route used by the Scandinavian Airlines

System and Canadian Pacific Airlines to link Canada with Norway. This true descendant of the Viking explorers died on October 17, 1973.

9. Although the buildings no longer exist—a school and boat club are now on the property—the scene has not changed dramatically. WCA's pilots would find the King George Hospital and many of the suburban streets familiar.

10. G.A. Thompson and Col. Arthur Tylee had flown the last leg of the first trans-Canada flight in October 1920.

11. In 1926 wealthy American J. Dalzell McKee flew across Canada in his own floatplane, a Douglas World Cruiser. The RCAF had squadron leader Earl Godfrey fly with him as co-pilot and arranged for all refuelling along the way. Over nine days and in thirty-five hours' flying time, McKee and Godfrey completed the first lone floatplane flight from Montreal to Vancouver. In gratitude, McKee endowed an award in his name to honour the person who had contributed the most during the preceding year to the advancement of Canadian aviation.

12. Richardson, with an eye to history, commissioned a book (*Arctic Pilot* [London: Thomas Nelson, 1939]) on WCA pilots' exploits on the Mackenzie River. The author was Kathleen Shackleton, sister of the British explorer Sir Ernest Shackleton. Unfortunately most copies were lost when the ship bringing them to Canada was torpedoed.

13. "Punch" Dickins, in *Canadian Aviation Anniversary Issue* (June 1978), p. 18.

14. Frank Ellis, "Airmen Adventurers to the Bay," *The Beaver* (Summer 1970), p. 56.

15. Francis Roy Brown (born in Stockton, Manitoba) was always mistaken for the Arthur Roy Brown (born in Carleton Place, outside Ottawa) who may have shot down the Red Baron in World War I. Francis Roy Brown left WCA to become a test pilot for MacDonald Brothers and in 1947 began his own bush carrier, Central Northern Airways, a predecessor of TransAir. Much later he was elected as a member of Parliament for Rupert's Land, Manitoba.

16. There was no better search and rescue pilot than Herbert Hollick-Kenyon. Besides participating in the MacAlpine Expedition, he also took part in the hunt for the Soviet flier Sigmund Levanevsky and was part of the Ellsworth expedition to Antarctica, where a plateau is named after him.

Chapter Two

1. From "Ballad of a Bush Pilot," by Captain C.R. Robinson:
In years gone by
I used to fly a Fairchild Eighty-Two
And was it fair or stormy air
We'd always muddle through
For hours I'd sit upon the bit
Of Kapok-padded seat
My knees tucked in beneath my chin
In comfort hard to beat

> The instruments, the cowling dents
> The grease spots on the glass
> I still recall them one and all
> As through the years I pass . . .
>
> I've carried boats and smelly goats
> In Junkers thirty-fours
> I froze my toes in Barkley-Grows
> On Great Bear's rocky shores . . .
>
> Throughout the years of sweat and tears
> This wish has come to me
> Before I die,
> I want to fly
> A Douglas DC-3.

2. J.R.K. Main, *Voyageurs of the Air* (Department of Transport Centennial Project) (Ottawa: Queen's Printer, 1967), p. 73.

3. Col. Redford Henry Mulock, C.B.E., D.S.O., A.D.C., had five "kills" over the Front and was one of the most decorated Canadian airmen in World War I. He could have risen to the rank of air vice-marshal but gave up his military career to concentrate on Canadian Vickers and WCA. His prestige proved invaluable to Richardson when representing WCA in Ottawa circles.

4. In defence of the Post Office, it should be noted that postal revenue alone at this time could never pay for the salaries of hundreds of rural mail carriers and clerks and the purchase of thousands of mailboxes. The operating costs could never be met, and the Post Office's annual deficits were prime targets in the House for members of the Opposition.

5. By October 27, the mail was being flown directly to Ottawa from Rimouski pier by a Canadian Transcontinental Airways FC-2.

6. Thornton had a well-deserved reputation for getting railroads out of bankruptcy: before he came to Canada, he was known in Britain as "the Yank who had saved the Great Eastern." The Grand Trunk Railway mess and Ottawa's bungling was beyond even his legendary talents.

7. After World War II, when both West and East Germany set up their own Lufthansas, each tried to claim legitimacy by appropriating the crane symbol. The "western" Lufthansa successfully sued its eastern cousin, forcing the latter to reappear as "Interflug."

8. K.M. Molson, *Pioneering in Canadian Air Transport* (Winnipeg: James Richardson & Sons, 1974), p. 264.

9. Aircraft were not equipped with landing lights or flares until 1932. Even then, pilots disliked using the flares to illuminate the runway because when dropped over an airfield, the flares caused shadows that made landings even more confusing. It was only when Curtiss, Douglas and Boeing began equipping their aircraft with landing lights in the nose and wings that pilots (and their passengers) had a reasonable chance of safety.

10. K.M. Molson, *Pioneering in Canadian Air Transport* (Winnipeg: James Richardson & Sons, 1974), p. 98.

11. As the first of the "quiet" aircraft, the Condors also had sleeping berths. Curtiss engineers discovered that using three-bladed propellers allowed for low engine speeds and permitted passengers to sleep through the night. Unfortunately because of their short range, the Condors had to land frequently and the passengers were woken up each time so that their ears wouldn't "pop."

12. As the Canadian Pacific liners were christened "Empresses," the smaller ships that the CNR built for Caribbean trade were appropriately called "Ladies."

13. It was Thornton's last hurrah. In 1930, wanting to compete with the Canadian Pacific oceanliners in the Far East, he ordered three ships from a British shipyard for a Prince Rupert–Alaska service that would eventually extend to the Orient. The ships were grossly overpowered and overpriced. Ottawa decided that in case of war they should be capable of being converted to armed merchantmen. Worse, Thornton discovered that no passengers wanted to go to Alaska. When the ships went to the breakers in 1936, the Canadian taxpayer was left with a bill of $82,176,699. Undismayed, the CNR entered the airline business the following year. Taxpayers were relieved that at least the first three aircraft (bought from Canadian Airways) cost a paltry $169,176.

14. Richardson's letter to Wilson, September 18, 1929. National Archives.

15. J.R.K. Main, *Voyageurs of the Air* (Ottawa: Queen's Printer, 1967), p. 103.

Chapter Three

1. Shirley Render, unpublished M.A. thesis, University of Manitoba, March 1984, p. 83. Post Office officials were said to be open to bribery, something that Canadian Airways' general manager, G.A. "Tommy" Thompson, who handled all contracts, refused to condone.

2. Walter Henry, Willowdale, Ontario, personal communication.

3. Alison Gardner, *Grant McConachie* (Don Mills: Fitzhenry & Whiteside Ltd., 1979), p. 22.

4. Russ Baker remembered taking a prospector to Usilika Lake one March day. After disembarking on the ice with his load, the prospector asked Russ if he ever came over this way in the summer. Upon being given the affirmative answer, he replied, "Well, how should I signal you if I want you to land? Should I shoot in front of you or behind you?" To humour him, Baker remarked that he doubted if he could hear the shot over the noise of the engine. The reply nearly caused him to fall out of the aircraft. "Oh, I think you would hear it; I can shoot pretty close!" For the remainder of the year, Russ never flew anywhere near that lake.

5. Shirley Render, unpublished M.A. thesis, University of Manitoba, March 1984, p. 65. When in 1936 a General Airways aircraft crashed because of overloading, killing the pilot and passengers, the government took no disciplinary action. The superintendent of air regulations is supposed to have said, "the

government had decided not to worry about overloading anymore but to let the operators find out the consequences by themselves."

6. George Lothian, *Flight Deck: Memoirs of an Airline Pilot* (Toronto: McGraw-Hill Ryerson, 1979), p. 33.

7. Ken Molson, *Pioneering in Canadian Air Transport* (Altona, MB: D.W. Friesen & Sons, 1974), p. 170.

8. Three Germans and one Irishman attempted the first east–west flight across the north Atlantic, from Baldonnel, Ireland, to New York on April 12, 1928. Compass problems and headwinds forced them down at Greenly Island, Labrador, 1,200 miles (1,920 km) from New York. A massive search operation was begun, with Vachon flying the first reporters to the site. Vachon was awarded the McKee Trophy in 1937 and later sat on the Air Transport Board.

9. Despite technological triumphs, transcontinental flying in the United States was still a "plane–train" affair; pilots carried special vouchers entitling passengers to train tickets in case of grounded flights.

10. In 1982 the Western Canada Airways Museum in Winnipeg bought a Spanish-built Junkers 52 trimotor from a dealer in Orlando, Florida. It had Bristol Aerospace remove the outboard engines to convert it into a replica of the "Flying Boxcar." Even in this age of Boeing 747s, it remains awe-inspiring.

11. David MacKenzie, *Canada and International Civil Aviation: 1932–1948* (Toronto: University of Toronto Press, 1989), p. 43.

12. On fair-weather days Joe Bertalino, who manned the Vancouver airport's radio, gave United's office in Seattle the local weather report by saying, "Yeah, the weather's pretty good, you can come in today." He remembered that they had a wonderful maxim that went, "If you can see Mount Baker, it's going to rain; if you can't see it, it's raining." British Columbia Archives & Records Service, *The Magnificent Distances: Early Aviation in British Columbia 1910–1940* (Sound Heritage Series No. 28), p. 45.

13. Don and Peggy Thompson, "Art Schade, Bush Pilot," *The Beaver* (Autumn 1979), pp. 23–24.

14. On May 20, 1932, Amelia Earhart had successfully crossed the Atlantic using a single-engine Lockheed Vega. The company was pleased with the publicity, and in 1935, when she chose the twin-engine Lockheed 10 for her "round the world" flight, it was ecstatic and allowed her to custom-design it. It is interesting to speculate what effect Earhart's choice had on sales of the 10, especially to James Richardson and C.D. Howe. The aviatrix disappeared in the Electra off Howland Island in the Pacific on July 2, 1937.

15. R. Bothwell, *C.D. Howe: A Biography* (Toronto: McClelland & Stewart, 1979), p. 107.

16. When the bound copies of the plan were returned to Winnipeg, CAL staff noted that one was missing. Despite all inquiries, it was never found.

17. J.R.K. Main, *Voyageurs of the Air: A History of Civil Aviation in Canada 1858–1967* (Ottawa: Department of Transport, 1967), p. 104.

18. James Richardson, "History of Canadian Airways," *Canadian Aviation* (July 1938), p. 11.

19. Walter Henry, "The Barkley-Grow T8P-1," *Western Canada Aviation Museum Aviation Review* (December 1994), pp. 19–20.

20. British Columbia Archives & Records Service, *The Magnificent Distances: Early Aviation in British Columbia 1910–1940* (Sound Heritage Series No. 28), p. 32.

21. *The Honker* (August 1939). This was the third edition of the revived newsletter, which was turned out "as a modest operation" by G.A. Thompson on CAL's Gestetner at the Winnipeg office. Thompson was very proud of the fact that one of the requests to be placed on the mailing list came from "no less a source than the Congressional Library" in Washington, DC.

Chapter Four

1. *The Monthly Honker* (October 1939). In the hysteria of the times, on November 11, 1939, Howard MacDonald was flying a CAL de Havilland Rapide out of Vancouver to Victoria when the aircraft sustained two rounds of buckshot from duck hunters. MacDonald returned to the airport, got the local policeman and drove off in search of the culprits. Their alibi was that they were shooting at birds and didn't notice the aircraft seventy feet above their heads. MacDonald had to endure days of ribbing from other CAL employees who greeted him with "Honk! Honk!"

2. British Columbia Archives & Records Service, *The Magnificent Distances: Early Aviation in British Columbia 1910–1940* (Sound Heritage Series No. 28), p. 67.

3. Ibid., pp. 68–69.

4. John Condit, *Wings over the West: Russ Baker and the Rise of Pacific Western Airlines* (Madeira Park, BC: Harbour Publishing, 1984), p. 23.

5. R.W. Ryan, *From Boxkite to Boardroom* (Moose Jaw: Moose Jaw Publications, 1987), pp. 45–48.

6. Ibid., p. 33.

7. At the movie's Broadway debut, an RCAF band provided the music on opening night. The success of the movie was slightly diminished by the Japanese attack on Pearl Harbor the following day.

8. The Avro Anson was a direct result of George Woods-Humphrey, then managing director of Imperial Airways, asking Avro in 1933 to design a short-haul airliner for the European routes. The first "652" was delivered to Imperial Airways on January 7, 1935, causing the Air Ministry to order 174 for Coastal Command and naming it the "Anson." Imperial Airways replaced their 652s with Lockheed Electras in 1938, and Coastal Command looked to do the same by 1940.

9. George Lothian, *Flight Deck: Memoirs of an Airline Pilot* (Toronto: McGraw-Hill Ryerson, 1979), pp. 75–76.

10. The movie *Captains of the Clouds* concluded with the chastened bush pilots ferrying their Hudsons to England and being attacked by a Luftwaffe fighter. Unable to get an authentic Messerschmitt for the movie, director Michael Curtiz substituted a Hurricane, covering it in swastikas. On the

strength of *Captains of the Clouds*, Curtiz was given *Casablanca* to direct the following year, for which he would win an Oscar.

11. Carl Burke, a CAL pilot who later flew for the Royal Air Force Ferry Command, used a ski-equipped de Havilland Rapide to bring the bodies of Banting, navigator William Bird, and pilot William Snailman out of the wreckage. Burke would begin Maritime Central Airways in 1941, which would become part of Eastern Provincial Airlines in 1963.

12. The figures are from D.M. Bain, *Canadian Pacific Air Lines: Its History and Its Aircraft* (Calgary: Cal/Oka Printing, 1987), p. 17.

13. George Lothian, "Flying with Ferry Commands," *Canadian Aviation 50th Anniversary Issue* (1978), p. 100.

14. D.M. Bain, *Canadian Pacific Air Lines: Its History and Its Aircraft* (Calgary: Cal/Oka Printing, 1987), p. 18.

15. *The Honker*, January 10, 1942.

16. John Condit, *Wings Over the West* (Madeira Park, BC: Harbour Publishing, 1984), pp. 27–28.

17. British Columbia Archives & Records Service, *The Magnificent Distances: Early Aviation in British Columbia 1910–1940* (Sound Heritage Series No. 28), p. 73.

18. Art Schade retired from active flying duties in 1948, having accumulated 10,500 hours, many of them flying over the bush. On August 6, 1974, the Government of Manitoba recognized his accomplishments by naming Schade Lake (lat. 52 degrees, long. 101 degrees) after him. Ontario also honoured him, with a Schade Lake and Schade River.

19. *The Monthly Honker* (August 1943), p. 2.

20. Cedric Mah was famous for having thrown two million dollars worth of gold bullion overboard to prevent a forced landing while flying over the Himalayas.

21. It was later revealed that the prime minister's office had sent a draft to London and that the CPR's old friend, Lord Beaverbrook, had somehow got his hands on it and gleefully leaked it to Coleman as part of the CPR (and newspaper) brotherhood.

22. Another of the stewardess's tasks was administering oxygen to the passengers. The Lodestars rarely flew higher than 10,000 feet because passengers became dizzy and even fainted due to the reduced level of oxygen at that altitude. The oxygen came from a portable cylinder, and the stewardess distributed it through masks to each passenger. She washed the masks on landing.

Chapter Five

1. Flo Whyard, Ernie Boffa's biographer, noted that this song was heard in the beer parlours of Yellowknife in the 1940s.

2. DC-3s inaugurated the Vancouver–Penticton route on September 8, 1947, and the Vancouver–Calgary run on September 22. Another DC-3 crashed at Okanagan Park, BC, on December 22, 1950. In all, the airline bought seventeen DC-3s. The last one, CF-CRX, remained in service until 1974. Probably the

most unusual cargo that any of the CPA DC-3s carried was for Max Ward in 1954. Max shipped three muskoxen from Yellowknife to Edmonton on a regular CPA DC-3 flight in the cabin with the human passengers. As the muskoxen were not toilet trained, the smell within the aircraft on the five-hour flight must have been memorable. Grant McConachie was *not* pleased. Max Ward, *The Max Ward Story: A Bush Pilot in the Bureaucratic Jungle* (Toronto: McClelland & Stewart, 1991), p. 125.

3. On September 9, 1949, one of the CPA Quebec DC-3s, CF-CUA, was blown up in mid-air in the notorious Albert Guay case—the first such incident in history. Albert Guay was a Quebec City jeweller who murdered his wife by having his girlfriend innocently plant a bomb on the Canadian Pacific flight— after taking out a $10,000 insurance policy two days before, on the unfortunate Mrs. Guay. Guay then tried to get his girlfriend to commit suicide. All twenty-three people on the aircraft were killed. The incident, according to the Montreal *Herald*, had everything: "sex, cash and explosives."

4. All the material on the flight attendants' struggle to organize is taken from N. Jill Newby, *The Sky's the Limit: The Story of the Canadian Air Line Flight Attendants Association* (Vancouver: CALFAA, 1986).

5. Stewardesses in Toronto, for example, had been getting to Malton airport, then in the wilderness, by taking the streetcar down Yonge Street to the Royal York Hotel, walking across to the Post Office terminal and then hitching a ride with the mail truck to the airport. Although the ride was free, since only two stewardesses could fit in the cab with the driver the others had to sit in the locked van, perched on the mail bags and spare tire. Looking clean and crisp for passengers after this journey must have been difficult. At the end of their flight, the women had to take the mail truck once more or beg pilots for a lift to the city.

6. Ernie Boffa's barnstorming experiences feature in Peter Pigott's *Flying Colours: The History of Commercial Aviation in Canada* (Vancouver: Douglas & McIntyre, 1997).

7. The stories of Ernie Boffa's career are supplied through the kind permission of Florence Whyard from her book *Ernie Boffa: Canadian Bush Pilot* (Anchorage: Alaska Northwest Publishing, 1986).

8. Alf Caywood was a prospector who joined Canadian Airways Ltd. as a mechanic's helper, graduating to pilot in 1938. During the war he flew for Eldorado Mining and Refining at Port Radium, on Great Bear Lake. He had been forced down in the Barrens for nine days in the winter of 1941, after his aircraft caught fire in mid-air. He will always be associated with use of the first DC-3 to be licensed commercially in Canada. Caywood was elected a member of Canada's Aviation Hall of Fame in 1988.

9. In 1956, using a DC-3 to work on the DEW Line, under Roland G. Le François, Boreal Airways changed its name to Nordair.

10. David Cruise and Alison Griffiths, *Lords of the Line* (Markham, ON: Penguin Books, 1988) pp. 367–68.

11. National Archives, Canadian Pacific Airlines, M.G. 28 III 48 Scrapbook No. 2.

12. In a personal communication, former president Ian Gray wrote that McConachie's philosophy could be characterized by the words, "If it's new, I'll take two!"

13. Unlike Cunard and other shipping lines, the CPR ships serviced Quebec City rather than New York, resulting in fewer and poorer customers.

14. When Walter Gilbert built the airstrip, entirely at CPA's expense, there was no road connecting it with Yellowknife. Passengers were required to take a floatplane that shuttled back and forth to town. These were called "elevator trips" because the shuttle went up and came down. A road was built in 1946, when the Department of Transport took over Yellowknife Airport from CPA.

15. A subsidiary of American Airlines, American Export Airlines, inaugurated the first wholly civilian postwar flights between New York and Bournemouth, England (via Gander and Shannon) on October 23, 1945, using DC-4s.

16. In 1954 the airline was bought by Qantas, which took over its Sydney–Vancouver service with Boeing 707s.

17. David MacKenzie, *Canada and International Civil Aviation* (Toronto: University of Toronto Press, 1989), p. 247.

18. McConachie also enjoyed the Australian beaches on the trip. The Sydney *Daily Mirror* noted, "The president of British Canadian Pacific Airlines, Mr. G.W.G. McConachie, is in town at the moment having the time of his life at Palm Beach. They don't have sea-going sharks in Canada, and G.W.G. was pretty nervous about them when he first took the waters. "You'll be all right. We'll blow a whistle," said the surf club boys reassuringly. The first time out he heard the whistle blow two hundred yards from shore. He thinks he clipped half a mile from the Olympic time, was dumped by a big roller and washed up on the beach "like a limp rag," as he says—to find it was only the 'bathe-between-the-flags' whistle after all."

19. R.W. Ryan, *From Boxkite to Boardroom* (Moose Jaw: Grand Valley Press, 1986), p. 84.

20. Apparently the Nationalist general in charge of the city's defence had the monopoly on timber.

21. Canadian nationals and diplomatic staff were later removed by the warship HMCS *Crescent*. The city would be completely overrun by May 20, and CF-TEP would be the last CPA aircraft to land in China until 1972.

22. Hugh Whittington, "Australian Odyssey," *Canadian Aviation* (August 1974), pp. 30–34.

23. Jill Newby, *The Sky's the Limit: The Story of the Canadian Air Line Flight Attendants Association* (Vancouver: CALFAA, 1986), p. 237.

Chapter Six

1. The Sydney *Daily Mirror* ran the following on September 28, 1949: "*Man Has 36 Seats in Plane All to Himself*. Mr. Max Henry Stampfli, a Swiss sheep farmer from Capertee (NSW) who left by Canadian Pacific Airlines Canadair yesterday for a holiday in Switzerland, will be a very pampered pas-

senger for the next four days. He will have 36 seats to himself and be attended by three beautiful Canadian air hostesses from Sydney to Vancouver. Mr. Stampfli, who is a 33-year-old bachelor, said he is 'thrilled' at the prospect of the trip."

2. P. Smith, *It Seems Like Only Yesterday: Air Canada, The First Fifty Years* (Toronto: McClelland and Stewart, 1986) p. 144.

3. Howe was determined to export his North Stars and sold them to BOAC for $670,000 each, $100,000 less than he had charged CPA. Rumours have persisted that the British gave Canada their colony of Newfoundland as part of the deal.

4. Eventually TCA overcame the engine noise with a crossover exhaust system, developed by a company engineer named Merlin MacLeod. After all the bugs had been worked out, the North Stars proved themselves to be safe and reliable, soldiering on in TCA service until 1961, BOAC until 1960 and the RCAF until 1966, as well as in the Royal Rhodesian Air Force.

5. The DC-6Bs had tiered bunks in the rear of the cabin, a favourite for newly married couples on their way to a honeymoon in Honolulu. It became, one flight attendant recalls, quite a task to wake them up on approach into Honolulu.

6. J.T. Radford, "Recollections of a DEW Line Pilot." Private papers, Canadian Airlines International Archives.

7. Don O'Grady, "DEWLINE," CP Air Internal Correspondence, Vancouver, April 15, 1982, pp. 7–11.

8. Bernt Balchen is one of the few foreigners named to Canada's Aviation Hall of Fame, for his flying in the Churchill airlift and pioneering polar flights in the 1950s.

9. Department of Transport Press Release #522. CAI DTD 7-55 R552.

10. Vancouver *Province*, June 21, 1956.

11. Not until June 17 did Governor General Vincent Massey accept Louis St. Laurent's resignation, so Sinclair and Company were within their rights to do so.

12. Denis Smith, *Rogue Tory: The Life and Legend of John G. Diefenbaker* (Toronto: Macfarlane Walter and Ross, 1995), p. 117.

13. John Condit, *Wings Over the West: Russ Baker and the Rise of Pacific Western Airlines* (Madeira Park, BC: Harbour Publishing, 1984), p. 150.

14. Ibid.

15. Quoted in S. Goldenberg, *Troubled Skies: Crisis, Competition and Control in Canada's Airline Industry* (Toronto: McGraw-Hill Ryerson, 1994), p. 19.

16. Ronald A. Keith, *Bush Pilot with a Briefcase: The Happy-Go-Lucky Story of Grant McConachie* (Toronto: Doubleday Canada, 1972), p. 298.

17. Contrary to John Condit's assertion in *Wings Over the West*, the flight attendants, not the airline, won the strike.

18. Capt. W.C. "Cam" Ross, "Flying the Britannia with Canadian Pacific," *CAHS Journal* (Spring 1990), pp. 24–25.

19. This was one of the "freedoms" adopted by the International Civil

Aviation Organization after the failure of the Freedom of the Skies at the Chicago Conference in 1944. It stated the privilege to take on passengers, mail and cargo destined for the territory of any other contracting state and the privilege to put down passengers, mail cargo coming from any such territory. Simply put, it allowed a foreign airline like BOAC to pick up passengers in Toronto and drop them off in San Francisco, thus competing with TCA.

Chapter Seven

1. Out of this foot-dragging came a proposal from two Calgary financiers, Max Bell and R.A. Brown, to buy out both airlines and offer the CPR and CNR 25 percent each of the shares in the new company—a return to the old Canadian Airways Ltd. arrangement. As much as he would have liked to have the airlines under one roof, Pickersgill was not prepared to allow a private company to monopolize Canada's air transport system.

2. It is unclear when the age-thirty restriction was adopted by CPA, but it appears that no one was ever forced out because of it.

3. Dick Ryan recorded an incident that reflects the attitude of both management and the stewardesses in the McConachie era. In 1950, Ken Dakin was in charge of labour disputes and was dealing with a request by the stewardesses to be given eight hours' rest after eight hours' duty in flight. "He pointed out the problem of trying to land an aircraft after eight hours in the air when there was nothing underneath it but water . . . and they replied, 'We don't care, that's what we want'."

4. R.W. Ryan, *From Boxkite to Boardroom* (Moose Jaw: Grand Valley Press, 1986), p. 84.

5. In 1975, CP Air spent $6 million worldwide ($850,000 in the US alone) to advertise its orange uniqueness. For that amount of money, the creative director of the agency came up with the slogan "Orange is Beautiful" and the even more memorable "Peel off to the South Pacific"—both incomprehensible in any other language.

6. PWA developed its own hotel and airfare package to the Grand Cayman Islands, but because the runway there was so short it had to buy a special model Boeing 707-138—the same type that CPA would lease from Standard Airways in 1967.

7. The first Canadian air freighter flight to China was a PWA Hercules 5 on March 3, 1973, sent from Calgary to Beijing with a load of electronic satellite equipment.

8. Seven years later the provincial government put PWA on the market in one of the biggest privatization deals in history, accomplished in two stages: from 100 percent to 15 percent, and by September 1984 from 15 percent to 4 percent.

9. Technically the airline became "all jet" on July 4, 1974, when it elected to take a group of air cadets in a DC-3 for a tour of the Fraser valley, showing them the airports at Vancouver and Abbotsford. The aircraft landed at Abbotsford, and after takeoff, the crew were unable to get the wheels up. It

returned to Abbotsford, and the passengers and crew were bussed to Vancouver. The DC-3 remained there overnight, and after much discussion and consultation the airline's chief pilots ferried it to Vancouver with the wheels down for its final flight. CP Air's last piston-engined airliner was sold to Harrison Airways Ltd. on October 23, 1974. The last DC-3 in Canadian Pacific colours is "CF-CPY," which rests on a pedestal outside the Whitehorse airport terminal, where it acts as a weathervane.

10. In October 1970, a duty day could not exceed 14 hours, which included checking in an hour before the flight and checking out an hour after. In any 30-day period a pilot could have up to 85 hours. Navigators had a maximum 10 1/2 hours, which they could spend on the flight deck in any 24-hour period; however, they could not average more than 75 hours a month during a block. Flight attendants averaged no more than 70 to 80 hours a month. All of the schedules were subject to many variations.

11. *CP Air News* (November 1970), p. 1.

12. Arthur Hailey's *Airport* was one of the bestsellers chosen, obviously by someone who hadn't read it.

13. In probably the most revealing press release ever, CP Air flew "Playboy Playmate" Roxanna Platt to Ottawa to pose in the Carleton University homecoming festivities. The press release had a photo of a fully clothed Ms. Platt with the advice, "For an 'inside' look at Miss Platt, please refer to the March *Playboy*." The airline discontinued stocking its Executive Jets with the magazine less than a year later, not on moral grounds or because its contents might be offensive to the stewardesses. Instead, it seems that the "serious and sophisticated businessmen" kept stealing so many of the *Playboys* en route that having to replace them so often became prohibitively expensive. In 1998, although Canadian Airlines International did stock *Sports Illustrated* on its flights, it didn't carry the "Swimsuit Issue."

14. CP Air's "Thigh in the Sky" episode was well managed by the airline and the stewardesses interviewed years later remember it fondly. It was nothing in comparison with the bitterness involved in PWA's celebrated "red panty" case. During the Calgary Stampede, in keeping with the "covered wagon and cowboy" theme, PWA stewardesses designed their own uniform: they wore cowboy boots, a blue western hat and vest, a white blouse, a short fringed skirt, which just about covered their buttocks, and red bloomers underneath. The "Stampeder" uniform was so popular that for the first time in airline history, during the demonstration of emergency procedures, the stewardess had the complete attention of all on board—for the wrong reason. It might have ended there, but, effective June 1, 1971, PWA decreed that all its female flight attendants were to wear the "Stampeder" outfit on all flights originating from Calgary and Vancouver. Several objected, particularly because the airline still served many logging and cowboy towns. The inevitable occurred, and a male passenger laid his hands on one of the stewardess's bloomers as she leaned over to serve another customer. As a result, two flight attendants showed up for work in their previous, more modest uniforms. The airline made a test case out of this and CALFAA took it up, whereupon the airline backed down. The "red

panty case" signalled a major victory for Canadian female flight attendants, who never had to put up with such humiliations again.

15. In 1974, the Chinese were granting visitor visas only to scientific groups, not to individual tourists.

16. Shirley Render, *No Place For A Lady* (Winnipeg: Portage & Main, 1992) p. 289. Via Transair and PWA, Rosella Bjornson would become the first female captain with Canadian Airlines International in 1990, and on June 6, 1997, she was inducted into Canada's Aviation Hall of Fame. At a banquet in her honour, she thanked all those who made it possible: "my father who instilled in me a love of flying at an early age, Transair management who gave me the opportunity that every pilot dreams of and . . . most of all, the employees of Canadian Airlines who are such a pleasure to work with. I am proud to be an employee of Canadian Airlines."

17. *CP Air News* (December 1973).

18. Only three were evident originally, because the fourth was leased to Braniff for four years. The 747-200s served until the last was sold off to Pakistan International Airlines in 1986.

19. The size of the 747 inspired awe from the very first, but also notable is the fact that when it has been fuelled up, the aircraft contorts and changes shape imperceptibly. Occasionally, Wardair flight attendants opened the doors of the 747 to cool it before the passengers boarded, only to discover that they couldn't be closed again—until the aircraft had been expensively defuelled.

20. CP Air pioneered the "bring your own lunch" concept twenty years before Freddie Laker's "Skytrain." Marketing came up with the idea of a "Thrift Service" on the Hawaii and Amsterdam flights in which passengers could buy a cheap boxed dinner before departure rather than paying for the regular in-flight meal with their tickets. A good idea in theory became a flight attendant's nightmare in practice. The passengers who had the boxed meal invariably found themselves nibbling their sandwiches beside someone enjoying a steak dinner. Many had even eaten their boxed meals while waiting in the departure lounge and now crazed with hunger at the scent of chicken or fish, filled themselves with alcohol. When told they couldn't order an in-flight meal "in flight"— because only the exact number had been put on board, they became abusive toward the cabin crew. The empty lunch containers were strewn around the cabin, all of it making for an unhappy flight for passengers and staff alike.

21. In 1974 the PWA Vancouver–Seattle flight had to stop technically at Victoria to satisfy government regulations, causing it to touch down at the airport and take off immediately, to the amusement of locals.

22. The 747s were due for expensive "section 41" maintenance checks, something that CP Air prided itself on. Rather than spend the money to do so, the airline took the opportunity to trade them in for the Pakistan International Airlines' DC-10s. It seemed like a good deal until it was discovered that once in CP Air service, they required a heavy expenditure as well.

23. For the life and times of Claude I. Taylor and the privatization of Air Canada, Peter Pigott's *Flying Canucks II* (Toronto: Dundurn Press, 1997), p. 121–132, is recommended.

24. Ronald Keith, "Air Canada and CP Air's Deregulated Dogfight," *Canadian Business* (October 1979), p. 56–60.

25. So incestuous was Gemini that the Federal Competition Tribunal investigated whether it should be dissolved as anticompetitive.

26. *CP News*, December 17, 1985. The name "CP Air" had always confused people in Hong Kong who associated "CP" with their own Cathay Pacific Airlines.

27. As the years have gone by, the DC-10 has attracted more complimentary names, such as the "Douglas Diner." This name comes about because it is the only three-pilot aircraft left in Canadian Airlines, and the second officer's seat, with its own table, has become a favourite place for all DC-10 pilots to "dine" rather than balance a meal tray on a pillow on one's lap.

28. David Cruise and Alison Griffiths, *Lords of the Line* (Markham, ON: Penguin Books, 1988) p. 456.

29. Jack Desmarais, "Canadian: Is this Brainchild of PWA Corp. Already Deregulation's Winner?" *Canadian Aviation* (November 1987) p. 25–28.

Chapter Eight

1. Carey French, "The Sky's The Limit," *Report on Business Magazine (The Globe and Mail)* (August 1987), pp. 18–22.

2. David Barrett, the former BC premier, who hadn't forgiven PWA for moving to Calgary, was then working as a radio talk-show host, and accused Eyton of union-busting tactics during the strike. Rhys immediately threatened him with a lawsuit, and Barrett backed down—on the air. "Respect," as Eyton said later, "is something you earn and you do not let irresponsible comments and individuals undermine it."

3. Gord Cope, "Calgary's Air Force," *Calgary Herald Sunday Magazine* (April 9, 1989), pp. 6–10.

4. Cecil Foster, in *Canadian Business* (February 1995), has it that although the Air Canada's directors really wanted Carty, "discussions faltered over a pay package for him," and they went for the second choice, Hollis Harris. Harris had been with Delta Airlines for thirty-six years and later had seen Continental Airlines into bankruptcy. He was running his own aviation consulting company in Atlanta when Air Canada called.

5. S. Goldenberg, *Troubled Skies: Crisis, Competition and Control in Canada's Airline Industry* (Toronto: McGraw-Hill Ryerson, 1994).

6. S.O. Fattedad, "Behind the Scenes at Canadian Airlines," *CGA Magazine* (June 1993), p. 32.

7. Capt. Rick Wiley, in conversation with the author.

8. A television commercial aired in Britain in 1982 showed passengers leaving one plane and walking along the wings to a Scandinavian Airlines System aircraft. But Chiat/Day/Mojo's Jack Neary and his partner, Duncan Milner, said they came up with the Canadian Airlines idea by themselves. The British agency withdrew its commercial when the pilots' association complained that it was in poor taste.

9. Michael Posner, "Flight into Oblivion," *Report on Business Magazine (The Globe and Mail)* (November 1996), p. 78.

10. Nothing displayed Harris's lack of knowledge of recent aviation history in Canada more than this. In 1978, Air Canada had received a $300 million loan from Ottawa at a favourable rate of 7.2 percent—3 percent below the prime lending rate at the time. When it was privatized in 1988, Air Canada continued to have a $130 million loan from the government; at the end of 1991, this was still $48 million.

11. Hugh Whittington, "West Wind Rising," *Air Transport World* (August 1994), p. 31.

12. The ad was so successful that staff received several inquiries from the public, who asked when "that aircraft with the signatures" would be visiting their city. Others thought that the airline should actually fly an aircraft like that. Unfortunately, Canadian Airlines' staff also took the spirit of the commercial literally and began signing their names to aircraft they had worked on—with felt markers and ball point pens—until management ordered them to stop.

13. The United States had granted similar antitrust immunity to KLM–Northwest, the purpose being to circumvent international air pacts that prevent mergers, restrict international ownership and prohibit cabotage.

14. In contrast, in April 1959 the monthly salary scale for flight attendant trainees was as follows: training period $200; first six months of service $275 (domestic), $315 (overseas); third year of service $335 (domestic), $375 (overseas); seventh year of service $401 (domestic), $446 (overseas). Deduction for the uniform upon completion of training was $20 a month until the total price of $160 was paid. Normal flying time on domestic lines was predicted to be 85 hours per month. No paid sick leave was granted in the first three months of service. For personnel not resident in Vancouver, the airline advised that accommodations be arranged through relatives prior to arrival, and the address of the YMCA at 997 Dunsmuir Street was also provided. On completion of training, the stewardesses were assigned to one of the following CPA bases: Vancouver, Edmonton, Prince Rupert or Montreal.

Conclusion

1. Peter C. Newman, "Gerry Schwartz has the right stuff." *Maclean's* September 13, 1999, p. 17.

2. Under code-sharing, two carriers share the same flight code numbers in reservation computers, enabling them to seamlessly load passengers on one another's planes. It makes it simpler to sell seats and enables airlines to offer more flights. For Canadian, the pact was beneficial because it could tap into American's marketing ability to sell seats in the US on Canadian's flights.

3. As part of the 1996 Operational Restructuring Plan, an agreement was reached with each of the airline's unions for the extension of all collective agreements for a four-year period ending December 31, 2000.

4. Steven Chase, "Canadian needed rules suspended to survive: CEO." *The Globe and Mail*, Sept. 15, 1999, p. B3.

5. Hugh Winsor, "Collenette tries to keep Canadian in the sky." *The Globe and Mail*, August 16, 1999.

6. Oliver Bertin, "On a wing and a prayer: The many routes that Canadian could take." *The Globe and Mail*, May 8, 1999, p. B3.

7. The reason why Alfonso Gagliano was included stemmed from a previous portfolio. As minister of labour in 1996 he had successfully resolved the Canadian Airlines flight attendants dispute with their airline by appointing conciliation mediator H. Allan Hope to settle it.

8. Orders-in-Council to enact a change in policy are a royal prerogative that Her Majesty the Queen in the person of the Governor General delegates to the Prime Minister and whomever he assigns. By using them policy can be implemented without parliamentary consultation.

9. *The Guardian*, Charlottetown, August 14, 1999.

10. It didn't take long for the media to dig up the dirt on how far Schwartz's political clout got him. In 1995 Collenette, then minister of defence, had arranged for Schwartz to go for a ride in a CF-18 fighter aircraft at CFB Bagotville.

11. "Report on Business," *The Globe and Mail*, August 25, 1999, p. B6.

12. Bruce Wallace, "Capital Solution." *Maclean's*, September 13, 1999, p. 20.

13. Giles Gherson and Rod McQueen, "Offer has nod of political approval," *National Post*, August 25, 1999.

14. Peter Lougheed, "Why I changed my mind." *The Globe and Mail* "Comment" section, October 22, 1999.

15. Under the Air Canada "poison pill," the airline issued one right for each Air Canada share then outstanding. It gave the directors the discretion to call the "separation time," that is, the rights were not exercisable until the separation time, which was the close of business on the tenth trading day after the date of the first public announcement of the acquisition of 10 percent or more of Air Canada shares and the commencement of an offer to acquire 10 percent or more of the shares. The "pill" stated that AirCo's August 24 announcement making the offer triggered the separation time.

16. Peter Kuitenbrouwer, "Firebrand Dave Ritchie in the eye of the storm." *National Post*, September 4, 1999, and Bill Tieleman, "Airline workers secret advantage: strong unions." *National Post*, September 13, 1999.

17. In 1998 when Loblaw Co. Ltd. tried to take over Quebec's Provigo chain of supermarkets the *Caisse* used its 35.7-percent stake in Provigo to make sure that the Ontario company would continue to buy from Quebec agriculture producers.

18. Heather Scoffield, "Ottawa won't say if airline bid may face merger review." *The Globe and Mail*, September 15, 1999, p. B3.

19. "Proxy Circular of AirCo in Respect of the Special Meeting of Shareholders of Air Canada," October, 1999, p. 13.

20. Onex AirCo letter accompanying pink proxy form dated October 13, 1999, "Dear Air Canada Shareholder," signed by Gerald W. Schwartz.

21. *Vancouver Sun*, September 29, 1999, p. A1.

22. Michael Colton reports from Ottawa. CBC Radio. October 26, 1999. 15:51.

23. Paul Farrar was until then best known for his 1992 role in bringing down the Reichmann's London Canary Wharf project. "This turkey won't fly," he said at a banker's meeting and cut off all CIBC funds.

24. James Rubin had honed his "vulture skills" when he represented the shareholders of toymakers Coleco in 1989. Bankrupt from making too many Cabbage Patch dolls, Coleco was about to default on its shareholders. Rubin not only won them enough money but engineered a takeover of Hasbro and got himself a 400-percent commission.

25. Lisa Wright, "Canadian on final course." *Toronto Star*, July 24, 2000.

26. Keith Macarthur. "Air Canada changes tune, plans to merge Canadian." *The Globe and Mail* Report on Business, p. B1.

27. *Toronto Star*, July 4, 2000.

A C K N O W L E D G E M E N T S

The material for this book came from many sources—archives, books and memoirs—and people, including employees, passengers and journalists. Wherever possible, contemporary material has been allowed to speak for itself, since to paraphrase it would have destroyed its flavour. The technology and politics of Canadian Airways, Canadian Pacific and Pacific Western Airlines have been admirably served by other writers, notably Ken Molson, Donald Bain, Ronald Keith and John Condit. My aim in this book was to narrate the events that took place between 1927 and 1997. Inevitably, a particular bias toward federal politics has crept in—a result of living too long in Ottawa, the self-acknowledged navel of the world. My special gratitude goes out to the following people and organizations.

In Canadian Airlines International: Steve Markey (vice-president, Corporate and Government Affairs), Jeffrey P. Angel (director, Corporate Communications), Andrew Geider (Corporate Archivist), Renee Smith-Valade, Rick P. Sloboda, Gary Bridgewater, Leiff Sollid, Gary Perdue, Sue Platts, Sandra Semple, Rick Wiley, Don Muir, Al Clark, Tony Archbold and Bob Randall, Jr.

People who still have the airline in their hearts: Rhys T. Eyton, Nina Morrison, Bob Randall, Sr. and Ian Alexander Gray.

Others whose information sent me along fruitful paths: Bev Tallon of the Western Canada Aviation Museum, Shirley Render, Florence Whyard and Patrick Gossage of "Media Profiles."

Organizations that provided help were: Canadian Pacific Archives/Bob Kennell, the Air Transport Association of Canada/Michael Skrobica, Transport Canada Corporate Services/Shaun Moran and the Staff at the Lester B. Pearson Library in the Department of Foreign Affairs.

While acknowledging the help and advice of all these people, I willingly exonerate them from blame for any errors and shortcomings, accepting these as my own responsibility.

I cannot close without expressing my deepest thanks to my wife, Avril, and my two daughters, Holly and Jade, to whom this work is dedicated.

B I B L I O G R A P H Y

Books

Bain, D.M. *Canadian Pacific Air Lines: Its History and Its Aircraft*. Calgary: Cal/Oka Printing, 1987.

Bothwell, R. *C.D. Howe: A Biography*. Toronto: McClelland & Stewart, 1979.

British Columbia Archives & Records Service. *The Magnificent Distances: Early Aviation in British Columbia 1910–1940* (Sound Heritage Series No. 28). 1980.

Christie, Carl A. *Ocean Bridge: The History of RAF Ferry Command*. Toronto: University of Toronto Press, 1995.

Condit, John. *Wings Over the West: Russ Baker and the Rise of Pacific Western Airlines*. Madeira Park, BC: Harbour Publishing, 1984.

Cruise, David and Alison Griffiths. *Lords of the Line*. Markham, ON: Penguin Books, 1988.

Ellis, Frank. *Canada's Flying Heritage*. Toronto: University of Toronto Press, 1954.

Goldenberg, S. *Troubled Skies: Crisis, Competition and Control in Canada's Airline Industry*. Toronto: McGraw-Hill Ryerson, 1994.

Hotson, Fred W. *The de Havilland Canada Story*. Toronto: CANAV Books, 1983.

Hudson, Kenneth, and Julian Pettifer. *Diamonds in the Sky: A Social History of Air Travel*. London: Bodley Head, 1979.

Keith, Ronald A. *Bush Pilot with a Briefcase: The Happy-Go-Lucky Story of Grant McConachie*. Toronto: Doubleday Canada, 1972.

Lothian, George. *Flight Deck: Memoirs of an Airline Pilot*. Toronto: McGraw-Hill Ryerson, 1979.

Mackenzie, David. *Canada and International Civil Aviation: 1932–1948*. Toronto: University of Toronto Press, 1989.

McGregor, Gordon. *The Adolescence of an Airline*. Montreal: Air Canada, 1970.

McLaren, Duncan D. *From Bush to Boardroom: A Personal View of Five Decades of Aviation History*. Winnipeg: Watson & Dwyer, 1992.

Main, J.R.K. *Voyageurs of the Air: A History of Civil Aviation in Canada, 1959–1967*. Ottawa: Queen's Printer, 1967.

Molson, Ken. *Pioneering in Canadian Air Transport*. Altona, MB: D.W. Friesen & Sons, 1974.

Newby, Jill. *The Sky's the Limit: The Story of the Canadian Air Line Flight Attendants' Association*. Vancouver: CALFAA, 1986.

Render, Shirley. *No Place for a Lady: The Story of Canadian Women Pilots 1928–1992*. Winnipeg: Portage & Main Press, 1992.

Ryan, R. *From Boxkite to Boardroom*. Moose Jaw: Grand Valley Press, 1986.

Sampson, Anthony. *Empires of the Sky*. London: Doubleday, 1988.

Smith, Denis. *Rogue Tory: The Life and Legend of John G. Diefenbaker*. Toronto: Macfarlane Walter & Ross, 1995.

Smith, Philip. *It Seems Like Only Yesterday: Air Canada, the First Fifty Years*. Toronto: McClelland & Stewart, 1986.

Solberg, Carl. *Conquest of the Skies: A History of Commercial Aviation in America*. Boston: Little, Brown & Co., 1979.

Stevenson, Garth. *The Politics of Canada's Airlines from Diefenbaker to Mulroney*. Toronto: University of Toronto Press, 1987.

Sutherland, Alice Gibson. *Canada's Aviation Pioneers: Fifty Years of McKee Trophy Winners*. Toronto: McGraw-Hill Ryerson, 1979.

Ward, Max. *The Max Ward Story: A Bush Pilot in the Bureaucratic Jungle*. Toronto: McClelland & Stewart, 1991.

Whyard, Florence. *Ernie Boffa, Canadian Bush Pilot*. Whitehorse: Beringian Books, 1994.

Magazine Articles

Cope, Gord. "Calgary's Air Force." *Calgary Herald Sunday Magazine*. April 9, 1989.

Dickins, Punch. "Canadian Aviation Anniversary Issue." *Canadian Aviation*. June 1978.

Ellis, Frank. "Airmen Adventurers to the Bay." *The Beaver*. Summer 1970.

Fattedad, Sid. "Behind the Scenes at Canadian Airlines." *CGA Magazine*. June 1993.

French, Carey. "The Sky's The Limit." *Report on Business Magazine (The Globe and Mail)*. August 1987.

Henry, Walter "The Barkley-Grow T8P-1." *Western Canada Aviation Museum Review*. December 1994.

Posner, Michael. "Flight Into Oblivion." *Report on Business Magazine (The Globe and Mail)*. November 1996.

Ross, Capt. W.C. (Cam). "Flying the Britannia with Canadian Pacific." *CAHS Journal*. Spring 1990.

Thompson, Don and Peggy. "Art Schade, Bush Pilot." *The Beaver*. Autumn 1979.

Whittington, Hugh. "Australian Odyssey." *Canadian Aviation*. August 1974.

_____. "West Wind Rising." *Air Transport World*. August 1994.

Archives

Canadian Airlines Archives.

National Archives of Canada. Canadian Pacific Airlines M.G. 28 III 48 Scrapbook No. 2.

I N D E X